MX 0714236 6

AF443467

Tuberculosis
The Microbe Host Interface

Edited by

Larry S. Schlesinger
The Ohio State University
Columbus, Ohio 43210, USA

and

Lucy E. DesJardin
University of Iowa
Iowa City, IA 52242, USA

Copyright © 2004
Horizon Bioscience
32 Hewitts Lane
Wymondham
Norfolk NR18 0JA
UK

www.horizonbioscience.com

British Library Cataloguing-in-Publication Data

A catalogue record for this book is available from the British Library

ISBN:0-9545232-1-0

Description or mention of instrumentation, software, or other products in this book does not imply endorsement by the author or publisher. The author and publisher do not assume responsibility for the validity of any products or procedures mentioned or described in this book or for the consequences of their use.

*Printed and bound in Great Britain
by Antony Rowe Ltd, Chippenham, Wiltshire, UK*

Contents

Books of Related Interest

For further information on these books contact:

Horizon Bioscience
32 Hewitts Lane, Wymondham
Norfolk
NR18 0JA, UK

Tel: +44(0)1953-601106
Fax: +44(0)1953-603068
Internet: www.horizonbioscience.com

Our Web site has details of all our books including full chapter abstracts, book reviews, and ordering information:

Contributors

Peter F. Barnes
CPIDC
Departments of Medicine
Microbiology and Immunology
University of Texas Health Center
11937 US Hwy. 271
Tyler
TX 75708
USA
e-mail: pbarnes@uthct.edu

Luiz E. Bermudez
Department of Biomedical Sciences
Oregon State University
Corvallis
OR 97331
USA
email: luiz.
bermudez@oregonstate.edu

Kristin A. Birkness
AIDS, STD and TB Lab Research
National Center for Infect Diseases
Centers for Disease Control & Prevent.
Atlanta
GA 30333
USA
email: kbirkness@cdc.gov

W. Henry Boom
Tuberculosis Research Unit
Case Western Reserve University
10900 Euclid Ave.
Cleveland
OH 44106-4984
USA
e-mail: whb@po.cwru.edu

M. Donald Cave
Dept. Anatomy and Neurobiology
University of Arkansas
Med. Sci. and Med. Res. Service
Central Arkansas Veterans Healthcare
System
Little Rock, AR
USA

John Chan
Albert Einstein College of Medicine
Depts. of Microbiol. and Immunol.
1300 Morris Park Ave.
Bronx
NY 10461
USA
e-mail: jchan@aecom.yu.edu

Josephine E. Clark-Curtiss
Dept. of Biology
Washington University
Campus Box 1137
St. Louis
MO 63130-4499
USA
e-mail: clarkcur@biology.wustl.edu

Daniel L. Clemens
Univ. of California at Los Angeles
Department of Medicine
37-121 CHS
UCLA School of Medicine
Los Angeles CA 90095
USA
e mail: dclemens@mednet.ucla.edu

Andrea Cooper
Trudeau Institute Inc.
100 Algonquin Ave.
Saranac Lake
NY 12983
USA
e-mail: acooper@trudeauinstitute.org

Lucy DesJardin
University Hygenic Laboratory
University of Iowa
102 Oakdale Hall H101-OH
Oakdale Campus
Iowa City
IA 52242
USA
e-mail: ldesjard@uhl.uiowa.edu

JoAnne Flynn
Depts of Mol. Genetics and Biochem.
University Pittsburg School Medicine
Pittsburg
PA 15261
USA

David Kusner
Univ. of Iowa Hospitals and Clinics
Division of Infectious Diseases
200 Hawkins Drive
SW54 General Hospital
Iowa City
Iowa 52242
USA
e-mail: David-kusner@uiowa.edu

Frederick D. Quinn
Dept. of Med. Microbiol. & Parasitol.
College of Veterinary Medicine
University of Georgia
Athens
GA 30602
USA
e-mail: fquinn@vet.uga.edu

Larry S. Schlesinger
Division of Infectious Diseases
Center for Microbial Interface Biology
The Ohio State University
456 W. 10th Avenue
Columbus
Ohio 43210
USA
e-mail: schlesinger-2@medctr.osu.edu

Zahra Toossi
Division of Infectious Disease
Department of Medicine
Case Western Reserve University
11100 Euclid Ave.
Cleveland
OH 44106
USA
e-mail: zxt2@po.cwru.edu

Preface

It has been approximately 120 years since Robert Koch identified the bacillus that causes tuberculosis (TB), *Mycobacterium tuberculosis*. Despite its early identification, fundamental questions remain regarding the pathogenesis of this microbe. Disturbingly, TB is reappearing in many countries as a public health crisis and continues as an epidemic worldwide. A staggering 2 to 3 million people around the globe die of TB each year, and estimated 2 billion are infected with *M. tuberculosis* and consequently are at risk for reactivation of disease.

In the 20th century, the United States made impressive strides in TB control. TB rates steadily declined from approximately 200 deaths per 100,000 per year to less than 1 death per 100,000 in 1985. Following 1985, however, the United States saw a reversal in the downward trend of TB incidence with increased rates of drug resistant bacterial strains, HIV co-infection, and a continued influx of immigrants from countries where TB is common. With a huge mobilization of public health resources, cases of TB have once again been declining since 1992. However, the rate of decline is slowing and recently several national agencies, including the Institutes of Medicine, have warned against complacency and neglect for this disease.

There are three major phases of TB in humans. The first is primary infection, in which innate immunity plays a major role. The second is latency, in which the individual remains healthy with a very low burden of viable, metabolically altered bacteria. The third phase is reactivation disease, which occurs in approximately 10% of healthy infected individuals. Beginning in the 1980's, enhanced awareness of the mounting TB burden resulted in an increase in the resources and research directed to TB. Our knowledge has increased markedly in the area of innate immunity and TB with improved models of primary infection. On the other hand, models to study latency are more limited, and in reactivation disease the host-microbe interactions are extraordinarily complex to model. To date, few new fundamental discoveries in TB have translated into clinical practice. However, new vaccine strategies are on the horizon that hold promise for increased efficacy, compared to the current BCG vaccine.

We have recently witnessed an explosion of information regarding the molecular genetics of *M. tuberculosis*. Knowledge gleaned from the genome has increased the pace and breadth of scientific discovery. We are now moving forward with investigation in the post genomics era. In order to more effectively acquire knowledge that is translatable into new diagnostics, therapies, and vaccines, there is now more need than ever for a multidisciplinary approach to study of the *M. tuberculosis*-host interaction. The title of this book "Tuberculosis: The Microbe Host Interface" emphasizes the requirement for input from scientists in a variety of scientific disciplines, such as Cell Biology, Molecular Biology, Microbiology, Pathology, Biochemistry, Pharmacology, Structural Biology and Bioinformatics, to successfully attack this complex interaction.

The major goal of this book is to present state-of-the-art technical strategies and emerging methodologies used for understanding the nature of the host response to infection with *M. tuberculosis*. Our current knowledge of the field, based on these methodologies, is also presented. Finally, different technical approaches are compared, and their strengths and weaknesses are highlighted. These chapters broadly encompass current and evolving *in vitro* assays, animal models, and approaches to the study of latency and molecular epidemiology.

It is hoped that in addition to providing current knowledge on TB pathogenesis, this book will enable new investigators from a variety of disciplines to develop more sophisticated models for the study of all three phases of TB.

The editors wish to thank Amanda MacFarlane, PhD for administrative assistance and Deb Nollen-Richter for editorial assistance. The concept of the book and its chapters originated at the University of Iowa during a number of discussions among members of the Schlesinger laboratory.

Larry Schlesinger
Lucy DesJardin

From: Tuberculosis: The Microbe Host Interface
Edited by: Larry S. Schlesinger and Lucy E. DesJardin

Chapter 1

M. tuberculosis Entry and Growth using Macrophage Models

Zahra Toossi and
Larry S. Schlesinger

Abstract

Although the central role for mononuclear phagocytes in the pathogenesis of tuberculosis has been known for a century, the resurgence of tuberculosis in the United States and worldwide over the past two decades has lead to an equally impressive resurgence of research aimed at further defining the molecular events underlying many aspects of the *M. tuberculosis* (MTB)-mononuclear phagocyte interaction. This chapter summarizes recent advances and compares various *in vitro* and *in vivo* models, pointing out strengths and weaknesses. It is clear that monocytes and macrophages differ phenotypically and functionally among mammals and between tissue compartments of the human host in ways that impact on the host response to MTB.

Introduction

Tuberculosis (TB) continues to plague mankind, with increasing worldwide mortality and morbidity. The intense interaction of TB with human immunodeficiency virus (HIV) infection over the last two decades has compounded the scene even further, and purports to more global devastation. Understanding the pathogenesis of infection with *Mycobacterium tuberculosis* (MTB) remains critical to development of new modalities to treat and prevent this disease. As macrophages are the cornerstone of MTB infection, a full understanding of phagocytosis and survival of MTB within this cellular target is needed to achieve this goal.

Humans are the sole natural host of MTB infection. The global infectious reservoir is immense with over 1/3 of the world population infected (Raviglione *et al.,* 1995). Maintenance of the infectious cycle among humans is dependent on the transmission of MTB from cases with active pulmonary TB to immunologically naïve subjects. The extreme susceptibility to MTB infection is reflected by the fact that aerosolization of a few (5-10) bacilli allows for a primary MTB infection (Smith *et al.,* 1965). The initial interaction of MTB with the host involves alveolar macrophages, which remain the only cell type shown to harbor MTB *in vivo* (Filley and Rook,1991). Phagocytosis and intracellular growth of MTB within alveolar macrophages initiate the events of primary infection. Activation of components of innate immunity, the recruitment of various classes of lymphocytes and monocytes to sites of infection, and the final development of specific immunity allow for the containment of infection.

The interaction of the host immune response with MTB can be divided into at least five types of events. 1. Events of primary MTB infection. 2. Events that are conducive to dissemination/progression of a primary infection. 3. Development of specific immunity that leads to containment of infection. 4. Maintenance of protective immunity versus immunologic events related to reactivation of MTB infection. 5. Immunologic events associated with active disease and immunopathology of TB. It is important to understand that involvement of components of the host immune system such as mononuclear phagocytes varies during the spectrum of MTB infection. Moreover, monocytes emigrate from the bone marrow and then differentiate to tissue macrophages in unique fashion in different organs. Thus, the molecular mechanisms underlying phagocytosis and intracellular growth of MTB in macrophages likely differ depending on the tissue site, degree of cell differentiation, presence of inflammatory mediators, etc. Various *in vitro* and *in vivo* models have been employed to understand these interactions more completely. Each of these models is best suited to assess only one or at most a few stages of the whole spectrum of MTB infection (Figure 1) and therefore any one model is inadequate in providing the full picture. As one example, whereas local

Spectrum of Host Immune Response to MTB

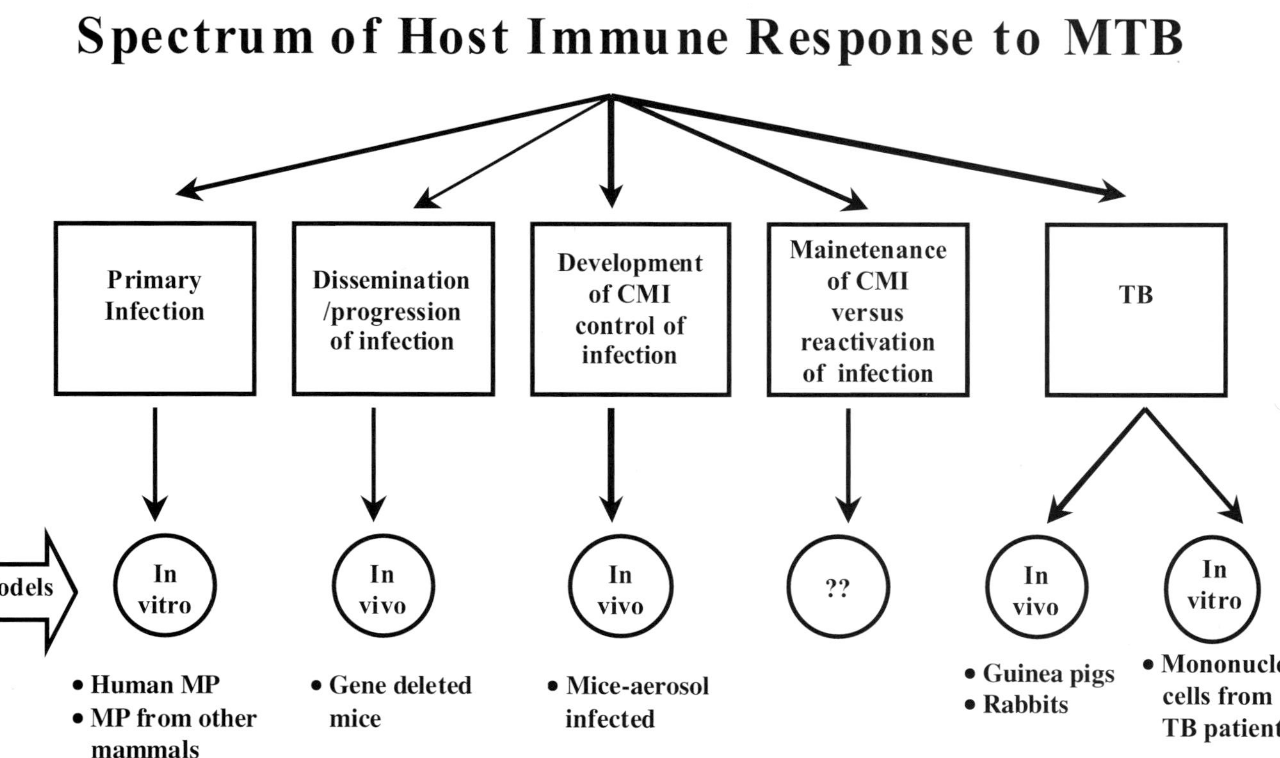

Figure 1. Models of interaction of the host immune system with MTB. *In vitro* and *in vivo* models for different stages of MTB infection.

production of tumor necrosis factor alpha (TNF-α) may activate macrophages allowing containment and prolonged maintenance of defenses against bacillary growth (Keane *et al.*, 2001), its excess production during active TB contributes to immunopathology (Bekker *et al.*, 2000).

This chapter will focus on the contribution of mononuclear phagocytes to MTB pathogenesis. How macrophages are targeted by bacilli and defeated in the battle against MTB to become safe-havens for intracellular growth will be discussed first. Then, both the *in vivo* and in *vitro models* of phagocytosis and growth of MTB within mononuclear phagocytes will be discussed. The focus will be primarily on the events related to primary MTB infection, with some reference to the remainder of the TB spectrum.

The Role of Macrophages in TB Pathogenesis: Phagocytosis and Post-phagocytic Events

Alveolar macrophages (AM) are the primary cell type targeted by MTB, and play a central role in the pathogenesis of this organism. However, the role of other mononuclear phagocytes such as the undifferentiated blood precursors of AM, monocytes (MN), that are heavily recruited to sites of MTB infection (Schwander *et al.*, 1996), and macrophages that have differentiated into dendritic cells (DC) *in situ* also need to be considered.

MTB enters mononuclear phagocytes by receptor-mediated phagocytosis (Schlesinger *et al.*, 1990) in which several major host cell receptors play a role. These include complement receptors (CRs), the mannose receptor (MR) and type A scavenger receptors (Schlesinger *et al.*, 1990, 1993; Stokes *et al.*, 1993; Hirsch *et al.*, 1994; Zimmerli *et al.*, 1996; Melo *et al.*, 2000). The CRs involved are CR1 (CD35) and the leukocyte integrins CR3 (CD11b/CD18) and CR4 (CD11c/CD18) (Harris *et al.*, 2000). The expression of CRs (particularly CR4) and the MR increases during monocyte differentiation into macrophages, and CR4 and the MR are highly expressed on AMs (Speert *et al.*, 1985; Myones *et al.*, 1988). Phagocytosis of MTB by human AMs is greater than that by monocytes and CR4 plays a particularly important role (Hirsch *et al.*, 1994). In the absence of specific antibody, Fcγ receptors do not play a role in phagocytosis (Schlesinger *et al.*, 1990), an important finding since entry via this receptor would be expected to generate a vigorous host response. Phagocytosis of MTB is enhanced in the presence of nonimmune serum as a result of complement component C3 opsonization of bacteria. However, there is evidence for the direct interaction between MTB surface components and CR3 during phagocytosis based on *in vitro* studies using human and murine macrophages as well as CR3-transfected Chinese hamster ovary (CHO) cell lines (Schlesinger *et al.*, 1993; Stokes *et al.*, 1993; Cywes *et al.*, 1996,1997).

The macrophage MR is a prototypic pattern recognition receptor (PRR) that binds with high affinity to mannose and fucose-containing glycoconjugates frequently found on the surface of a variety of microbes [reviewed in (Medzhitov *et al.*, 2000)]. The MR is a member of a family of C-type lectins that is expressed on MDMs, tissues macrophages, and dendritic cells but not monocytes (Speert *et al.*, 1985; Stahl *et al.*, 1990, 1998). AMs demonstrate high MR activity (Wileman *et al.*, 1986). In contrast to CRs, the macrophage MR mediates phagocytosis of the virulent MTB strains, Erdman and H37Rv, but not the attenuated H37Ra strain (Schlesinger *et al.*, 1993) The MR also mediates uptake of other mycobacteria (Astarie-Dequeker *et al.*, 1999). The linear α1-2-linked oligomannosyl "caps" of MTB cell wall lipoarabinomannan (LAM) serve as ligands for the MR during bacterial phagocytosis (Schlesinger *et al.*, 1994, 1996; Kang *et al.*, 1998). Subtle differences exist in the ability of LAM from different MTB strains to bind to the MR and the inositol phosphate-capped AraLAM from *M. smegmatis* does not bind to this receptor (Schlesinger *et al.*, 1994). Other MTB surface molecules accessible to the MR are arabinomannans, mannans, and mannoproteins (Ortalo-Magne *et al.*, 1995; Dobos *et al.*, 1996).

The high MR activity on AMs is noteworthy for TB pathogenesis. MR activity is increased by IL-4, IL-13 and glucocorticoids and inhibited by IFN-γ. It has been postulated that induction by these mediators as well as by TGF-β produces an alternative activation state of macrophages with many attributes characteristic of AMs [reviewed in (Goerdt *et al.*, 1999)]. The phenotypic and molecular characteristics of these macrophages differ considerably from those of classically activated macrophages. For example, alternatively activated macrophages express high levels of PRRs such as the MR and scavenger receptors but do not display enhanced killing functions towards microbes (Munder *et al.*, 1998; Becker *et al.*, 2000;Stein *et al.*, 1992). In this regard, the abundant surfactant associated protein, surfactant protein A (SP-A), produced in the lung interacts with macrophages to enhance MR activity (Beharka *et al.*, 2002). In contrast, surfactant protein D (SP-D) blocks the LAM-MR pathway (Ferguson *et al.*, 1999). Furthermore, nitric oxide (NO) and oxidant production in response to stimuli are reduced in these cells (Oren *et al.*, 1963; Fels *et al.*, 1986). There is recent evidence for potential involvement of the MR in mycobacterial infection in humans (Siddiqui *et al.*, 2001; Hill *et al.*, 2001).

Although CRs and the MR are the major receptors that mediate phagocytosis on mononuclear phagocytes, other receptors may also participate in MTB phagocytosis either alone or in conjunction with CRs and/or the MR. CD14 has been found to mediate uptake of nonopsonized MTB by human microglia, the resident macrophage in the brain (Peterson *et al.*, 1995), and uptake of *M. bovis* by porcine alveolar macrophages (Khanna *et al.*, 1996). Class A scavenger receptors participate in the uptake of nonopsonized

M. tuberculosis by MDMs (Zimmerli *et al.,* 1996). Potential MTB ligands for mononuclear phagocytes and nonprofessional phagocytes include a mammalian cell entry protein (Arruda *et al.,* 1993), a heparin-binding hemagglutinin (Menozzi *et al.,* 1998), glucan (Cywes *et al.,* 1997; Ehlers *et al.,* 1998; Schwebach *et al.,* 2002), PE_PGRS proteins (Brennan *et al.,* 2001), phosphatidylinositol mannoside (PIM) (Hoppe *et al.,* 1997), and antigen 85 (Hetland *et al.,* 1994).

Recent literature indicates that the nature of receptor-ligand interactions for MTB can regulate the early host cell response and the fate of the bacterium. The C3-CR entry pathway for MTB is postulated to provide the bacterium safe passage into mononuclear phagocytes. To the potential advantage of the pathogen, ligation of CR3 does not uniformly trigger toxic host cell responses (Wilson *et al.,* 1980) and has recently been shown to selectively suppress interleukin 12 (IL-12) production, an important mediator of the cellular immune response to MTB (Marth *et al.,* 1997). There is little or no oxidative response to MTB by non-activated human macrophages during phagocytosis (Wayne *et al.,* 1995), a result similar to that obtained with *M. kansasii* using a human myeloid cell line (Le Cabec *et al.,* 2000).

The binding sites for MTB on CR3 may impact on the host cell response and bacterial fate. That there are fundamental differences in the binding interactions to CR3 between C3-opsonized and non-opsonized MTB is supported by *in vitro* studies in which the phagocyte membrane is depleted of cholesterol to ascertain the importance of cholesterol-rich microdomains (rafts). Cholesterol depletion results in significant reduction in mycobacterial uptake (Gatfield *et al.,* 2000). Cholesterol depletion of neutrophils decreased the uptake of non-opsonic mycobacteria but not serum-opsonized organisms, indicating that nonopsonic uptake via CR3 selectively involves a GPI-anchored protein in membrane rafts (Peyron *et al.,* 2000). Different molecular mechanisms for binding involving distinct epitopes on CR3 lead to transduction of different cellular responses depending on the state of bacterial opsonization (Le Cabec *et al.,* 2002).

The LAM-MR pathway also appears to be preferable for MTB. MR-dependent phagocytosis is not coupled to activation of the NADPH oxidase in nonactivated phagocytes (Ezekowitz *et al.,* 1985; Astarie-Dequeker *et al.,* 1999) and the LAM-MR pathway appears to be important in limiting P-L fusion events (Kang and Schlesinger, 1998;Fratti *et al.,* 2001). MTB LAM is reported to inhibit IL-12 production via the MR by generating a negative signal in the cell (Nigou *et al.,* 2001). Thus, like CRs, involvement of LAM and the MR during MTB phagocytosis may enhance intracellular survival.

The Macrophage Response Following MTB Phagocytosis: Oxidant Production, Apoptosis, and Iron (Fe) Acquisition

Once within phagocytes, MTB survives several host anti-mycobacterial defenses. Upon activation by phagocytosis or cytokines, macrophages produce the microbicidal molecules reactive oxygen and nitrogen intermediates (ROI and RNI). However, MTB gene products confer resistance to ROI (Manca *et al.*, 1999). Also, mycobacterial components, such as LAM and phenolic glycolipid are potent scavengers of oxygen radicals (Chan *et al.*, 1989, 1991). Therefore ROI may not play an important role in containment of MTB. However, patients with chronic granulomatous disease are predisposed to TB to some extent (Lau *et al.*, 1998), and mice with deletion of components of the NADPH oxidase have a slight enhancement in the development of TB upon infection (Adams *et al.*, 1997). Nonetheless, Cooper *et al.* have shown that the increased susceptibility to aerosolized MTB in mice lacking $p47^{phox}$, is only transient during the early innate immune response (Cooper *et al.*, 2000).

On the other hand, MTB is sensitive to RNI (MacMicking *et al.*, 1997). Deletion of the gene for the inducible nitric oxide synthase (iNOS), which is a critical enzyme in the RNI pathway, increases the susceptibility of mice to both laboratory and clinical strains of MTB (Scanga *et al.*, 2001). Further, administration of inhibitors of iNOS decreases the survival time of infected animals remarkably (Chan *et al.*, 1995). However the relevance or adequacy of nitric oxide (NO) production in human MTB infection is still not clear. Studies that examine the ability of human mononuclear phagocytes to produce NO demonstrate that whereas both mature and immature cells express iNOS mRNA upon MTB infection (Rich *et al.*, 1997), AM from some but not all subjects produce moderate to high amounts of NO. However, both iNOS mRNA (Nicholson *et al.*, 1996), and some NO activity (Wang *et al.*, 1998) have been demonstrated in lung macrophages of patients with active TB. Thus in humans, only low levels of NO appear to be produced and these levels may contribute to the inflammatory response in addition to effects on MTB growth. In addition, MTB may be genetically resistant to RNI to some extent (Ruan *et al.*, 1999).

The control of intracellular growth of mycobacteria by macrophages has also been linked to the ability of infected cells to undergo apoptosis (Molloy *et al.*, 1994). Further it has been suggested that MTB virulence is associated with the inability of infected macrophages to undergo apoptosis (Balcewicz-Sablinska *et al.*, 1998). Stimulation by ATP through its receptor, $P2X_7$ induces apoptosis and thereby killing of MTB (Kusner and Barton, 2001). Activation through this pathway enhances phagosome-lysosome fusion (Fairbairn *et al.*, 2001), and appears to be independent of both ROI (Fairbairn *et al.*, 2001) and RNI (Sikora *et al.*, 1999; Fairbairn *et al.*, 2001). In a recent study, induction of apoptosis in macrophages by MTB was associated with increased rather than decreased growth of the pathogen (Santucci *et al.*, 2000). Thus, the role of macrophage apoptosis in MTB containment remains unsettled.

Intracellular growth of MTB in macrophages depends in part on competition between the host cell and the pathogen for nutrients such as Fe. Whereas macrophages require Fe for activation of antimicrobial pathways, the intracellular survival of MTB is also dependent on its ability to acquire Fe. Several mycobacterial components are involved in Fe acquisition (Calder and Horwitz,1998). Mycobactins and exochelins (exomycobactins) are siderophores that allow for adequate Fe acquisition for MTB growth (Gobin *et al.*, 1995; De Voss *et al.*, 1999). That intracellular MTB resides in the recycling endosomal pathway, may enhance access to transferrin-bound Fe (Clemens *et al.*, 1996). However, intracellular MTB can acquire Fe by more than one route (Olakanmi *et al.*, 2002).

The Macrophage Response Following MTB Phagocytosis: Cytokine Production

Macrophage cytokine responses are initiated by the interaction of MTB with macrophages prior to, at the time of, and subsequent to phagocytosis. The armamentarium of macrophage defenses against pathogens includes cytokine responses that activate cellular defense mechanisms (TNF-α, IL-1-β, IL-6) or deactivate macrophages (IL-10, TGF-β). Regardless of the timing of release or nature of the cytokine produced, a critical issue is that exposure to cytokines determines subsequent macrophage responses through autocrine/paracrine pathways (Erwig *et al.*, 1998). The production and interaction of cytokines may amplify or counteract beneficial or damaging circuits *in situ* (Toossi *et al.*, 1997; Othieno *et al.*, 1999).

Prior to phagocytosis, induction of cytokines by MTB occurs predominantly through Toll-like receptors (TLR) two and four (Means *et al.*, 1999). TLR signaling may be responsible for the majority of TNF-α (Underhill *et al.*, 1999) that is induced at sites of MTB infection (Schwander *et al.*, 2000; Toossi *et al.*, 2001) . Induction of macrophage IL-12 has also been shown to be TLR-mediated (Brightbill *et al.*, 1999). While induction of TNF-α and IL-12 initiate the innate immune response, MTB or its components are able to modulate acquired T-cell responses through TLR-mediated alteration of expression of histocompatibility antigens (Noss *et al.*, 2001). In addition, cytokine responses induced by mycobacterial cell wall LAM (Barnes *et al.*, 1992; Dahl *et al.*, 1996) that are mediated through surface CD14 or TLRs may contribute to excess cellular apoptotic responses (Aliprantis *et al.*, 2000) that are at sites of MTB infection (Hirsch *et al.*, 2001).

Phagocytosis of MTB induces pro-inflammatory cytokines including TNF-α. Virulent MTB is capable of inducing higher amounts of TNF-α at a lower multiplicity of infection (MOI) as compared to avirulent mycobacteria (Toossi, unpublished observations). During intracellular growth, a number of proteins and non-protein components are produced and secreted (Beatty *et al.*, 2000) that may act on infected or uninfected macrophages to produce cytokines. This scenario is best exemplified by MTB 85 antigen (alpha antigen), which is abundantly secreted by MTB (Salata *et al.*, 1991), is fibronectin-binding (Abou-Zeid *et al.*, 1998), and immunodominant (Huygen *et al.*, 1988). Recently, the expression of MTB 85B mRNA has been found to be increased during early infection of monocytes and macrophages (Graham *et al.*, 1999; Wilkinson *et al.*, 2001). Further, the expression of 85B correlates with the amount of secreted TNF-α, and with subsequent intracellular mycobacterial growth (Wilkinson *et al.*, 2001). Of note, the expression of MTB 85B mRNA was induced by both endogenous and exogenous TNF-α in alveolar macrophages and monocytes (Toossi, Unpublished data). MTB 85B immunoreactivity can be identified in macrophage culture supernatants within hours after infection and correlates with its gene expression (Toossi, Unpublished data). Importantly, induction of TNF-α is augmented when mononuclear phagocytes are stimulated with mycobacterial 85B in complex with fibronectin (Aung *et al.*, 1996). Thus, the interaction of MTB 85B and host TNF-α may create a vicious cycle in which each induces the production of the other. Another abundant MTB protein that may contribute to TNF-α circuits is glutamine synthetase (Wallis *et al.*, 1993), a protein implicated in the immunopathogenesis of MTB (Harth and Horwitz *et al.*, 1999).

In Vitro Assays of MTB- Mononuclear Phagocyte Interactions

Cell Type

In vitro studies to address interactions between MTB and monocytes and macrophages have utilized a variety of cell types. These include primary monocytes and macrophages from humans and a variety of other mammals such as mice, guinea pigs, rabbits, rats, primates, cows; and from fish. In addition, a number of human myeloid cells (e.g. THP-1, U937, Mono-Mac-6, HL-60) and murine phagocyte cell lines (e.g. J774, RAW, MH-S) have been utilized to a variable extent (Lederman *et al.*, 1994; Coppolino *et al.*, 1995; Lee & Horwitz, 1995; Wright *et al.*, 1996; Rockett *et al.*, 1998; Stokes *et al.*, 1998; Stokes and Doxsee, 1999; Melo and Stokes, 2000; Passmore *et al.*, 2001; Sato *et al.*, 2002). Numerous studies have demonstrated significant differences between human and other mammalian mononuclear phagocytes in a variety of biological functions including expression and activity of surface receptors, regulation of antigen presenting molecules, cytokine responses, signaling, oxidant responses, resistance to mycobacterial growth, etc. In addition, cells from different tissue compartments vary greatly in their biological activity. For example, there are significant differences between alveolar macrophages and macrophages from other sources (e.g. peritoneal, bone marrow, or blood). Alveolar macrophages are alternately activated, which is likely to have important implications for TB pathogenesis (Goerdt *et al.,* 1999).

All known myeloid cell lines vary in the expression and activity of surface receptors, including those involved in MTB phagocytosis, and differ from primary human mononuclear phagocytes. Also, it is important to consider that a relatively large proportion of metabolic activity in myeloid cell lines is devoted to cell division. Cells driven to differentiation by phorbol esters add complexity to the interpretation of cellular responses (e.g. signaling responses). In contrast, primary mononuclear phagocytes are terminally differentiated cells. Since the human is the natural reservoir for MTB and the bacterium is uniquely adapted to this host, the use of primary human monocytes and macrophages, particularly alveolar macrophages, is most relevant for understanding the pathogenesis of TB in humans. *In vitro* assay conditions have been established for the use of such cells that allow for a variety of experimental questions to be asked. Having said this, there are also notable limitations to the use of primary human cells such as cost, availability of human donors, number of cells that can be obtained (particularly limited for alveolar macrophages in which approximately 10^7 cells can be harvested from a single human donor) and biohazard. For specific types of studies such as biochemical assays, isolated phagosome studies, and protein purification assays, the use of other primary mammalian cells and cell lines has advantages. A particularly strong approach to investigation is to utilize primary human monocytes and macrophages for initial fundamental observations and then to compare these observations with cells obtained from other mammals and various myeloid cell lines. If the particular process being studied is similar in the different cell types, then more mechanistic type studies can be pursued with more easily accessible cells. Newer genetically engineered myeloid cells that express proteins that are otherwise reduced in activity and/ or expression offer promise for even better *in vitro* models.

Assay Conditions

In vitro studies of MTB and human mononuclear phagocytes are most informative about the events of primary MTB infection. Assays that simulate complex *in vivo* settings are difficult and have been utilized less frequently.

Studies employing MTB-infected blood monocytes, monocyte-derived macrophages (MDM), and alveolar macrophages demonstrate that phagocytosis is efficient and intracellular growth is unrestrained (Schlesinger *et al.,* 1990, 1993; Hirsch *et al.,* 1994a, 1994b).

Assay conditions impact on the successful use of primary human cells. In one study, a significant percentage of *in vitro* cultured macrophages underwent apoptotic cell death over short periods of time (Santucci *et al.,* 2000). This was attributed to the culture of macrophages on polystyrene surfaces, the absence of human serum, or density of cultured cells. *In vitro* culture of MTB-infected mononuclear phagocytes beyond 10-14 days is difficult except when using a very low multiplicity of infection (MOI) (Olakanmi *et al.,* 2000). The MOI, length of infection, and virulence properties of the particular strain of MTB utilized are all important determinants in assays of intracellular growth in macrophages. Another important variable of intracellular mycobacterial growth is the donor source of macrophages (Hoal-van Helden *et al.,* 2001) . Studies using high MOIs (50-100: 1, bacteria/ cell) of the attenuated MTB strain H37Ra indicate that the rate of replication within AM and monocytes is similar over the first 4 days in culture (Hirsch *et al.,* 1994a), however, during days 4-7, growth in monocytes continues to increase (Hirsch *et al.,* 1994a, 1994b), whereas growth in AM plateaus (Hirsch *et al.,* 1994a). Conditions of infection that may mimic the more "physiological" encounter of the pathogen with mononuclear phagocytes *in vivo,* i.e. the virulent MTB strains H37Rv and Erdman used at a low MOI (0.1-1: 1), reveal that MTB establishes itself rapidly in monocytes (Byrd *et al.,* 1997; Silver *et al.,* 1998; Schlesinger *et al.,* 1990) or macrophages (Douvas *et al.,* 1986; Schlesinger, 1993), growing with a doubling time of approximately 36 hr (Silver *et al.,* 1998). *In vitro* -derived macrophages appear to control the intracellular growth of MTB more efficiently when cultured at higher densities (Boechat *et al.,* 2001). Intracellular growth of clinical isolates of MTB is either faster (Hoal-van Helden *et al.,* 2001), or at the same rate as H37Rv (Manca *et al.,* 1999), although results vary (Hoal-van Helden *et al.,* 2001). Earlier work had shown a faster intracellular growth of another epidemic-associated strain compared to non-epidemic strains (Zhang *et al.,* 1999). The conditions of the assays differ in these studies making direct comparisons somewhat difficult.

Newer studies have challenged the dogma that macrophage activating cytokines uniformly reduce, while macrophage deactivating cytokines enhance, intracellular growth of MTB. For example, TNF-α has been shown to enhance intracellular growth (Byrd *et al.,* 1997), whereas IL-10 appears to convert dendritic cells to cells with characteristics of macrophages that are better able to control MTB growth (Fortsch *et al.,* 2000). On the other hand, TGF-β enhances the intracellular growth of MTB, and neutralization of endogenous TGF-β restores the ability of monocytes to contain bacillary growth (Hirsch *et al.,* 1994b, 1997).

In Vivo Models of TB Pathogenesis:
The Role For Macrophages

The murine model of MTB infection has been helpful in deciphering the development and maintenance of protective immunity against MTB infection. Aerosolized MTB infection in mice, which mimics the natural route of infection by targeting the lungs with small numbers of organisms, leads to a chronic infection maintained by acquired immunity (North *et al.*, 1995). A controlled and limited growth of MTB for prolonged periods of time (weeks) is followed by increased local pathology and rapid bacterial growth and finally death of the animals (Orme, 1997). In this model, depletion of AM abrogates the capacity of MTB to establish infection (Leemans *et al.*, 2001) underscoring the importance of these cells in the establishment of an MTB infection.

A recent *in vivo* study using CR3 deficient mice does not show a difference in bacterial burden or pathology between these animals and their controls (Hu *et al.*, 2000). However, in this study, intravenous inoculation of bacteria was used rather than the aerosol route, which is the natural route of infection. As compartmentalization of the immune response is well established and the lung is unique in this respect, the relative role of the C3-CR3 pathway in TB pathogenesis remains unresolved.

TNF-α from macrophages is essential for formation of bactericidal granulomas (Kindler *et al.*, 1989), and mice with a disrupted TNF receptor gene develop rapidly disseminated infection (Jacobs *et al.*, 2000; Ehlers *et al.*, 1999). Further, TNFR1 knock out mice develop more necrotic granulomas than wild type mice (Flynn *et al.*, 1995). Transgenic mice with over-expression of TNFR1, which neutralizes TNF-α, showed overgrowth and marked inhibition of macrophage differentiation within granulomas after BCG infection resulting in extensive caseation necrosis in lesions within the lung (Garcia *et al.*, 1997). Disruption of NF-IL-6 also results in severe MTB infection. Whereas the expression of cytokines such as IL-12, TNF-α, and IFN-γ is not altered, production of GM-CSF and oxidative killing by polymorphonuclear cells is impaired (Sugawara *et al.*, 2001). However, the role of GM-CSF in MTB containment has not been fully investigated.

Deletion of the gene for IFN-γ markedly increases the susceptibility to tuberculosis (Cooper *et al.*, 1993, Flynn *et al.*, 1993). The effects of IFN-γ are in part mediated through induction of iNOS and thereby RNI. In fact, mice lacking the iNOS gene also develop fatal TB indicating a requirement for NO in the control of MTB infection (MacMicking *et al.*, 1997). Cooperativity between IFN-γ and TNF-α in the control of MTB is suggested by studies of mice that are transgenic for sTNFR1 and have deletion of the IFN-γ gene (Garcia *et al.*, 1997). The strength of the IFN-γ response is to some extent dependent on the heterodimeric IL-12 which is produced by macrophages

(Trinchieri *et al.,* 1995). IL-12 p40 chain knock out mice are unable to control mycobacterial growth (Garcia *et al.,* 1997). Recent studies suggest that in addition to their role as mycobacteriocidal agents, both IFN-γ and NO control the host inflammatory response and regulate the progression of an organized granulomatous response (Cooper *et al.*, In press).

Summary

Mononuclear phagocytes play a central role in TB pathogenesis by serving both as host cell reservoirs for the bacterium and important modulators of the immune response, linking innate and adaptive immune responses. Our knowledge has increased greatly in defining the major ligands and receptors for bacterial phagocytosis as well as host and bacterial components that play a role in determining bacterial fate. However, these interactions are complex. Results from *in vitro* assays vary depending on the source of the mononuclear phagocyte utilized, i.e. humans versus other mammals, or lung versus other tissue compartments. The use of myeloid cell lines has advantages for specific assays. However, they are known to be altered in a number of biological activities that have important impact on the MTB-macrophage interaction. The identification of similar responses in primary human mononuclear phagocytes and cell lines is particularly useful for elucidating pathogenic mechanisms. It is clear that there are multiple regulatory loops for determining the effectiveness of the host cell response to MTB. These impact on oxidative responses, the degree of phagosome-lysosome fusion, and regulation of apoptosis and complex cytokine networks. *In vivo* studies confirm the important role for macrophages in MTB pathogenesis both directly and indirectly. Further definition of the molecular events underlying MTB-mononuclear phagocyte interactions is necessary and will be aided by newer experimental techniques including the ability to regulate the expression of proteins in primary macrophages (Elbashir *et al.,* 2001).

References

Abou-Zeid, C., T.L. Ratliff, H.G. Wiker, M. Harboe, J. Bennedsen, and G.A. Rook. 1988. Characterization of fibronectin-binding antigens released by *Mycobacterium tuberculosis* and *Mycobacterium bovis* BCG. Infect. Immun. 56: 3046-3051.

Adams, L.B., M.C. Dinauer, D.E. Morgenstern, and J.L. Krahenbuhl. 1997. Comparison of the roles of reactive oxygen and nitrogen intermediates in the host response to *Mycobacterium tuberculosis* using transgenic mice. Tuber. Lung Dis. 78: 237-246.

Aliprantis, A.O., R.B. Yang, D.S. Weiss, P. Godowski, and A. Zychlinsky. 2000. The apoptotic signaling pathway activated by Toll-like receptor-2. EMBO J. 19: 3325-3336.

Armstrong, J.A., and P.D. Hart. 1975. Phagosome-lysosome interactions in cultured macrophages infected with virulent tubercle bacilli. Reversal of the usual nonfusion pattern and observations on bacterial survival. J. Exp. Med. 142: 1-16.

Arruda, S., Bomfim, G., Knights, R., Huima-Byron, T., and Riley, L.W. 1993. Cloning of an *M. tuberculosis* DNA fragment associated with entry and survival inside cells. Science. 261: 1454-1457.

Astarie-Dequeker, C., N'Diaye, E.N., Le Cabec, V., Rittig, M.G., Prandi, J., and Maridonneau-Parini, I. 1999. The mannose receptor mediates uptake of pathogenic and nonpathogenic mycobacteria and bypasses bactericidal responses in human macrophages. Infect. Immun. 67(2): 469-477.

Aung, H., Z. Toossi, J.J. Wisnieski, R.S. Wallis, L.A. Culp, N.B. Phillips, M. Phillips, L.E. Averill, T.M. Daniel, and J.J. Ellner. 1996. Induction of monocyte expression of tumor necrosis factor alpha by the 30-kD alpha antigen of *Mycobacterium tuberculosis* and synergism with fibronectin. J. Clin. Invest. 98: 1261-1268.

Balcewicz-Sablinska, M.K., J. Keane, H. Kornfeld, and H.G. Remold. 1998. Pathogenic *Mycobacterium tuberculosis* evades apoptosis of host macrophages by release of TNF-R2, resulting in inactivation of TNF-alpha. J. Immunol. 161: 2636-2641.

Barnes, P.F., D. Chatterjee, J.S. Abrams, S. Lu, E. Wang, M. Yamamura, P.J. Brennan, and R.L. Modlin. 1992. Cytokine production induced by *Mycobacterium tuberculosis* lipoarabinomannan. Relationship to chemical structure. J. Immunol. 149: 541-547.

Beatty, W. L., Rhoades, E. R., Ullrich, H. J., Chatterjee, D., Heuser, J. E., and Russell, D. G. 2000. Trafficking and release of mycobacterial lipids from infected macrophages. Traffic. 1: 235-247.

Becker, S., and Daniel, E.G. 2000. Antagonistic and additive effects of IL-4 and interferon-gamma on human monocytes and macrophages: effects on Fc receptors HLA-D antigens, and superoxide production. Cell. Immunol. 129: 351-362.

Beharka, A.A., Gaynor, C.D., Kang, B.K., Voelker, D.R., McCormack, F.X., and Schlesinger, L.S.. 2002. Pulmonary surfactant protein A up-regulates activity of the mannose receptor, a pattern recognition receptor expressed on human macrophages. J. Immunol. 169: 3565-3573.

Bekker, L.G., A.L. Moreira, A. Bergtold, S. Freeman, B. Ryffel, and G. Kaplan. 2000. Immunopathologic effects of tumor necrosis factor alpha in murine mycobacterial infection are dose dependent. Infect. Immun. 68: 6954-6961.

Boechat, N., F. Bouchonnet, M. Bonay, A. Grodet, V. Pelicic, B. Gicquel, and A.J. Hance. 2001. Culture at high density improves the ability of human macrophages to control mycobacterial growth. J. Immunol. 166: 6203-6211.

Braun, J., J. Brandt, J. Listing, A. Zink, R. Alten, W. Golder, E. Gromnica-Ihle, H. Kellner, A. Krause, M. Schneider, H. Sorensen, H. Zeidler, W. Thriene, and J. Sieper. 2002. Treatment of active ankylosing spondylitis with infliximab: a randomised controlled multicentre trial. Lancet. 359: 1187-1193.

Brennan, M.J., Delogu, G., Chen, Y., Bardarov, S., Kriakov, J., Alavi M. *et al.* 2001. Evidence that mycobacterial PE-PGRS proteins are cell surface constituents that influence interactions with other cells. Infect Immun. 69(12): 7326-7333.

Brightbill, H.D., D.H. Libraty, S.R. Krutzik, R.B. Yang, J.T. Belisle, J.R. Bleharski, M.Maitland, M.V. Norgard, S.E. Plevy, S.T. Smale, P.J. Brennan, B.R. Bloom, P.J. Godowski, and R.L. Modlin. 1999. Host defense mechanisms triggered by microbial lipoproteins through toll-like receptors. Science. 285: 732-736.

Byrd, T.F. 1997. Tumor necrosis factor alpha (TNFalpha) promotes growth of virulent *Mycobacterium tuberculosis* in human monocytes iron-mediated growth suppression is correlated with decreased release of TNFalpha from iron- treated infected monocytes. J. Clin. Invest. 99: 2518-2529.

Calder, K.M. and M.A. Horwitz. 1998. Identification of iron-regulated proteins of *Mycobacterium tuberculosis* and cloning of tandem genes encoding a low iron-induced protein and a metal transporting ATPase with similarities to two-component metal transport systems. Microb. Pathog. 24: 133-143.

Chan, J., T. Fujiwara, P. Brennan, M. McNeil, S.J. Turco, J.C. Sibille, M. Snapper, P. Aisen, and B.R. Bloom. 1989. Microbial glycolipids: possible virulence factors that scavenge oxygen radicals. Proc. Natl. Acad. Sci. USA. 86: 2453-2457.

Chan, J., K. Tanaka, D. Carroll, J. Flynn, and B.R. Bloom. 1995. Effects of nitric oxide synthase inhibitors on murine infection with *Mycobacterium tuberculosis*. Infect. Immun. 63: 736-740.

Chan, J., X.D. Fan, S.W. Hunter, P.J. Brennan, and B.R. Bloom. 1991. Lipoarabinomannan, a possible virulence factor involved in persistence of *Mycobacterium tuberculosis* within macrophages. Infect. Immun. 59: 1755-1761.

Clemens, D.L., and M.A. Horwitz. 1995. Characterization of the *Mycobacterium tuberculosis* phagosome and evidence that phagosomal maturation is inhibited. J. Exp. Med. 181: 257-270.

Clemens, D.L., and M.A. Horwitz. 1996. The *Mycobacterium tuberculosis* phagosome interacts with early endosomes and is accessible to exogenously administered transferrin. J. Exp. Med. 184: 1349-1355.

Cooper, A.M., D.K. Dalton, T.A. Stewart, J.P. Griffin, D.G. Russell, and I.M. Orme. 1993. Disseminated tuberculosis in interferon gamma gene-disrupted mice. J. Exp. Med. 178: 2243-2247.

Cooper, A.M., J. Magram, J. Ferrante, and I.M. Orme. 1997. Interleukin 12 (IL-12) is crucial to the development of protective immunity in mice intravenously infected with *Mycobacterium tuberculosis*. J. Exp. Med. 186: 39-45.

Cooper, A.M., B.H. Segal, A.A. Frank, S.M. Holland, and I.M. Orme. 2000. Transient loss of resistance to pulmonary tuberculosis in p47(phox-/-) mice. Infect. Immun. 68: 1231-1234.

Coppolino, M., Leung-Hagesteijn, C., Dedhar, S., and Wilkins, J. 1995. Inducible interaction of integrin $\alpha_2\beta_1$ with calreticulin - Dependence on the activation state of the integrin. J.Biol.Chem. 270: 23132-23138.

Cywes, C., Hoppe, H.C., Daffe, M., and Ehlers, M.R.W. 1997. Nonopsonic binding of *Mycobacterium tuberculosis* to complement receptor type 3 is mediated by capsular polysaccharides and is strain dependent. Infect.Immun. 65[10], 4258-4266.

Cywes C., Godenir N.L., Hoppe H.C., Scholle R.R., Steyn L.M., Kirsch R.E., *et al.* 1996. Nonopsonic binding of *Mycobacterium tuberculosis* to human complement receptor type 3 expressed in Chinese hamster ovary cells. Infect. Immun. 64: 5373-5383.

Dahl, K.E., H. Shiratsuchi, B.D. Hamilton, J.J. Ellner, and Z. Toossi. 1996. Selective induction of transforming growth factor beta in human monocytes by lipoarabinomannan of *Mycobacterium tuberculosis*. Infect. Immun. 64: 399-405.

Desjardins, M., L.A. Huber, R.G. Parton, and G. Griffiths. 1994. Biogenesis of phagolysosomes proceeds through a sequential series of interactions with the endocytic apparatus. J. Cell Biol. 124: 677-688.

De Voss, J.J., K. Rutter, B.G. Schroeder, and C.E. Barry, 3rd. 1999. Iron acquisition and metabolism by mycobacteria. J. Bacteriol. 181: 4443-4451.

Dobos, K.M., Khoo, K.H., Swiderek, K.M., Brennan, P.J., and Belisle, J.T. 1996. Definition of the full extent of glycosylation of the 45- kilodalton glycoprotein of *Mycobacterium tuberculosis*. J. Bacteriol. 178: 2498-2506.

Douvas, G.S., E.M. Berger, J.E. Repine, and A.J. Crowle. 1986. Natural mycobacteriostatic activity in human monocyte-derived adherent cells. Am. Rev. Respir. Dis. 134: 44-48.

Ehlers, S., J. Benini, S. Kutsch, R. Endres, E.T. Rietschel, and K. Pfeffer. 1999. Fatal granuloma necrosis without exacerbated mycobacterial growth in tumor necrosis factor receptor p55 gene-deficient mice intravenously infected with *Mycobacterium avium*. Infect. Immun. 67: 3571-3579.

Ehlers, M.R.W., and Daffe, M. 1998. Interactions between *Mycobacterium tuberculosis* and host cells: are mycobacterial sugars the key? Trends Microbiol. 6(8): 328-335.

Elbashir, S.M., Harborth, J., Lendeckel,W., Yalcin,A., Weber, K., and Tuschl, T. 2001. Duplexes of 21-nucleotide RNAs mediate RNA interface in cultured mammalian cells. Nature. 411: 494-498.

Erwig, L.P., D.C. Kluth, G.M. Walsh, and A.J. Rees. 1998. Initial cytokine exposure determines function of macrophages and renders them unresponsive to other cytokines. J. Immunol. 161: 1983-1988.

Fairbairn, I.P., C.B. Stober, D.S. Kumararatne, and D.A. Lammas. 2001. ATP-mediated killing of intracellular mycobacteria by macrophages is a P2X(7)-dependent process inducing bacterial death by phagosome-lysosome fusion. J. Immunol. 167: 3300-3307.

Fels, A., and Cohn, Z.A. 1986. The alveolar macrophage. J. Appl. Physiol. 60: 353-369.

Ferguson, J.S., Voelker, D.R., McCormack, F.X., and Schlesinger, L.S. 1999. Surfactant protein D binds to *Mycobacterium tuberculosis* bacili and lipoarrabinomannan via carbohydrate-lectin interactions resulting in reduced phagocytosis of the bacteria by macrophages. J. Immunol. 163: 312-321.

Filley, E.A., and G.A. Rook. 1991. Effect of mycobacteria on sensitivity to the cytotoxic effects of tumor necrosis factor. Infect. Immun. 59: 2567-2572.

Flynn, J.L., J. Chan, K.J. Triebold, D.K. Dalton, T.A. Stewart, and B.R. Bloom. 1993. An essential role for interferon gamma in resistance to *Mycobacterium tuberculosis* infection. J. Exp. Med. 178: 2249-2254.

Flynn, J.L., M.M. Goldstein, J. Chan, K.J. Triebold, K. Pfeffer, C.J. Lowenstein, R. Schreiber, T.W. Mak, and B.R. Bloom. 1995. Tumor necrosis factor-alpha is required in the protective immune response against *Mycobacterium tuberculosis* in mice. Immunity. 2: 561-572

Fortsch, D., M. Rollinghoff, and S. Stenger. 2000. IL-10 converts human dendritic cells into macrophage-like cells with increased antibacterial activity against virulent *Mycobacterium tuberculosis*. J. Immunol. 165: 978-987.

Fratti, R.A., Backer, J.M., Gruenberg, J., Corvera, S., and Deretic, V. 2001. Role of phosphatidylinositol 3-kinase and Rab5 effectors in phagosomal biogenesis and mycobacterial phagosome maturation arrest. J. Cell. Biol. 154(6): 631-644.

Garcia, I., Y. Miyazaki, G. Marchal, W. Lesslauer, and P. Vassalli. 1997. High sensitivity of transgenic mice expressing soluble TNFR1 fusion protein to mycobacterial infections: synergistic action of TNF and IFN- gamma in the differentiation of protective granulomas. Eur. J. Immunol. 27: 3182-3190.

Gatfield, J., and Pieters, J. 2000. Essential role for cholesterol in entry of mycobacteria into macrophages. Science . 288(5471): 1647-1650.

Gobin, J., Moore, C. H., Reeve, J. R., Jr., Wong, D. K., Gibson, B. W., and Horwitz, M. A. 1995. Iron acquisition by *Mycobacterium tuberculosis*: Isolation and characterization of a family of iron-binding exochelins. Proc. Natl. Acad. Sci. USA. 92: 5189-5193.

Goerdt, S., and Orfanos, C.E. 1999. Other functions, other genes: alternative activation of antigen-presenting cells. Immunity. 10: 137-142.

Goren, M.B., P. D'Arcy Hart, M.R. Young, and J.A. Armstrong. 1976. Prevention of phagosome-lysosome fusion in cultured macrophages by sulfatides of *Mycobacterium tuberculosis*. P roc. Natl. Acad. Sci. USA. 73: 2510-2514.

Graham, J.E., and J.E. Clark-Curtiss.1999. Identification of *Mycobacterium tuberculosis* RNAs synthesized in response to phagocytosis by human macrophages by selective capture of transcribed sequences (SCOTS). Proc. Natl. Acad. Sci. USA. 96: 11554-11559.

Harris, E.S., McIntyre, T.M., Prescott, S.M., Zimmerman, G.A. 2000. The leukocyte integrins. J. Biol. Chem. 275: 23409-23412.

Harth, G., and M.A. Horwitz. 1999. An inhibitor of exported *Mycobacterium tuberculosis* glutamine synthetase selectively blocks the growth of pathogenic mycobacteria in axenic culture and in human monocytes: extracellular proteins as potential novel drug targets. J. Exp. Med. 189: 1425-1436.

Hetland, G., and Wiker, H.G. 1994. Antigen 85C on *Mycobacterium bovis*, BCG and *M. tuberculosis* promotes monocyte-CR3-mediated uptake of microbeads coated with mycobacterial products. Immunol. 82: 445-449.

Hill, A.V.S. 2001. Genetic susceptibility to mycobacterial disease. American Society for Microbiology General Meeting, May 21, 2001.

Hirsch, C.S., J.J. Ellner, R. Blinkhorn, and Z. Toossi. 1997. *In vitro* restoration of T cell responses in tuberculosis and augmentation of monocyte effector function against *Mycobacterium tuberculosis* by natural inhibitors of transforming growth factor beta. Proc. Natl. Acad. Sci. USA. 94: 3926-3931.

Hirsch, C.S., J.J. Ellner, D.G. Russell, and E.A. Rich. 1994a. Complement receptor-mediated uptake and tumor necrosis factor-alpha- mediated growth inhibition of *Mycobacterium tuberculosis* by human alveolar macrophages. J. Immunol. 152: 743-753.

Hirsch, C.S., T. Yoneda, L. Averill, J.J. Ellner, and Z. Toossi. 1994b. Enhancement of intracellular growth of *Mycobacterium tuberculosis* in human monocytes by transforming growth factor-beta 1. J. Infect. Dis. 170: 1229-1237.

Hirsch, C.S., Z. Toossi, J.L. Johnson, H. Luzze, L. Ntambi, P. Peters, M. McHugh, A. Okwera, M. Joloba, P. Mugyenyi, R.D. Mugerwa, P. Terebuh, and J.J. Ellner. 2001. Augmentation of apoptosis and interferon-gamma production at sites of active *Mycobacterium tuberculosis* infection in human tuberculosis. J. Infect. Dis. 183: 779-788.

Hirsch, C.S., Ellner, J.J., Russell, D.G., and Rich, E.A. 1994. Complement receptor-mediated uptake and tumor necrosis factor- alpha-mediated growth inhibition of *Mycobacterium tuberculosis* by human alveolar macrophages. J. Immunol. 152: 743-753.

Hoal-van Helden, E.G., D. Hon, L.A. Lewis, N. Beyers, and P.D. van Helden. 2001. Mycobacterial growth in human macrophages: variation according to donor, inoculum and bacterial strain. Cell Biol. Int. 25: 71-81.

Hu, C., Mayadas-Norton, T., Tanaka, K., Chan, J., and Salgame, P. 2000. *Mycobacterium tuberculosis* infection in complement receptor 3-deficient mice. J.Immunol. 165:2596-2602.

Huygen, K. 1998. DNA vaccines: application to tuberculosis. Int. J. Tuberc. Lung Dis. 2: 971-978.

Jacobs, M., N. Brown, N. Allie, and B. Ryffel. 2000. Fatal *Mycobacterium bovis* BCG infection in TNF-LT-alpha-deficient mice. Clin. Immunol 94: 192-199.

Kang, B.K., and Schlesinger, L.S. 1998. *M. tuberculosis* lipoarabinomannan is a microbial determinant that reduces phagosome-lysosome fusion following phagocytosis events in human macrophages. Keystone Symposia TB: Molecular Mechanisms and Immunologic Aspects.

Keane, J., S. Gershon, R. P. Wise, E. Mirabile-Levens, J. Kasznica, W. D. Schwieterman, J. N. Siegel, and M. M. Braun. 2001. Tuberculosis associated with infliximab, a tumor necrosis factor alpha-neutralizing agent. N. Engl. J. Med. 345: 1098-104.

Khanna, K.V., Choi, C.S., Gekker, G., and Peterson, P.K., Molitor T.W. 1996. Differential infection of porcine alveolar macrophage subpopulations by nonopsonized *Mycobacterium bovis* involves CD14 receptors. J. Leukocyte. Biol. 60: 214-220.

Kindler, V., A.P. Sappino, G.E. Grau, P.F. Piguet, and P. Vassalli. 1989. The inducing role of tumor necrosis factor in the development of bactericidal granulomas during BCG infection. Cell. 56: 731-740.

Kusner, D.J., and J.A. Barton. 2001. ATP stimulates human macrophages to kill intracellular virulent *Mycobacterium tuberculosis* via calcium-dependent phagosome-lysosome fusion. J. Immunol. 167: 3308-3315.

Lau, Y.L., G.C. Chan, S.Y. Ha, Y.F. Hui, and K.Y. Yuen. 1998. The role of phagocytic respiratory burst in host defense against *Mycobacterium tuberculosis*. Clin. Infect. Dis. 26: 226-227.

Le Cabec, V., Carreno, S., Moisand, A., Bordier, C., Maridonneau-Parini, I. 2002. Complement receptor 3 (CD11b/CD18) mediates type I and type II phagocytosis during nonopsonic and opsonic phagocytosis, respectively. J. Immunol. 169: 2003-2009.

Lederman, M. M., Georges, D. L., Kusner, D. J., Mudido, P., Giam, C.-Z., and Toossi, Z. 1994. *Mycobacterium tuberculosis* and its purified protein derivative activate expression of the human immunodeficiency virus. J. Acquir. Immune Defic. Syndr. 7: 727-733.

Lee, B.-Y. and Horwitz, M. A. 1995. Identification of macrophage and stress-induced proteins of *Mycobacterium tuberculosis*. J. Clin. Invest. 96: 245-249.

Leemans, J.C., N.P. Juffermans, S. Florquin, N. van Rooijen, M.J. Vervoordeldonk, A. Verbon, S.J. van Deventer, and T. van der Poll. 2001. Depletion of alveolar macrophages exerts protective effects in pulmonary tuberculosis in mice. J. Immunol. 166: 4604-4611.

MacMicking, J.D., R.J. North, R. LaCourse, J.S. Mudgett, S.K. Shah, and C.F. Nathan. 1997. Identification of nitric oxide synthase as a protective locus against tuberculosis. Proc. Natl. Acad. Sci. USA. 94: 5243-5248.

MacMicking, J., Q.W. Xie, and C. Nathan. 1997. Nitric oxide and macrophage function. Annu. Rev. Immunol. 15: 323-350.

Manca, C., S. Paul, C.E. Barry, 3rd, V.H. Freedman, and G. Kaplan. 1999. *Mycobacterium tuberculosis* catalase and peroxidase activities and resistance to oxidative killing in human monocytes *in vitro*. Infect. Immun. 67: 74-79.

Manca, C., L. Tsenova, C.E. Barry, 3rd, A. Bergtold, S. Freeman, P.A. Haslett, J.M. Musser, V.H. Freedman, and G. Kaplan. 1999. *Mycobacterium tuberculosis* CDC1551 induces a more vigorous host response *in vivo* and *in vitro*, but is not more virulent than other clinical isolates. J. Immunol. 162: 6740-6746.

Marth, T., and Kelsall, B.L. 1997. Regulation of interleukin-12 by complement receptor 3 signaling. J. Exp. Med. 85: 1987-1995.

Means, T.K., S. Wang, E. Lien, A. Yoshimura, D.T. Golenbock, and M.J. Fenton. 1999. Human toll-like receptors mediate cellular activation by *Mycobacterium tuberculosis*. J. Immunol. 163: 3920-3927.

Medzhitov, R., and Janeway, C., Jr. 2000. Innate Immunity. N. Engl. J. Med. 343(5): 338-344.

Melo, M. D. and Stokes, R. W. 2000. Interaction of *Mycobacterium tuberculosis* with MH-S, an immortalized murine alveolar macrophage cell line: a comparison with primary murine macrophages. Tuber. Lung Dis. 80: 35-46.

Menozzi, F.D., Bischoff, R., Fort, E., Brennan, M.J., and Locht, C. 1998. Molecular characterization of the mycobacterial heparin-binding hemagglutinin, a mycobacterial adhesin. Proc. Natl. Acad. Sci. USA. 95: 12625-12630.

Molloy, A., P. Laochumroonvorapong, and G. Kaplan. 1994. Apoptosis, but not necrosis, of infected monocytes is coupled with killing of intracellular bacillus Calmette-Guerin. J. Exp. Med. 180: 1499-1509.

Munder, M., Eichmann, K., Modolell, M. 1998. Alternative metabolic states in murine macrophages reflected by the nitric oxide synthase/arginase balance: competitive regulation by CD4+ T cells correlates with Th1/Th2 phenotype. J Immunol 1998; 160: 5347-5354.

Myones, B.L., Dalzell, J.G., Hogg, N., and Ross, G.D. 1988. Neutrophil and monocyte cell surface p150,955 has iC3b-receptor (CR4) activity resembling CR3. J. Clin. Invest. 82: 640-651.

Nicholson, S., G. Bonecini-Almeida Mda, J.R. Lapa e Silva, C. Nathan, Q.W. Xie, R. Mumford, J.R. Weidner, J. Calaycay, J. Geng, N. Boechat, and *et al*. 1996. Inducible nitric oxide synthase in pulmonary alveolar macrophages from patients with tuberculosis. J. Exp. Med. 183: 2293-2302.

North, R.J. 1995. *Mycobacterium tuberculosis* is strikingly more virulent for mice when given via the respiratory than via the intravenous route. J. Infect. Dis. 172: 1550-1553.

Noss, E.H., R.K. Pai, T.J. Sellati, J.D. Radolf, J. Belisle, D.T. Golenbock, W.H. Boom, and C.V. Harding. 2001. Toll-like receptor 2-dependent inhibition of macrophage class II MHC expression and antigen processing by 19-kDa lipoprotein of *Mycobacterium tuberculosis*. J. Immunol. 167: 910-918.

Olakanmi, O., Britigan, B. E., and Schlesinger, L. S. 2000. Gallium disrupts iron metabolism of mycobacteria residing within human macrophages. Infect. Immun. 68: 5619-5627.

Olakanmi, O., Schlesinger, L. S., Ahmed, A. and Britigan, B. E., 2002. Intraphagosomal *Mycobacterium tuberculosis* acquires iron from both extracellular transferring and intracellular iron pools. Impact of interferon-gamma and hemochromatosis. J. Biol. Chem. 277: 49727-49734.

Oren, R., Farnham, A.E., Saito, K., Milofsky, E., and Karnovsky, M.L. 1963. Metabolic patterns in three types of phagocytizing cells. J. Cell. Biol. 17: 487-501.

Orme, I. 1997. In: TB pathogenesis Protection and Control. Bloom, B.R., ed. ASM PRESS, Rhoades, ER. p.113-134.

Ortalo-Magné, A., Dupont, M.-A., Lemassu, A., Andersen, A.B., Gounon, P., and Daffé, M. 1995. Molecular composition of the outermost capsular material of the tubercle bacillus. Microbiology 141: 1609-1620.

Othieno, C., C.S. Hirsch, B.D. Hamilton, K. Wilkinson, J.J. Ellner, and Z. Toossi. 1999. Interaction of *Mycobacterium tuberculosis*-induced transforming growth factor beta1 and interleukin-10. Infect. Immun. 67: 5730-5735.

Peterson, P.K., Gekker, G., Hu, S., Sheng, W.S., Anderson, W.R., Ulevitch, R.J., *et al.* 1995. CD14 receptor-mediated uptake of nonopsonized *Mycobacterium tuberculosis* by human microglia. Infect. Immun. 63: 1598-1602.

Peyron, P., Bordier, C., N'Diaye, E.N., and Maridonneau-Parini, I. 2000. Nonopsonic phagocytosis if Mycobacterium kansasii by human neutrophils depends on cholesterol and is mediated by CR3 associated with glycosylphosphatidylinositol-anchored proteins. J. Immunol. 165(9): 5186-5191.

Raviglione, M.C., D.E. Snider, Jr., and A. Kochi. 1995. Global epidemiology of tuberculosis. Morbidity and mortality of a worldwide epidemic. JAMA. 273: 220-226.

Rich, E.A., M. Torres, E. Sada, C.K. Finegan, B.D. Hamilton, and Z. Toossi. 1997. *Mycobacterium tuberculosis* (MTB)-stimulated production of nitric oxide by human alveolar macrophages and relationship of nitric oxide production to growth inhibition of MTB. Tuber. Lung Dis. 78: 247-255.

Rockett, K. A., Brookes, R., Udalovi, R., Vidal, V., Hill, A. V., and Kwiatkowski, D. 1998. 1,25-Dihydroxyvitamin D3 induces nitric oxide synthase and suppresses growth of *Mycobacterium tuberculosis* in a human macrophage-like cell line. Infect.Immun. 66: 5314-5321.

Ruan, J., G. St John, S. Ehrt, L. Riley, and C. Nathan. 1999. *noxR3*, a novel gene from *Mycobacterium tuberculosis*, protects *Salmonella typhimurium* from nitrosative and oxidative stress. Infect. Immun. 67: 3276-3283.

Salata, R.A., A.J. Sanson, I.J. Malhotra, H.G. Wiker, M. Harboe, N.B. Phillips, and T.M. Daniel. 1991.Purification and characterization of the 30,000 dalton native antigen of *Mycobacterium tuberculosis* and characterization of six monoclonal antibodies reactive with a major epitope of this antigen. J. Lab. Clin. Med. 118: 589-598.

Santucci, M.B., M. Amicosante, R. Cicconi, C. Montesano, M. Casarini, S. Giosue, A. Bisetti, V. Colizzi, and M. Fraziano. 2000. *Mycobacterium tuberculosis*-induced apoptosis in monocytes/macrophages: early membrane modifications and intracellular mycobacterial viability. J. Infect. Dis. 181: 1506-1509.

Sato, K., Tomioka, H., Shimizu, T., Gonda, T., Ota, F., and Sano, C. 2002. Type II alveolar cells play roles in macrophage-mediated host innate resistance to pulmonary mycobacterial infections by producing proinflammatory cytokines. J. Infect.Dis. 185: 139-1147.

Scanga, C.A., V.P. Mohan, K. Tanaka, D. Alland, J.L. Flynn, and J. Chan. 2001. The inducible nitric oxide synthase locus confers protection against aerogenic challenge of both clinical and laboratory strains of *Mycobacterium tuberculosis* in mice. Infect. Immun. 69: 7711-7717.

Schlesinger, L.S. 1993. Macrophage phagocytosis of virulent but not attenuated strains of *Mycobacterium tuberculosis* is mediated by mannose receptors in addition to complement receptors. J. Immunol. 150: 2920-2930.

Schlesinger, L.S., Bellinger-Kawahara, C.G., Payne, N.R., and Horwitz, M.A. 1990. Phagocytosis of *Mycobacterium tuberculosis* is mediated by human monocyte complement receptors and complement component C3. J. Immunol. 144: 2771-2780.

Schlesinger LS, Hull SR, Kaufman TM. 1994. Binding of the terminal mannosyl units of lipoarabinomannan from a virulent strain of *Mycobacterium tuberculosis* to human macrophages. J. Immunol. 152: 4070-4079.

Schlesinger, L.S., Kaufman, T.M., Iyer, S., Hull, S.R., and Marciando, L.K. 1996. Differences in mannose receptor-mediated uptake of lipoarabinomannan from virulent and attenuated strains of *Mycobacterium tuberculosis* by human macrophages. J. Immunol. 157: 4568-4575.

Schwander, S.K., E. Sada, M. Torres, D. Escobedo, J.G. Sierra, S. Alt, and E.A. Rich.1996. T lymphocytic and immature macrophage alveolitis in active pulmonary tuberculosis. J. Infect. Dis. 173: 1267-1272.

Schwander, S.K., M. Torres, C.C. Carranza, D. Escobedo, M. Tary-Lehmann, P. Anderson, Z. Toossi, J.J. Ellner, E.A. Rich, and E. Sada. 2000. Pulmonary mononuclear cell responses to antigens of *Mycobacterium tuberculosis* in healthy household contacts of patients with active tuberculosis and healthy controls from the community. J. Immunol. 165: 1479-1485.

Siddiqui, M.R., Meisner, S., Tosh, K., Balakrishnan, K., Ghei, S., Fisher, S.E., *et al.* 2001. A major susceptibility locus for leprosy in India maps to chromosome 10p13. Nature Genetics. 27: 439-441.

Sikora, A., J. Liu, C. Brosnan, G. Buell, I. Chessel, and B.R. Bloom. 1999. Cutting edge: purinergic signaling regulates radical-mediated bacterial killing mechanisms in macrophages through a P2X7-independent mechanism. J. Immunol. 163: 558-561.

Silver, R.F., Q. Li, and J.J. Ellner. 1998. Expression of virulence of *Mycobacterium tuberculosis* within human monocytes: virulence correlates with intracellular growth and induction of tumor necrosis factor alpha but not with evasion of lymphocyte- dependent monocyte effector functions. Infect. Immun. 66: 1190-1199.

Smith, D.W., E. Wiegeshaus, R. Navalkar, and A.A. Grover. 1965. Host-parasite relationships in experimental airborne tuberculosis. I. Preliminary studies in BCG-vaccinated and nonvaccinated animals. J. Bacteriol. 91: 718-724.

Speert, D.P., and Silverstein S.C. Phagocytosis of unopsonized zymosan by human monocyte-derived macrophages: Maturation and inhibition by mannan. J. Leukocyte Biol. 38: 655-658.

Stahl, P.D. 1990. The macrophage mannose receptor: current status. Am. J. Respir. Cell. Mol. Biol. 2: 317-318.

Stahl, P.D., and Ezekowitz, R.A. 1998. The mannose receptor is a pattern recognition receptor involved in host defense. Curr. Opin. Immunol. 10: 50-55.

Stein, M., Keshav, S., Harris, N., and Gordon S. Interleukin 4 potently enhances murine macrophage mannose receptor activity: A marker of alternative immunologic macrophage activation. J Exp Med. 176: 287-292.

Stokes, R. W. and Doxsee, D. 1999. The receptor-mediated uptake, survival, replication, and drug sensitivity of *Mycobacterium tuberculosis* within the macrophage-like cell line THP- 1: a comparison with human monocyte-derived macrophages. Cell.Immunol. 197: 1-9.

Stokes, R.W., Haidl, I.D., Jefferies, W.A., and Speert, D.P. 1993. Mycobacteria-macrophage interactions: Macrophage phenotype determines the nonopsonic binding of *Mycobacterium tuberculosis* to murine macrophages. J. Immunol. 151: 7067-7076.

Stokes, R. W., Thorson, L. M., and Speert, D. P. 1998. Nonopsonic and opsonic association of *Mycobactium tuberculosis* with resident alveolar macrophages is inefficient. J.Immunol. 160: 5514-5521.

Sugawara, I., S. Mizuno, H. Yamada, M. Matsumoto, and S. Akira. 2001. Disruption of nuclear factor-interleukin-6, a transcription factor, results in severe mycobacterial infection. Am. J. Pathol. 158: 361-366.

Toossi, Z., J.L. Johnson, R.A. Kanost, M. Wu, H. Luzze, P. Peters, A. Okwera, M. Joloba, P. Mugyenyi, R.D. Mugerwa, H. Aung, J.J. Ellner, and C.S. Hirsch. 2001. Increased replication of HIV-1 at sites of *Mycobacterium tuberculosis* infection: potential mechanisms of viral activation. J. Acquir. Immune Defic. Syndr. 28: 1-8.

Toossi, Z., M. Mincek, E. Seeholtzer, S.A. Fulton, B.D. Hamilton, and C.S. Hirsch. 1997. Modulation of IL-12 by transforming growth factor-beta (TGF-beta) in *Mycobacterium tuberculosis*-infected mononuclear phagocytes and in patients with active tuberculosis. J. Clin. Lab. Immunol. 49: 59-75.

Trinchieri, G. 1995. In: Annual Review of Immunology. Paul, W.E., Fathman, C.G. and Metzger, H., eds. Annual Reviews Inc. p. 251-76.

Underhill, D.M., A. Ozinsky, K.D. Smith, and A. Aderem. 1999. Toll-like receptor-2 mediates mycobacteria-induced proinflammatory signaling in macrophages. Proc. Natl. Acad. Sci. USA. 96: 14459-14463.

Wallis, R.S., R. Paranjape, and M. Phillips. 1993. Identification by two-dimensional gel electrophoresis of a 58-kilodalton tumor necrosis factor-inducing protein of *Mycobacterium tuberculosis*. Infect. Immun. 61: 627-632.

Wang, C.H., C.Y. Liu, H.C. Lin, C.T. Yu, K.F. Chung, and H.P. Kuo. 1998. Increased exhaled nitric oxide in active pulmonary tuberculosis due to inducible NO synthase upregulation in alveolar macrophages. Eur. Respir. J. 11: 809-815.

Wayne, S., Denning, G., Kusner, D.J., Kaufman, T.M., and Schlesinger, L.S. 1995. Phagocytosis of virulent and attenuated strains of *M. tuberculosis* by human macrophages does not generate detectable superoxide anion or hydrogen peroxide. Clin. Res. 43: 219A.

Wileman, T.E., Lennartz, M.R., and Stahl, P.D. 1986. Identification of the macrophage mannose receptor as a 175-kDa membrane protein. Proc. Natl. Acad. Sci. USA. 83: 2501-2505.

Wilkinson, R.J., L.E. DesJardin, N. Islam, B.M. Gibson, R.A. Kanost, K.A. Wilkinson, D. Poelman, K.D. Eisenach, and Z. Toossi. 2001. An increase in expression of a *Mycobacterium tuberculosis* mycolyl transferase gene (*fbpB*) occurs early after infection of human monocytes. Mol. Microbiol. 39: 813-821.

Wilson, C.B., Tsai, V., and Remington, J.S. 1980. Failure to trigger the oxidative metabolic burst by normal macrophages: Possible mechanism for survival of intracellular pathogens. J. Exp. Med. 1980: 151: 328-346.

Wright, E. L., Quenelle, D. C., Suling, W. J., and Barrow, W. W. 1996. Use of Mono Mac 6 human monocytic cell line and J774 murine macrophage cell line in parallel antimycobacterial drug studies. Antimicrob.Agents Chemother. 40: 2206-2208.

Zhang, M., J. Gong, D.H. Presky, W. Xue, and P.F. Barnes. 1999. Expression of the IL-12 receptor beta 1 and beta 2 subunits in human tuberculosis. J. Immunol. 162: 2441-2447.

Zimmerli, S., Edwards, S., and Ernst, J.D. 1996. Selective receptor blockade during phagocytosis does not alter the survival and growth of *Mycobacterium tuberculosis* in human macrophages. Am. J. Respir. Cell. Mol. Biol. 15: 760-770.

From: Tuberculosis: The Microbe Host Interface
Edited by: Larry S. Schlesinger and Lucy E. DesJardin

Chapter 2

Analysis of Post-Phagocytic Events: Membrane trafficking and the *Mycobacterium tuberculosis* phagosome

Daniel L. Clemens

Abstract

Mycobacterium tuberculosis subverts the normal membrane trafficking pathway of the host cell: it prevents the maturation and acidification of its phagosome, and thereby achieves a phagosome that is hospitable for its growth. The composition of the phagosome and the interactions between the phagosome and other host cell organelles have been studied by a variety of different techniques, including transmission electron microscopy, immunoelectron microscopy, immunofluorescence microscopy, and biochemical and immunological analysis of isolated phagosomes. In addition, the pH of the phagosome has been studied both by fluorescence and electron microscopy based techniques. Each of these methods for characterizing the properties and the membrane trafficking interactions of the *M. tuberculosis* phagosome has advantages and disadvantages, and each technique has the capacity to complement the other techniques in providing a more complete picture of the composition and membrane trafficking interactions of the *M. tuberculosis* phagosome.

Introduction

Membrane Trafficking and the Endocytic and Phagocytic Pathways

Macrophages play an important role in host defense by phagocytosing invading microbes and degrading the microbes in an acidified, hydrolase rich phagolysosome. *M. tuberculosis* is a successful pathogen because it is able to subvert the host cell membrane trafficking machinery and prevent the maturation of its phagosome into a phagolysosome. Methods to study the membrane trafficking events of the *M. tuberculosis* phagosome and to delineate how this pathogen evades the host cell defenses are active areas of research and are the subject of this review.

The membrane trafficking patterns of endosomes and phagosomes containing inert material have been studied extensively. Endocytosed material proceeds through a series of membrane bound compartments – early endosomes, late endosomes, and lysosomes - with distinct membrane markers at each stage. Cartoon illustrations from two decades ago depict a simple concept of membrane trafficking of phagosomes in which, at some point following phagocytosis, the phagosome fuses with lysosomes and becomes a phagolysosome. We now understand that this concept is in error and that phagosomes (Pitt, *et al.* 1992a, 1992b; Desjardins, *et al.* 1994b; and Desjardins, 1995) and macropinosomes (Racoosin and Swanson, 1993) undergo a gradual maturation process involving sequential interactions with endocytic vesicles, leading to exchange of vesicle contents and membrane material between the phagosome or macropinosome and the vesicles of the endocytic pathway. This process is also accompanied by progressive acidification of the phagosome. Thus, the maturation and the membrane trafficking of phagosomes and macropinosomes containing inert particles closely mirror the membrane trafficking events of the endosomal-lysosomal pathway.

Immediately after phagocytosis, early phagosomes interact with early endosomes and acquire markers of early endosomes, such as Rab5, transferrin receptor, and early endosomal antigen-1 (EEA1). With further maturation, the phagosome interacts more with late endosomes and less with early endosomes and acquires late endosomal markers, such as the mannose-6-phosphate receptor, Rab7, lysosome associated membrane glycoproteins (LAMPs: e.g.

LAMP-1, LAMP-2, and CD63) and acid hydrolases. Ultimately the phagosome fuses with lysosomes. Lysosomes are distinguished from late endosomes by the absence of recycling receptors, such as the mannose-6-phosphate receptor (M6PR). The phagosome acquires vacuolar-proton ATPase as it matures, which acts to acidify the maturing phagosome. Whereas early endosomes and early phagosomes have a pH of 5.5 – 6.5, phagolysosomes and lysosomes have a pH of 5 or less.

Considerable progress has been made in understanding the cellular machinery that mediates membrane docking and fusion events in eukaryotic cells, including the importance of molecules such as NSF (N-ethylmaleimide sensitive factor), SNAP (soluble NSF attachment protein), SNAREs (SNAP attachment proteins), and Rab-GTPases (Ferro-Novick and Jahn, 1994; Rothman, 1994). Each membrane-bound compartment within a cell is thought to have a unique set of Rab-GTPases that regulate vesicle docking and the assembly of SNARE complexes (Hall, 1990; Nuoffer and Balch, 1994; Pfeffer, 1994). The SNARE proteins, in turn, assemble into fusion-competent, oligomeric core complexes that form tight connections between membranes (Hanson *et al.*, 1997). In the cases studied to date, SNARE core complexes consist of one R (arginine)-SNARE and 3 Q (glutamine)-SNARE domains (Antonin *et al.*, 2000a). The core complexes are then reversibly dissociated by NSF, a chaperon ATPase, acting in conjunction with α-SNAP (Sollner *et al.*, 1993).

Rab-GTPases

Over 50 small GTPases of the Rab family have been identified and several have been demonstrated to be key regulators of vesicular docking and fusion events in the endocytic and phagocytic pathways. The Rab proteins are synthesized on free ribosomes in the cytosol and are post-translationally modified by the addition of a geranyl-geranyl lipid group to each of two cysteines located near the carboxy terminus. Rab-GTPases bind GTP and hydrolyze it to GDP. The GDP-bound form of geranylated-Rab proteins are bound by a chaperon protein, GDI (guanine nucleotide dissociation inhibitor) and are inserted onto the cytoplasmic aspect of their target membranes. Whereas Rab proteins are inserted only onto specific organelle membranes, GDI is promiscuous and can escort any Rab protein to its target membrane. The Rab-GTPases on the membrane exchange GDP for GTP and hydrolyze GTP to GDP. This GTP/GDP cycle is important both in the shuttling of Rab proteins between their target membranes and the cytosol, and also in regulation of vesicular docking and fusion by Rab proteins. The membrane associated, GTP-bound forms of Rab proteins promote vesicular docking and fusion by recruiting SNARE proteins. The GDP-bound forms of Rab proteins do not promote fusion and can be bound by GDI and recycled into the cytosolic pool of Rab proteins. Rab-GTPases that are present on early and late endosomes include Rab4, 5, 7,

9 and 11. Rab4, 5, and 11 define endocytic compartments that are accessible to exogenously added transferrin. Rab5 is present on the early endosomal compartment and regulates membrane trafficking events involving this compartment (Chavrier *et al.*, 1990; Barbieri *et al.*, 1994; Bucci *et al.*, 1995; Stenmark *et al.*, 1995). Rab11 is present on a pericentriolar recycling compartment and regulates recycling through this compartment (Ullrich *et al.*, 1996). Rab4 is found on peripherally located recycling vesicles and is involved in regulation of membrane recycling between early endosomes and the plasma membrane (Van der Sluijs *et al.*, 1992; Bottger *et al.*, 1996). Consistent with this function, it is also found in patches on early endosomes and the pericentriolar recycling vesicles, where it overlaps with Rab5 and Rab11, respectively (Sonnichsen *et al.*, 2000). Rab9 is present on late endosomes and the trans-Golgi network (TGN), regulates protein transport from the TGN to late endosomes, and is required for lysosome biogenesis (Lombardi *et al.*, 1993; Shapiro *et al.*, 1993.). Rab7 is present on late endosomes and to a limited extent, on lysosomes (Chavrier *et al.*, 1990; Meresse *et al.*, 1995.), and is believed to regulate vesicular trafficking events involving these compartments (Meresse *et al.*, 1995.). The importance of Rab-GTPases in regulation of membrane trafficking has been demonstrated in part by transfection of mammalian cells with genes for mutant Rab proteins that are defective either in GTP hydrolysis or in GDP-GTP exchange. These mutations are associated with either dominant fusion competent or dominant fusion incompetent phenotypes, respectively. For example, cells expressing a dominant positive Rab5 mutant (defective in GTP hydrolysis and therefore GTP-bound) develop large, swollen early endosomes (Bucci *et al.*, 1992). Cells expressing a dominant negative Rab5 mutant (defective in GDP-GTP exchange, and therefore GDP-bound) show impaired endocytosis (Bucci *et al.*, 1992), and this mutant form of Rab5 does not support fusion between early endosomes in an in vitro assay (Gorvel, 1991). Similarly, expression of a dominant-negative Rab9 impaired lysosomal biogenesis by impairing transport from TGN to late endosomes (Riederer *et al.*, 1994), and expression of a dominant negative Rab7 impaired transport from early to late endosomes (Press *et al.*, 1998).

Rab Effectors and SNARE Proteins

The Rab effectors and SNARE proteins that act downstream of activated Rab-proteins are beginning to be understood. In the case of Rab5 and early endosomes, the activated form of Rab5 (Rab5-GTP) recruits phosphoinositol-3-kinases (PI3K) to the endosomal membrane (Christoforidis *et al.*, 1999a). The PI3K in turn converts membrane phosphoinositol-2 phosphate (PI(2)P) to phosphoinositol-3-phosphate (PI(3)P). Membrane PI(3)P is required for stable binding of the EEA1 to the endosomal membrane (Christoforidis *et al.*, 1999b). EEA1 then interacts with the transmembrane endosomal SNARE protein, syntaxin 13, in the formation of an active, fusion competent core

Table 1. Markers and properties of the endosomal - lysosomal pathway

	Early Endosome	Peripheral Recycling Endosome	Pericentriolar Recycling Endosome	Late Endosome	Lysosome	*M. tb* phagosome
Transferrin Receptor	+	+	+	-	-	+
Transferrin	+	+	+	-	-	+
Rab5	+	±	-	-	-	+
Rab4	+	+	+	-	-	?
Rab11	-	-	+	+	-	?
Vacuolar Proton Pump	+	+	+	++	++	-
Rab7	-	-	-	+	±	+
Rab9	-	-	-	+	-	?
M6PR	±	±	±	+	-	-
Cathepsins B and D	-	-	-	+	+	±
LAMP$_{1,2}$ and CD63	-	-	-	+	+	±
pH	5.9- 6.2	6.5	6.5	5.5	< 5	> 5.5

complex that can dock and fuse with a corresponding active SNARE core complex on another membrane (McBride *et al.*, 1999). In addition to syntaxin 13, additional SNARE proteins that are required for homotypic early endosomal fusion and likely to participate in formation of the early endosomal SNARE core complex are endobrevin, (also known as VAMP-8) (Antonin *et al.*, 2000b), vti1a (also known as Vti1-rp2) (Antonin *et al.*, 2000a), and syntaxin 4 (Ward *et al.*, 2000). For other membranes and other Rab-GTPases, different effector proteins and SNARE proteins are involved. In the case of late endosomes, the fusion promoting core complex consists of the R-SNARE, endobrevin (this SNARE is common to both early and late endosomes) and the three Q-SNAREs: syntaxin 7, vti1b, and syntaxin 8 (Antonin *et al.*, 2000a). At least two lysosomal SNARE proteins have been identified: fusion between late endosomes and lysosomes and homotypic fusion between lysosomes in human alveolar macrophages has been shown to require syntaxin 7 and VAMP-7 (Ward *et al.*, 2000).

Table 1 summarizes the distribution of Rab-GTPases and other important molecules and markers of the endocytic pathway. In the cases in which it has been reported, the presence or relative exclusion of the marker on *M. tuberculosis* phagosomes is also indicated.

Table 2 shows the roles of several known Rab effectors in the endocytic pathway. Some of these effectors are soluble and are recruited from the cytosol, and others are present on other membranes in addition to those for which they have been shown to play a physiological role. The symbols "+", "-", and "?" indicate whether an effector is known to be required, known not to be required, or not reported to be required, respectively, for homotypic fusion of the indicated compartment. Although the distribution of some of these markers on *M. bovis* BCG phagosomes has received some study (Fratti *et al.*, 2001, 2002) their distribution on phagosomes of virulent *M. tuberculosis* has not been reported.

Membrane Trafficking and Vacuoles Containing Intracellular Parasites

The preceding descriptions of membrane trafficking pertain to vacuoles containing inert particles. Intracellular parasites have been shown to subvert the host cell's normal membrane trafficking pathways to achieve intracellular compartments that are more hospitable for their growth and multiplication. After internalization by the host cell, intracellular parasites follow one of three general pathways inside the host cell (Horwitz, 1988): (1) the extraphagosomal pathway, in which the pathogen lyses the phagosomal membrane and resides freely in the host cell cytoplasm, (2) the phagolysosomal pathway, in which the pathogen resides and multiplies in an acidified phagosome that does fuse

Table 2. Rab effectors

	Early Endosome	Late Endosome	Lysosome
PIPK and PI(3)P	+	?	?
EEA1	+	-	-
syntaxin 4	+	-	-
syntaxin 13	+	-	-
vti1a (Vti1-rp2)	+	-	-
endobrevin (VAMP-8)	+	+	?
syntaxin 8	-	+	?
vti1b	-	+	?
syntaxin 7	-	+	+
VAMP-7	-	-	+

with lysosomes, and (3) a phagosomal pathway in which the pathogen resides in a phagosome that does not fuse with lysosomes. Examples of intracellular pathogens that have been reported to follow the extraphagosomal pathway by lysing their phagosome include *Trypanosoma cruzi* (Tanowitz *et al.*, 1975; Noguiera and Cohn, 1976), *Listeria monocytogenes* (Gaillard *et al.*, 1987), *Shigellae* (Sansonetti *et al.*, 1986; Clerc *et al.*, 1987), and some species of *Rickettsia* (Winkler, 1990). Examples of pathogens that reside in acidified phagolysosomes include *Coxiella burnetii* (Burton *et al.*, 1971) and *Leishmania amazonensis* (Antoine *et al.*, 1990). Pathogens that have been shown to reside in phagosomes that do not fuse with lysosomes include *Legionella pneumophila* (Horwitz, 1983) and *Chlamydia psittaci* (Wyrick and Brownridge, 1978). Within these three general pathways, individual parasites exhibit additional unique variations that can be distinguished ultrastructurally and in terms of the interactions of the parasite or its vacuole with other host cell organelles. These unique variations that characterize the intracellular compartment and its host cell interactions provide the fundamental basis for understanding the cell biology and pathogenetic mechanisms of intracellular parasites.

As noted above, even within the three general pathways of intracellular parasites, there is considerable variation that makes each pathogen's pathway unique. For example, among intracellular parasites that reside in phagosomes that do not fuse with lysosomes, there is considerable diversity of membrane trafficking interactions with the host cell (Sinai and Joiner, 1997; Clemens, 1996). Although *L. pneumophila* and *M. tuberculosis* both reside in phagosomes that are only mildly acidified and that do not fuse with lysosomes, their membrane trafficking patterns are markedly different (Clemens and Horwitz, 1995; Clemens *et al.*, 2000a). Whereas *L. pneumophila* never acquires markers of the endocytic pathway, *M. tuberculosis* shows an arrested maturation at a stage with early endosomal markers, such as transferrin, transferrin receptor and Rab5 (Table 1) (Xu *et al.*, 1994; Clemens and Horwitz, 1995; Clemens, 1996; Clemens and Horwitz, 1996; Clemens *et al.*, 2000a). Thus, whereas the *M. tuberculosis* phagosome resides in an arrested maturation along the endocytic pathway, the *L. pneumophila* phagosome appears to reside in a pathway that does not interact with the endocytic pathway.

This review will describe the methods that have been used to study the intracellular pathway of *M. tuberculosis*; it will examine the strengths and weaknesses of these methods, and it will describe some promising new methods of examining membrane trafficking.

Selection and Verification of a Model System

Choice of a Host Cell

Ideally, the host cell to be studied should be selected on the basis of its properties and its biological relevance rather than on the basis of convenience or habit. Different species and different cell types have different properties and interact differently with *M. tuberculosis*. Humans are naturally susceptible to *M. tuberculosis* and the bacterium replicates primarily within macrophages. Mice are relatively resistant to *M. tuberculosis* and their macrophages are also less permissive to mycobacterial growth. This difference between mouse and human macrophages in permissiveness to mycobacterial growth may reflect the facility with which mouse macrophages, but not human macrophages, generate reactive nitrogen intermediates (Chan *et al.*, 1992; Arias *et al.*, 1997).

One could argue that the most biologically relevant host cell for the study of tuberculosis is the alveolar macrophage. These cells can be obtained by bronchioalveolar lavage and several laboratories have studied their interactions with *M. tuberculosis*. Unfortunately, obtaining alveolar macrophages exposes human subjects to significantly more discomfort and risk of injury than is posed by a simple blood draw. A close approximation to the alveolar macrophage is the human peripheral blood monocyte and monocyte derived macrophage. As peripheral blood monocytes are progenitors of tissue resident macrophages (including alveolar macrophages and macrophages of liver and spleen that are natural host cells in human tuberculosis), they are clearly a biologically relevant cell type. A variety of different cell lines have also been studied in models of *M. tuberculosis* infection. Cell lines offer convenience and unique properties – particularly cell lines that are defective in various properties or that have been engineered to express specific genes, but their properties may differ from those of natural host cells. The human monocyte-like cell lines, THP-1, HL-60, and U937, have been shown to support the growth of *M. tuberculosis* and have been used as models to study the host-pathogen interaction (Lee and Horwitz, 1995). Mouse J774 macrophage-like cell line has also been used as a model cell to study *M. tuberculosis* and

M. bovis BCG (Gatfield and Pieters, 2000). As noted above, cell lines may have different properties than primary monocytes and macrophages. For example, none of the cell lines listed above has functional mannose receptors, and the J774 cell line lacks the mannose-6-phosphate receptor. Deficiencies such as these can either be fortuitous or detrimental to the model system. If cell lines are studied, then it is important to attempt to verify that the conclusions drawn from them also apply to more natural host macrophages, ultimately including the most relevant cell: human alveolar macrophages.

Whichever system is used to study the interaction between *M. tuberculosis* and its host cells, it is important to measure and document (a) the viability of the bacteria and (b) the intracellular growth of the bacteria.

Assessment of Bacterial Viability

Bacterial viability has been assessed by two methods: (1) comparison of Petroff-Hausser counts with CFUs obtained by plating and (2) evaluation of the bacteria with fluorescent probes that distinguish between dead and live bacteria. Comparison of bacterial counts and CFUs is straightforward, but has the disadvantage of requiring a delay of 2 - 3 weeks before one knows the viability of the bacteria that were used in an experiment. However, we have found that 7 – 8 day old colonies of *M. tuberculosis* scraped from recently prepared plates, consistently yield a mycobacterial viability of > 70%. Use of dual fluorescent probes, such as fluorescein diacetate (converted to fluorescent fluorescein by live bacteria) and ethidium homodimer (which stains damaged bacteria red) can be used to obtain an immediate assessment of bacterial viability (McDonough and Kress, 1995), but should be checked by the first method to assure the sensitivity of the fluorescent stains.

Following Intracellular Growth of *M. tuberculosis* Within Host Cells

M. tuberculosis is a pathogenic bacterium that causes disease by growing within host cells. Model systems in which the mycobacteria do not grow productively within the host cells may not reflect the pathogenic properties that are relevant to mycobacterial disease. The growth of *M. tuberculosis* within cell monolayers can be assessed by lysing the monolayer with hypotonic detergent at sequential times after infection and plating serial dilutions of the cell lysate and the culture medium supernatant (Hirsch *et al.*, 1993). We have found that prolonged exposure of the samples to hypotonic detergent decreases the mycobacterial viability; therefore the exposure of the mycobacteria to hypotonic detergent should be minimized by immediate dilution in plating buffer containing albumin (Tullius, Clemens, and Horwitz, unpublished data).

Ultrastructural Analysis of the Phagosome

Considerable information about the interaction between a parasite and its host cell can be obtained by standard electron microscopic ultrastructural analysis. Transmission electron microscopy (TEM) can provide information regarding integrity of the phagosomal membrane, interaction of the phagosome with other host cell organelles, and the degree of spaciousness of the phagosome after entry and as the bacteria multiply within the host cell. Ultrastructural analysis can also provide vital information regarding the host cell, as certain features are pathognomonic of apoptosis or necrosis.

In 1971, Armstrong and Hart reported ultrastructural analysis of *M. tuberculosis* infected cells by transmission electron microscopy. These investigators obtained excellent ultrastructural preservation and contrast by simultaneous fixation and staining of cells in osmium tetroxide and glutaraldehyde, post-staining with uranyl acetate, and embedding in epoxy resin. They observed that the *M. tuberculosis* bacteria were consistently within membrane bound vacuoles and that the phagosomal wall was relatively tightly juxtaposed to the bacteria (Armstrong and Hart, 1971). In most cases, only a single bacillus was observed per phagosome in the plane of the section, suggesting that the phagosome grows, divides, and is remodeled in concert with the bacteria. In contrast to the situation with *Legionella pneumophila* and *Toxoplasma gondii*, the mycobacterial phagosome has not been observed to associate with other host cell organelles, such as endoplasmic reticulum or mitochondria.

Artifacts and Shortcomings of Ultrastructural Analysis by Standard TEM

Examination of biological structures by standard TEM requires that the sample be fixed, dehydrated with organic solvents, infiltrated and embedded in a plastic resin, thin sectioned, and stained with electron dense metals to provide contrast. Each of these steps has a potential to introduce artifacts that alter the ultrastructural features of the sample. One of the most obvious artifacts seen in thin sections of resin embedded mycobacteria is the tendency of the resin to pull away from the edges of the mycobacteria and for the mycobacteria to literally pop out of the resin during thin sectioning. This has been attributed to exclusion of the resin from the thick, waxy coat of the mycobacteria (Armstrong and Hart, 1971).

Whereas some investigators have observed the *M. tuberculosis* phagosome to remain intact (Armstrong and Hart, 1971; Clemens and Horwitz, 1995), others have reported that a variable percentage of the mycobacteria escape into the cytoplasm or reside in phagosomes with incomplete membranes

(McDonough *et al.*, 1993; Myrvik *et al.*, 1984). The capacity of membranes to be stained and visualized by transmission EM varies with the composition of the membrane, the staining techniques employed, the extent to which membrane molecules are extracted during dehydration and infiltration, and the thickness of the section (thicker sections blur the membranes). With traditional TEM staining techniques, membranes with low protein content may be difficult to visualize. In this respect, negative staining of ultrathin cryosections can offer advantages for the staining of membranes: the sample is not exposed to organic solvents and the negative staining technique is well suited to visualize protein poor membranes.

The thick waxy coat of mycobacteria may also contribute to the difficulties in visualizing phagosomal membranes during standard TEM. Extraction and smearing of mycobacterial wax into adjacent areas of resin may obscure staining of phagosomal membranes. Regions of membranes will also be lost during thin sectioning if the section fractures and skips rather than cuts evenly at the mycobacterial cell wall/phagosome interface. Incomplete infiltration of the mycobacterial cell wall by resin increases the likelihood that the sample will fracture and skip out of the plane of the section. This artifact can yield an intact phagosomal membrane on one side of the mycobacterium and no discernible membrane on the other side.

Analysis of Interactions of the *M. tuberculosis* Phagosome With the Endosomal-Lysosomal Pathway

Transmission EM and Analysis of Phagosome-Lysosome Fusion

Interaction of the M. tuberculosis Phagosome with Secondary Lysosomes

Historically, the reduced capacity of the *M. tuberculosis* phagosome to interact with lysosomes was recognized before its interaction with other elements of the endocytic pathway was appreciated. This is because it is technically easier to label and study primary and secondary lysosomes than early and late endosomes. Lysosomes can be labeled by allowing macrophages to internalize electron dense material, such as ferritin (Armstrong and Hart, 1971), colloidal thorium dioxide (Sibley *et al.*, 1987), or colloidal gold (Clemens and Horwitz, 1995), followed by a chase period of 1 h or more to ensure that the majority of the electron dense material has left the early and late endosomes and has entered the lysosomal compartment. The capacity of a phagosomal compartment to fuse with secondary lysosomes can then be monitored by transmission electron microscopy by determining whether or not the electron dense marker enters the phagosomal compartment. Using this technique, it has been shown that live, morphologically intact *M. tuberculosis* (Armstrong and Hart, 1971; Clemens and Horwitz, 1995), and *M. leprae* (Sibley *et al.*, 1987) resist fusion with secondary lysosomes labeled with the electron dense probes. Phagosomes containing killed or morphologically damaged bacteria, on the other hand, fused extensively with the labeled lysosomes (Armstrong and Hart, 1971; Clemens and Horwitz, 1995; Sibley *et al.*, 1987).

Ferritin has an electron dense iron-laden core and a uniform size of 5.5 nm. However, it is not as electron dense as thorium dioxide or colloidal gold, and therefore can be more difficult to recognize if the samples have been stained too intensely with lead or osmium. Colloidal gold is readily prepared (Roth, 1983; Slot and Geuze, 1985; Slot *et al.*, 1988) or purchased in a range of monodisperse sizes (1 nm to 25 nm are most useful). Colloidal gold particles of 1 nm size can be distinguished in lightly stained sections, but are more easily visualized after silver intensification. (Note, however, that enhancement by silver intensification precludes the use of different sizes of gold in the same cells or the same section.) The colloidal gold can be prepared with a variety of different proteins coating the surface of the gold particles (Roth, 1983), thereby directing the endocytosis of the particles via different receptors. In addition to the above electron dense endocytic markers, horseradish peroxidase (HRP) has also been used frequently as an endocytic probe to label lysosomes. While HRP is not electron dense, its enzymatic activity can be developed with diaminobenzidine and hydrogen peroxide to yield an osmiphilic precipitate. For example, the gradual acquisition of HRP by maturing latex bead phagosomes has been demonstrated by Desjardins *et al.* (1994b).

The interplay between phagosomes and endocytic compartments can involve either sequential symmetric fusion/fission ("kiss and run") interactions (Desjardins, 1995) or asymmetric fusion (kiss and fuse) interactions. The *Coxiella burnetii* vacuole appears to engage in asymmetric fusion interactions with lysosomal compartments, thereby yielding enormous parasitophorous vacuoles that contain essentially all the lysosomal contents of the host cell. The *M. tuberculosis* phagosome, on the other hand, appears to engage in symmetric fusion/fission interactions with host cell endocytic compartments, thereby yielding a tightly fitting phagosome and allowing the host cell endocytic compartments to retain their structural integrity. While fusion of a secondary lysosome with a phagosome is expected to lead to transfer of solute from the secondary lysosome to the phagosome, the efficiency of this transfer will vary depending upon whether the interaction between the phagosome and the vesicle is asymmetric (kiss and fuse) or symmetric ("kiss and run"). Smaller electron dense markers will be transferred more efficiently than larger markers from vesicle to phagosome if the fusion pore is limited in size or if the duration of the fusion interaction is limited. Large clumps of particulate matter may not be able to pass through a fusion pore at all. Thus smaller endocytic markers may be more sensitive markers of symmetric fusion interactions between phagosomes and secondary lysosomes. Whichever endocytic probe is employed to label secondary lysosomes, it is important to include a control particle that is targeted to phagolysosomes (e.g. latex beads or killed bacteria) in order to confirm that the electron dense material is indeed transferred to phagolysosomes.

The M. tuberculosis Phagosome and Lysosomal Acid Phosphatase

Endocytic probes report fusion interactions between the phagosome and endocytic compartments. However, it is conceivable that a parasite vacuole could mature and acquire endosomal or lysosomal proteins and lysosomal characteristics by interaction with vesicles of the biosynthetic pathway or by interacting with primary lysosomes, without ever interacting with endosomes or secondary lysosomes, and therefore without acquiring endocytosed probes. Acid hydrolase enzymatic activity can be examined by cytochemical detection of electron dense products by transmission EM. Acid phosphatase cytochemistry has a venerable history in this regard, having been employed in the early electron microscopic descriptions of lysosomes (de Duve, 1963; Novikoff, 1963). At acidic pH, acid phosphatase catalyzes the hydrolysis of substrates such as beta-glycerophosphate and the enzymatic activity is localized by capturing the phosphate as the electron dense lead phosphate precipitate. Armstrong and Hart (1971) reported that acid phosphatase cytochemistry yielded similar results to those obtained with ferritin labeling of lysosomes (i.e. acid phosphatase was largely excluded from the *M. tuberculosis* phagosome), but found the results difficult to quantify because of inferior ultrastructure and membrane preservation associated with the acid phosphatase staining technique. Whereas Sibley *et al.* (1987) found that *M. leprae* phagosomes did not fuse with ferritin labeled lysosomes in mouse macrophages, Steinhoff *et al.* (1989) reported that some *M. leprae* containing phagosomes in mouse Schwann cells did acquire staining for the lysosomal enzyme acid phosphatase; however, the percentage of phagosomes acquiring staining was not quantitated and the electron micrographs presented do suggest heterogeneity. In studies of mouse bone marrow derived macrophages, Frehel *et al.* (1986) demonstrated that phagosomes containing killed *M. avium* acquired abundant staining with acid phosphatase, but that phagosomes containing live *M. avium* acquired only limited amounts of acid phosphatase staining.

Immunoelectron Microscopic Detection of Specific Molecules of the Endocytic Pathway

Examining the interaction of the *M. tuberculosis* phagosome with early and late endosomes has required the localization of specific probes and the use of immuno-electron microscopy techniques distinct from those employed with standard ultrastructural TEM. Available methods for examining the localization of specific antigens by electron microscopy are: pre-embedding immunostaining and post-embedding immunostaining of ultrathin resin embedded sections or of ultrathin cryosections.

Pre-embedding Immunolocalization

Antigens can be immunolocalized in infected cells either before embedding and sectioning (pre-embedding immunolocalization) or after sectioning (post-embedding immunolocalization). Pre-embedding immuno-localization requires relatively gentle fixation of the infected host cells, in order that the cells can be permeabilized to allow antibody penetration. Therefore fixation is accomplished with low concentrations and short durations of formaldehyde (e.g. 2% for 30 - 60 min). Formaldehyde is a monomeric fixative which does relatively little cross-linking and does very little harm to antigenicity. Relatively gentle permeabilization can be accomplished with low concentrations (0.05 - 0.1%) of detergents such as saponin, Tween-20, or Triton X-100. Antigens can be immunolocalized either by horseradish peroxidase-conjugated antibodies, or by nanogold-conjugated antibodies. The HRP is developed with diaminobenzidine/hydrogen peroxide followed by osmium tetroxide, and the immuno-nanogold can be facilitated by silver intensification. After the antibody staining steps are completed, considerably more vigorous fixation can be employed (osmium and glutaraldehyde) prior to dehydration and embedding in plastic resins.

Pre-embedding immunostaining has the potential advantage of allowing antigen-antibody interaction prior to antigen denaturation by more vigorous fixation and embedding. Another advantage is that both HRP and silver-intensified nanogold can be visualized both by light and by electron microscopy. Nevertheless, there are several important caveats regarding pre-embedding staining. First, it is important to include appropriate controls to insure that the sample is adequately permeabilized. We have observed that conditions sufficient to permeabilize the plasma membrane are often insufficient to permeabilize the *M. tuberculosis* phagosome. In other words, we have found that conditions of fixation and detergent or solvent permeabilization that allow excellent immunostaining of microtubules and early endosomal antigen (EEA1) within the host cell cytoplasm can be insufficient to allow immunostaining of the majority of *M. tuberculosis* bacteria. While this reinforces the concept that *M. tuberculosis* resides within a membrane bound phagosome, it also emphasizes the potential for invalid conclusions from pre-embedding immunostaining if appropriate controls are not included to insure that all compartments are accessible to antibody. A second caveat for the pre-embedding technique is that fixation and permeabilization conditions that are adequate to allow antibody-antigen interaction in the compartment of interest may cause solubilization and extraction of the antigens and loss of cellular ultrastructure. Because of these problems, pre-embedding immunostaining should not be the first choice for ultrastructural immunolocalization studies.

Post-embedding Immunolocalization

Post-embedding immunolocalization eliminates the above noted problems of incomplete permeabilization of the compartments of interest since all compartments are sliced open and are accessible to antibody. For the most part, post-embedding immunolocalization is done either in plastic resin embedded samples or in cryosections.

Post-embedding Immunolocalization on Resin Embedded Sections

While some immunolocalizations have been done successfully in samples embedded in the extremely hydrophobic resins (such as Epon or Araldite) that are traditionally used for ultrastructural analysis, these immunolocalizations tend to be successful only with antigens that are extremely abundant. It is likely that these resins denature protein antigens and bury the antigenic epitopes within the hydrophobic plastic matrix. As antibodies are hydrophilic molecules within an aqueous medium, they have little chance of penetrating the hydrophobic plastic resin and interacting with antigens present in the section.

Several plastic resins are compatible with immunodetection of host and parasite antigens. These include London Research (LR) White, LR Gold, and the Lowicryl series of resins (the polar Lowicryl K4M and non-polar Lowicryl HM20). LR Gold and the Lowicryl resins are polymerized by UV light at temperatures of –20°C or below. It is thought that this low temperature embedding helps to preserve water associated with polar molecules in the section, thereby minimizing denaturation and ultrastructural changes otherwise associated with embedding in resins that do not accept water. It is also likely that, by retaining some water in the sample, these resins are more porous and that the pores in the resin improve the accessibility of the antigenic epitopes to immunostaining.

The apolar Lowicryl HM20 does not accept water, whereas the K4M does tolerate water, however, Lowicryl HM20 can be polymerized at lower temperatures than K4M. Embedding in LR White is done by thermal curing at 58°C, which is more convenient than the UV polymerization at low temperatures. In my experience, the preservation of antigenicity with LR White embedding is equal or superior to that which I have achieved with the Lowicryl resins. Whereas the Lowicryl resins are multicomponent resins that must be mixed in the proper ratios, LR White resin is ready to use straight from the bottle. Nevertheless, LR White embedding still requires dehydration at least through 70% ethanol, which can still lead to denaturation and extraction of lipids from the sample. A deficiency of LR White is that membranes tend to be difficult to see with LR White embedding. Organelles are recognized primarily by contrast staining of their protein and nucleic acid content, rather than by staining of their lipids. Thus, phagosomal membranes are not apparent but are instead surmised by the interface of phagosomal space and cytoplasmic protein. Likewise, mitochondrial membranes are not seen, and the mitochondria are instead recognized by the characteristic staining pattern of proteins of the christae. Smooth ER is extremely difficult to visualize in LR White samples that are intended for immunostaining. Some membrane structure can be seen if the samples are stained with osmium tetroxide, but osmium tetroxide staining is not compatible with immunolabeling.

The low temperature embedding resins can accommodate the "freeze substitution" technique of cryopreservation, in which conventional chemical fixation is replaced by rapid freezing of the sample, followed by low temperature dehydration and embedding. Rapid freezing is accomplished by slamming the sample against a polished metal mirror that has been cooled with liquid propane. The frozen sample is immersed in an ultracold mixture of alcoholic uranyl acetate, infiltrated with the low temperature embedding resin, and polymerized by UV light. This method has yielded good ultrastructure and immunostaining in other systems, but the technique requires specialized equipment and has not yet been applied to the study of mycobacteria or mycobacteria infected macrophages.

Post-embedding Immunolocalization on Cryosections

In cryo-immunoelectron microscopy, the sample is embedded in ice rather than in a plastic resin. Whereas a plastic resin can bury epitopes and make them inaccessible to antibody, ice melts and many more antigenic epitopes are accessible to antibody in cryosections. In addition, since there is no dehydration in organic solvents or infiltration in plastic resin, denaturation of protein antigens and extraction of lipid from the sample is completely avoided.

The technique of immunoelectron microscopy of ultra thin frozen sections was pioneered by Tokuyasu, Geuze, Slot and co-workers (Geuze *et al.*, 1981; Tokuyasu, 1986). It has been applied to the study of mycobacterial trafficking by Russell and co-workers (Russell *et al.*, 1994; Sturgill-Koszycki *et al.*, 1994; Xu *et al.*, 1994) and by Clemens *et al.* (Clemens and Horwitz, 1995, 1996;

Clemens *et al.*, 2000a, 2000b). The technique of cryo-immunoelectron microscopy has been reviewed by Griffiths (1993) and by Russell (1994).

Fixation

When fixing samples for post-embedding immunoelectron microscopy, the strength of the fixative is determined by the sensitivity of the antigens of interest. In general, 2 – 4% formaldehyde for 1 – 2 h at room temperature is safe for all antigens and provides some ultrastructural preservation. Formaldehyde concentrations as high as 8%, and as long as 12 – 18 h have also been used successfully. Glutaraldehyde provides a greater degree of cross-linking, thereby providing better ultrastructural preservation. Therefore, whenever possible, some glutaraldehyde (e.g. 0.05 – 0.5%) should also be included in the fixative, in combination with 2 – 4% formaldehyde. Unfortunately, the antigenicity of many proteins is destroyed even with low concentrations and brief exposures to glutaraldehyde. On the other hand, some antigenic epitopes are very resistant to glutaraldehyde; typically these are epitopes that lack amino groups capable of reacting with aldehydes. For example, immunostaining of dinitrophenol-tagged molecules (e.g. the pH marker, DAMP, discussed below), digoxigenin-tagged molecules, and carbohydrate antigens (e.g. lipoarabinomannan) are unaffected by glutaraldehyde fixation, even when used at relatively high concentrations of 2%. Periodate-lysine-paraformaldehyde (PLP) fixation (McLean and Nakane, 1974) is an alternative method of cross-linking that avoids glutaraldehyde and is tolerated by most antigenic epitopes. However, it is possible that some carbohydrate epitopes might be harmed by the periodate oxidation.

Cryoprotection

After fixation, the monolayers can be scraped into PBS with 0.1% BSA (to minimize adherence of the cells to the wall of the centrifuge tube), centrifuged, resuspended and pelleted in warm 10% gelatin. The gelatin is solidified on ice, trimmed into small cubes (0.5 – 1 mm^3), and infiltrated with 2.3 M sucrose containing 20% polyvinylpyrolidone (PVP), pH 7 – 7.4. This cryoprotects the samples, preventing formation of ice crystals during freezing, and the PVP acts as a plasticizer to improve sectioning properties. The cryoprotected blocks are positioned on the heads of aluminum pegs and frozen in liquid nitrogen.

Cryosectioning and Section Pick-up

Cryosectioning is usually done at −100 to −120 °C with glass or diamond knives. Colder temperatures result in harder blocks, permitting thinner sections. Glass knives should be prepared by the technique of symmetric breaks (Griffiths *et al.*, 1983) and diamond knives should be those manufactured specifically for cryosectioning. The frozen sections are moved down, away from the edge of knife, and are collected on a drop of sucrose in a wire loop suspended over the sections. Section pick-up is a critical step. In the technique developed by Tokuyasu, frozen sections are collected on drops of 2.3 M sucrose. This has the advantage of stretching the sections and eliminating wrinkles that are introduced during sectioning. However, in some samples, even those that are strongly fixed, there can be considerable expansion artifact induced as a result of the stretching (Liou and Slot, 1994; Liou *et al.*, 1996). This problem can be alleviated by increasing the viscosity of the sucrose pick-up solution by including methylcellulose in the sucrose pick-up solution (Liou and Slot, 1994; Liou *et al.*, 1996). The disadvantage of increasing the viscosity of the pick-up solution is that the sections may be more wrinkled. Thawed sections on the bottom of the sucrose droplet are transferred to Formvar coated grids (with or without carbon coating) that have been freshly glow discharged. Glow discharging makes the surface of the grids more hydrophilic and decreases non-specific staining. The grids are placed face down on a wet plate of 2% gelatin on ice to allow the sucrose to diffuse away from the section gradually. The plate of gelatin is liquefied at 37°C for 5 – 10 min, and the grids are floated, section side down, on drops of blocking solution. If the sections are to be stained with protein A gold reagents, then the blocking solution contains albumin and fish gelatin. If the sections are to be stained with antibody-gold reagents, then the blocking solution can also contain 5 – 10% non-immune serum of the same species used to prepare the immunogold reagent. The grids are stained on drops of primary antibody, washed on drops of buffer, and stained with the colloidal gold reagent diluted in the same blocking buffer.

Protein A gold reagents have several advantages and a few disadvantages when compared with immunogold reagents. For rabbit primary antibodies, the staining tends to be more abundant than that observed with gold-labeled secondary antibodies. The use of the sequential protein A gold technique (Geuze *et al.*, 1981) allows localization of three consecutive antigens by rabbit primary antibodies in combination with 3 different sizes of protein A gold (PAG). Glutaraldehyde (0.5%) is used as a blocking step between the sequential immunogold labeling steps. (On ultrathin sections, treatment with 0.5% usually has no effect on antigenicity or antigen-antibody accessibility.) The same could be achieved with the secondary antibody technique only by using different species for the primary antibody. A disadvantage of PAG is that Protein A binds poorly to many mouse immunoglobulin isotypes and to goat IgG. Therefore, a secondary rabbit anti-mouse or anti-goat antibody must be used for protein A gold staining with these species of primary antibodies.

Even with cryosections, there are clearly steric factors that limit the immunostaining. Therefore, smaller colloidal gold particles typically give more abundant immunostaining than larger particles.

In immunostaining of mycobacterial infected host cells, it is important to be aware of the problem of anti-mycobacterial reactivity of antiserum raised with Freund's complete adjuvant. Alternative adjuvants (e.g. Titermax (Bennett *et al.*, 1992)) should be used instead. Unfortunately, many commercial polyclonal antibodies are prepared with complete Freund's adjuvant. In addition, some rabbits may have natural exposure to mycobacteria leading to humoral reactivity against mycobacterial antigens. In some cases the anti-mycobacterial reactivity can be overcome by adsorption of the antiserum to acetone treated *M. tuberculosis*. Affinity purification of antiserum is also a valuable method for assuring the specificity of polyclonal antisera. The protein antigen is resolved by polyacrylamide gel electrophoresis, transferred to polyvinylidene diflouride (PVDF), stained with Coomassie blue, and strips of PVDF bearing the band are trimmed and blocked with buffer containing albumin. Polyclonal antiserum diluted in blocking buffer is incubated with the strips and monospecific antibody is eluted with 100 mM glycine, pH 2.5, with 0.1% BSA carrier protein, immediately neutralized with 1.0 M Tris, pH 8, and dialyzed against PBS.

The Final Staining and Embedding of Cryosections

The staining of cryosections is achieved with a mixture of positive and negative staining by uranyl acetate that is included in the final embedding solution. This method provides good staining of membranes, but ribosomes and some fibrillar material may be obscured by the negative staining background (Tokuyasu, 1986).

Inert Particles as Internal Controls

In the evaluation of immunostaining of *M. tuberculosis* infected cells, it is useful to analyze and quantitate the immunostaining of phagolysosomes containing inert control particles. The control particles should be easily identified and should remain morphologically distinguishable or immunochemically identifiable for the required duration of the experiment. Dead bacteria are good control particles for short term experiments, but may be relatively rapidly degraded and not reliably identified at times longer than 2 h. Mycobacteria are an exception and at least some of them are morphologically recognizable for several days (Armstrong and Hart, 1971). Zymosan particles (autoclaved yeast cell walls) are readily phagocytosed via the mannose receptor, are targeted to phagolysosomes, and are only slowly digested by macrophages (Schlesinger, 1994). Latex beads are readily phagocytosed by macrophages and are morphologically distinguishable within macrophages indefinitely and have served as a model phagolysosomal compartment in many studies (Desjardins *et al.*, 1994a, 1994b; Garin *et al.*, 2001). While it is clear that latex bead phagosomes do mature to phagolysosomes, Chantal de Chastellier and colleagues (de Chastellier and Thilo, 1997) have expressed concern that these particles differ considerably in morphology and surface properties from bacterial pathogens. Therefore, control particles with size, shape, and surface properties similar to live mycobacteria should also be studied as controls, when feasible.

Immunostaining Controls

Isotypic control antibodies and preimmune or non-immune serum are essential controls to assure the specificity of the immunostaining results and to evaluate the degree of non-specific staining. In some cases, biological controls are the most ideal. For example, when studying the distribution of an epitope-tagged recombinant protein in transduced or transfected host cells, the parental cells that do not express the epitope-tagged protein are the ideal negative immunostaining control.

Overview of Post-embedding Immunoelectron Microscopy

For most purposes, ultrathin cryosections offers superior preservation of antigenicity, better accessibility of antibody to antigen, and equal or better ultrastructural preservation than is achieved with plastic resin embedding techniques. However, it has the disadvantage of requiring more specialized equipment and requiring liquid nitrogen for storage of the samples. For some purposes, resin embedded sections may be preferable because the rigid support provided by the resin eliminates the risk of spatial distortions that can occur during sample pick-up of cryosections. Because of the differences in staining and embedding methods, whereas membranes are more easily visualized on ultrathin cryosections, fibrillar material (e.g. actin filaments) are more easily visualized in resin embedded sections.

Examination of the Interaction of the M. tuberculosis Phagosome with Early and Late Endosomes by Post-Embedding Immunoelectron Microscopy

Early endosomes can be identified by exogenously added endocytic probes either kinetically, by fixing cells within 5 min of endocytosis, or, more conveniently, by an endocytosed probe that is confined to early endosomes, such as transferrin. Transferrin is internalized by receptor-mediated endocytosis and traffics to early endosomes, but not to late endosomes or lysosomes. The demonstration of delivery of endocytosed transferrin to *M. tuberculosis* phagosomes by immunoelectron microscopy (Figure 1) revealed the capacity of these phagosomes to interact with early endosomes (Clemens and Horwitz, 1996).

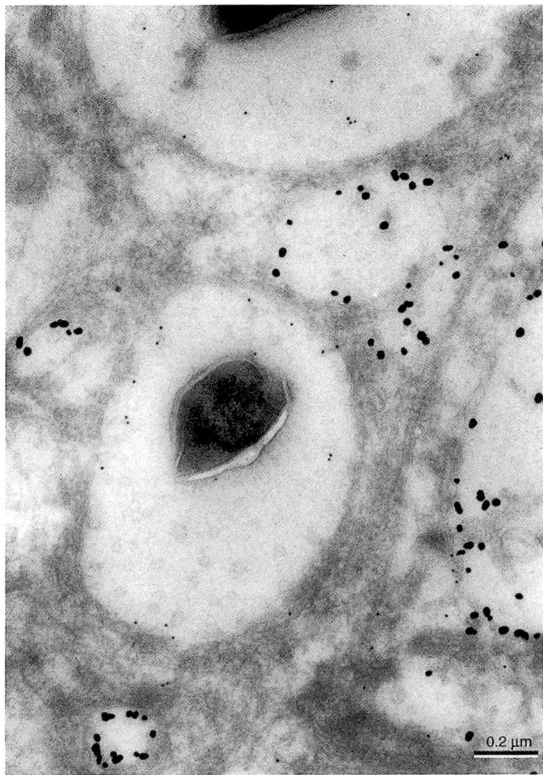

Figure 1. Cryosection immunogold staining for transferrin in a human macrophage infected with *M. tuberculosis*. Human monocyte derived macrophages were infected for 3 d in medium lacking human transferrin. washed with media lacking transferrin, and incubated for 1 h with culture medium containing human transferrin, fixed, and processed for cryo-immunoelectron microscopy. Lysosomes were prelabeled with 20 nm-BSA gold followed by an overnight chase before the addition of transferrin. The *M. tuberculosis* phagosome has acquired abundant transferrin immunogold staining (10 nm gold particles) but has not acquired any of the larger BSA-gold particles used to label lysosomes. In addition, the transferrin immunogold staining is absent from the BSA-gold labeled compartments.

The capacity of *M. tuberculosis* and *M. avium* phagosomes to interact with the endocytic pathway has also been investigated by Russell *et al.* (1996). These authors demonstrated by immunoelectron microscopy that cholera toxin B subunit that had bound to GM1 gangliosides on the surface of *M. tuberculosis* and *M. avium* infected mouse macrophages subsequently was internalized and within 5 minutes was detected within the mycobacterial vacuoles. This time course is consistent with delivery from early endosomes. Treatment with brefeldin A did not prevent the delivery of cholera toxin to the mycobacterial vacuoles, further supporting the concept that delivery of the cholera toxin was by interaction with the endocytic pathway rather than by traffic of the toxin-bound gangliosides through the Golgi apparatus (Russell *et al.*, 1996). Similarly, Russell *et al.* (1994) have biotinylated the surface proteins of macrophages, and demonstrated by immunogold staining of cryosections that the biotinylated proteins traffic to *M. avium* phagosomes.

The acquisition of other numerous endogenous markers of the endocytic pathway by *M. tuberculosis* phagosomes in mouse (Xu *et al.*, 1994) and human cells (Clemens, 1996; Clemens and Horwitz, 1996, 1995; Clemens *et al.*, 2000a, 2000b) has been examined by immunogold staining of ultrathin cryosections and the results are summarized in Table 1 above (see also Figure 2).

Analysis of Post-Phagocytic Events by Light Microscopy

While light microscopy will never offer the level of ultrastructural detail provided by electron microscopy and immunoelectron microscopy, it offers several advantages over electron microscopy in the evaluation of post-phagocytic events: (1st) the capacity to observe live cells, (2nd) a greater sensitivity for many antigens, (3rd) facile examination of a larger sample size, and (4th) a more facile examination of the cells in 3 dimensions. Whereas standard TEM of thin sections is restricted to the plane of the individual sections and therefore may miss structures and interactions that are out of the plane of the sections; light microscopy readily views the entire 3-dimensional phagosome/host cell interaction. For example, Rab-GTPases and Rab-effectors have been shown to accumulate in patches on endosomal structures (Sonnichsen *et al.*, 2000). Therefore, a thin section through a section may entirely miss these patches of the Rab protein and their effectors. Immunofluorescence microscopy, on the other hand, allows a visualization of the patchy distribution of these proteins (Sonnichsen *et al.*, 2000). The capacity of light microscopy to capture the entire thickness of the cell, rather than a single plane, is also advantageous in evaluating the interaction of a parasite vacuole with the host cell cytoskeletal apparatus.

The fluorescence microscopy field has evolved rapidly in recent years, both in terms of the number of new fluorescent probes for cellular organelles

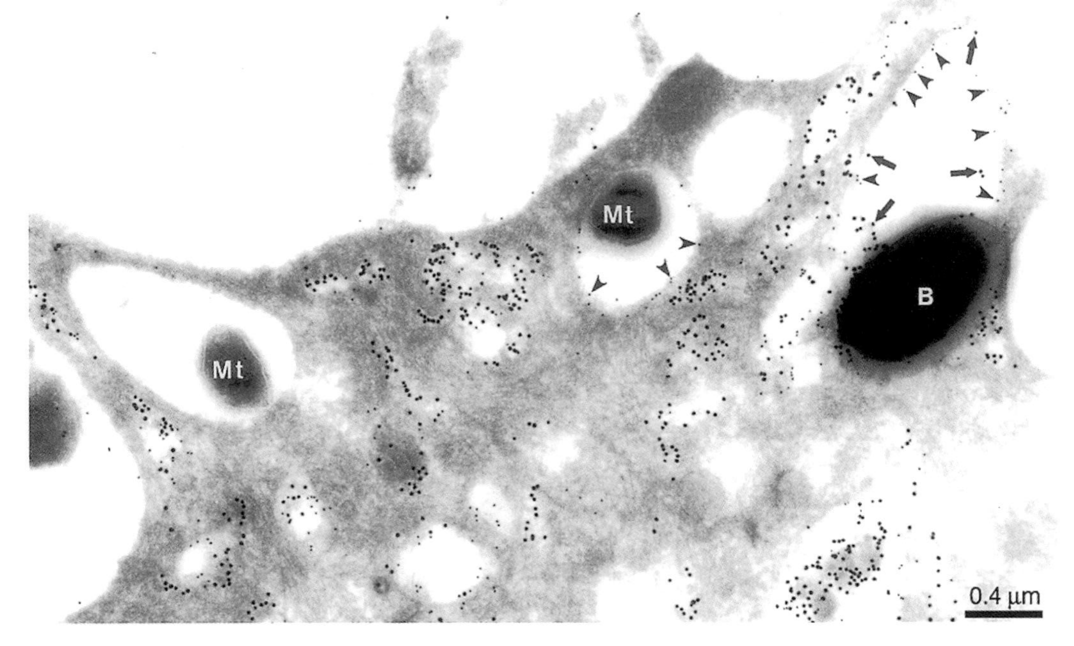

Figure 2. Cryosection immunogold staining for the lysosome associated membrane glycoprotein, CD63, in human monocytes infected with *M. tuberculosis*. Human monocytes were cultured for 5 days and lysosomes were labeled with large BSA-gold particles. The monocytes were coincubated with 1 µm latex beads and live *M. tuberculosis* and fixed and processed for cryo-immunoelectron microscopy one day after infection. The cryosection is stained for CD63, which is absent from one *M. tuberculosis* (Mt) phagosome and present at an intermediate level in a second phagosome (arrowheads). The latex bead (B) phagosome has abundant staining for CD63 (arrowheads) and has also acquired the large BSA-gold particles (arrows) that were used to prelabel lysosomes.

and cellular processes, and in terms of the microscopy equipment used to visualize the interaction of these probes with fixed and living cells. Fluorescence microscopy methods include (1) conventional wide-field epifluorescence microscopy, (2) confocal laser scanning microscopy, (3) multiphoton fluorescence microscopy, (4) fluorescence microscopy with image deconvolution, and (5) fluorescence video microscopy of living cells by each of the preceding 4 microscopy methods. More detailed reviews of these various techniques are available (Inoué, 1995; Inoué and Spring, 1997; McNally *et al.*, 1999; Straub *et al.*, 2000; White *et al.*, 2001). A brief description of each of these microscopic techniques follows:

Conventional Wide-field Epifluorescence Microscopy

In conventional wide-field epifluorescence microscopy, light from a source such as a mercury or xenon lamp passes through the objective lens to illuminate the entire field of view of the specimen. Light emitted by excited fluorescent molecules in the specimen is focused and the image magnified by the microscope objective and ocular lenses. A major limitation of conventional wide-field epifluorescence microscopy is out-of-focus blur from regions of the specimen above and below the focal plane. Image deconvolution, confocal scanning laser microscopy, and multiphoton fluorescence microscopy are techniques that address this problem.

Fluorescence Microscopy with Image Deconvolution

Although image deconvolution can be applied to any digitized images, in biological applications this technique is most often applied to images collected with a standard wide-field fluorescence microscope with a sensitive charge coupled device (CCD) camera. The digital images are processed by computer algorithms that take into account the laws of wave optics and the fact that an infinitely small point of light will be focused by a lens (even a perfect lens) into a 3-dimensional diffraction pattern or point spread function (PSF). The algorithms process the data from the raw image with theoretical or empirical mathematical models of the 3-dimensional PSF to remove blur and to provide an estimate of the original structure (the deconvoluted image). This technique works well with thin, fixed samples but less well with living cells, in which movement within the plane of the section complicates successful deconvolution. Because the deconvolution algorithms make assumptions about the underlying structure (e.g. smoothness), there is a danger that some features of the structure will be artifactually removed. In addition, the algorithms can cause the disappearance of dim features and cause bright features to expand beyond their actual size (Wallace *et al.*, 2001). In colocalization studies, these artifacts could cause false negatives and false positives. Nonetheless, this technique can provide much greater contrast and resolution than conventional wide-field epifluorescence microscopy and is comparable to that which is achieved with confocal laser scanning microscopy.

Confocal Laser Scanning Microscopy

With confocal laser scanning microscopy (CLSM) an excitatory laser beam is focused through the microscope objective lens onto the specimen and is scanned over the specimen. The fluorescence signal from the specimen returns via the same objective and is filtered through a pinhole aperture before passing to a photodetector. The pinhole is in a conjugate focal plane to the focal plane being observed in the specimen. Therefore, fluorescence from the point being illuminated passes unobstructed through the pinhole, but fluorescence from above or below the plane of focus will be spread out and will, predominantly, be blocked by the pinhole aperture. Thus CLSM can be used to collect optical sections, free from out-of-focus blur, from fluorescent probes distributed within either fixed or living cells. CLSM provides greater contrast and resolution than is achieved with conventional wide-field fluorescence microscopes. With CLSM, 3-D views of the cells can be reconstructed from a series of sections collected at defined increments in focal plane. A disadvantage of CLSM is that the intense laser light tends to cause even more photobleaching and phototoxicity to the specimen than wide-field fluorescence microscopy.

Multiphoton Fluorescence Microscopy

Multiphoton fluorescence microscopy is a relatively new technique that uses a laser and a scanning microscope system very similar to that of CLSM, except that much longer wavelengths of light are selected and the pinhole aperture is omitted. Multiphoton fluorescence microscopy offers several advantages in the study of living cells (White *et al.*, 2001). Wide-field fluorescence microscopy and CLSM excite the entire volume of the living cell at the wavelength at which the fluorescent probe is maximally excited, which can cause quenching of the fluorescent probe and phototoxicity to the cell. MFM only excites in the region of interest, since the unfocused photons are of too low an energy to excite the fluorophore. Multiphoton fluorescence microscopy focuses on the region of interest a beam of photons of lower energy (longer wavelength) than that needed to excite the fluorescent probe. Only in the region of interest where the beam is focused is the photon density high enough that fluorescent probe molecules can be excited simultaneously by two photons of lower energy. In areas above and below the region of interest, the photon density is too low for excitation, thereby minimizing bleaching and phototoxicity. Since the problem of out of focus blur is eliminated by the absence of excitation in planes above or below the plane of focus, the pin-hole aperture is eliminated, thus allowing the detection system to collect more of the excited light.

To summarize these techniques, the main problem with standard wide-field fluorescence microscopy is out of focus blur caused by excitation of fluorescent molecules above and below the region of interest. The other three imaging techniques employ different strategies to remove or avoid the out of focus blur: Image deconvolution uses computer programs to remove the blur from digitized images; CLSM uses a pinhole aperture to eliminate the majority of light emitted from planes that are not conjugate to the plane of focus in the specimen; and mutliphoton fluorescence microscopy uses photons of lower energy that only excite molecules in the plane of focus in the specimen.

Fluorescence Video Microscopy of Living Cells

Fluorescence video microscopy of living cells can be employed with any of the above fluorescence microscopy methods and allows the investigator to follow and record membrane trafficking events in living cells. The capacity to view processes in live cells allows unique observations that cannot be made on fixed cells, including the direct observation of vesicle-phagosome fusion events (including determining whether the interactions are of the "kiss and fuse" or the "kiss and run" variety), and observing the relative motility of a phagosome and adjacent organelles. The use of quantitative fluorescence video microscopy is essential to the ratio fluorescence measurements to determine phagosomal pH (discussed below) and to follow intracellular calcium levels (Malik *et al.*, 2000, 2001). Fluorescence video microscopy has the potential to be particularly powerful when combined with endogenous expression of GFP-tagged proteins and mutant-GFP-tagged proteins, as this allows non-invasive studies of the trafficking of virtually any host protein in living cells infected with *M. tuberculosis*. To date, however, very few video microscopy studies have been reported with virulent *M. tuberculosis* and most of the studies have involved less pathogenic mycobacteria.

In addition to recording events in living cells, video microscopy allows image enhancement and amplification of fixed and living cells so that objects that are too faint to be seen by eye or recorded on film can be observed and recorded electronically (Inoué and Spring, 1997).

Immunofluorescence Microscopy Protocols and Applications

Many different protocols have been published for immunofluorescence examination of host cells infected with mycobacterial and other pathogens. A valuable resource for fluorescent and immunofluorescent cell biology experiments is the Molecular Probes "Handbook of Fluorescent Probes and Research Chemicals" (Haugland, 1996).

Fluorescent Probes of the Endocytic Pathway

The endocytic pathway can be labeled readily by the use of fluorescent tracers that are added to the cell culture medium. Fluorescent-labeled transferrin can be used to label early endosomes, and fluorescent labeled dextrans can be employed in pulse-chase fashion to label the endocytic pathway. Fluorescent-dextrans are not fixed or immobilized by aldehyde fixation or by fixation in organic solvents, therefore, if the host cells are to be permeabilized prior to viewing, then the lysine-fixable forms of the fluorescent-dextrans should be used, otherwise the fluorescent probe will be lost from the cell with permeabilization.

Immunofluorescent Staining of Fixed Cells

The techniques employed for immunofluorescent staining of fixed cells are similar to those described above for pre-embedding staining for immunoelectron microscopy. However, since ultrastructure will not be viewed, the permeabilization is usually harsher. (It is instructive to examine, by TEM, cells that have been fixed and permeabilized under light microscopy conditions and to observe how little ultrastructure remains!)

Fixatives for Immunofluorescence

Fixation techniques for immunofluorescence microscopy are markedly different from those employed for electron microscopy. Protein cross-linking that is employed for preservation of ultrastructure in EM is not desired in fluorescence microscopy, as it would prevent adequate permeabilization of the cell. Whereas it is desirable to maintain membrane ultrastructure for EM, with immunofluorescence it is vital to permeabilize and destroy membranes to ensure that all compartments are accessible to antibody immunostaining.

Organic Solvents

Water miscible organic solvents, such as acetone and methanol, fix antigens by precipitation and simultaneously permeabilize the cells by destroying membrane structure. Fixation in acetone or methanol preserves the antigenicity of most proteins, even to monoclonal antibodies. In addition, in our experience, the fluorescence of GFP and GFP-fusion proteins is not reduced by fixation in acetone or methanol. A low temperature ($-20°C$) is usually employed to reduce the extraction of cellular components. Employing an air-drying step after immersion of the monolayer in acetone or methanol can assist further in disrupting membranes and permeabilizing the cell.

Formaldehyde

Formaldehyde is commonly used in combination with detergent permeabilization for immunofluorescence microscopy. For example, 2% formaldehyde in 0.075 M sodium phosphate for 30 – 60 min, followed by permeabilization for 10 min with 0.1% Triton X-100 in PBS, or 0.05% saponin in PBS containing 0.2% BSA (McCafferey and Farquhar, 1995). In general, glutaraldehyde is not employed as a fixative in fluorescence microscopy because it causes autofluorescence and because it cross-links proteins so well that adequate permeabilization of the sample to antibodies is difficult.

Periodate-lysine-paraformaldehyde

This has also been used in combination with detergent permeabilization for immunofluorescence (e.g. Roy *et al.*, 1998). This fixative may reduce lateral mobility of proteins in membranes and decrease the degree to which some antigens are extracted after permeabilization. However, because it provides superior cross-linking and immobilization of proteins, there may also be a greater danger that some compartments will not be adequately permeabilized and that antigens within them will be inaccessible to antibody. No matter which fixation and permeabilization technique is employed, it is vitally important to verify that the permeabilization of relevant intracellular compartments has been adequate. The immunostaining pattern obtained with aldehyde fixation and detergent permeabilization can be compared to that obtained with acetone or methanol fixation/permeabilization. To assess whether lysosomal compartments have been adequately permeabilized, one should verify that immunostaining of lysosomal antigens (e.g. cathepsin D, LAMP-1, or LAMP-2) reveals the expected tubulovesicular fluorescent pattern. If one wishes to draw conclusions regarding antigens within *M. tuberculosis* or *M. bovis* BCG phagosomes, then it is important to verify that these phagosomes have been adequately permeabilized to immunostaining. One method is to determine whether GFP-expressing bacteria can be stained by rhodamine- or Texas red-conjugated antibodies (using either direct or indirect antibody staining). Unfortunately, controls such as these are not often performed, and, many protocols, particularly those that involve formaldehyde fixation combined with detergent permeabilization, do not permeabilize the majority of *M. tuberculosis* or *M. bovis* BCG phagosomes, although they do permeabilize the plasma membrane.

Quenching, Blocking, and Immunostaining

Following aldehyde fixation, any remaining reactive aldehyde groups should be quenched by incubation for 10 min at room temperature either with 10 mM glycine in PBS or 50 mM ammonium chloride in PBS. Non-specific antibody binding sites are blocked by incubation in non-immune serum of the species that will be used for the secondary antibody steps (e.g. 5% goat or donkey serum in PBS). Primary and secondary antibodies should be diluted (usually to 0.2 – 10 µg/ml of IgG) in buffer containing BSA (e.g. PBS with 0.2% BSA). If problems with non-specific staining are encountered then the non-immune blocking serum can also be included.

When monolayers are grown on cover slips, it is convenient to conduct the fixation, blocking, washing, and staining procedures in 24 well tissue culture plates. However, if the antibody solution is expensive or scarce, then it is desirable to employ a staining technique that requires a smaller volume. Following the fixation, permeabilization, quenching, and washing steps, blot the back of the cover slips on tissue or filter paper and then float the cover slips section side down on drops of blocking solution and then on drops of antibody solution (30 – 50 µl is sufficient). The drops of blocking buffer or antibody solution are placed on sheets of paraffin secured to glass plates. The antibody incubations are done in a humid chamber to prevent evaporation and drying of the blocking or antibody solutions. Simultaneous or sequential double immunofluorescent staining is possible by using primary antibodies from different species (e.g. rat and mouse) followed by secondary antibodies conjugated to different fluorescent probes (e.g. fluorescein conjugated goat anti-mouse IgG and rhodamine conjugated goat anti-rabbit IgG). Cover slips are washed in buffer between staining steps and after a final wash, are mounted on glass slides with a mounting medium. A mixture of glycerol: PBS (3: 1) containing 1 mg/ml p-phenylenediamine can be prepared as an anti-bleaching mounting medium. The cover slip can be secured by coating the edges with clear nail polish. Several commercial mounting media, e.g. Molecular Probes Antifade Mounting Medium, are also available and are even more effective at reducing photobleaching during viewing. Antifade Mounting Medium solidifies in 30 – 60 min, thus eliminating the nail polish step.

Fluorescence Microscopy of Cells and Bacteria Expressing Recombinant Green Fluorescent Protein

Analysis of either fixed or living host cells and mycobacteria expressing green fluorescent protein (GFP) and mutant forms of GFP represents a powerful technique for the study of host parasite interaction. Multiple enhanced GFP constructs have been prepared and are available commercially as vectors that can be used to prepare fusion proteins. The cDNA of essentially any gene of interest can be fused to the coding sequence of GFP (or a mutant GFP of a different color) and expressed within the host cells (or within mycobacteria), thereby allowing the distribution of the particular protein to be followed. Even dual-color fluorescence imaging can be done with GFP constructs of different colors (Ellenberg *et al.*, 1999). At present, the combination of cyan-fluorescent protein (CFP) and yellow-fluorescent protein (YFP), are the most suitable for dual-color imaging in living cells (Ellenberg *et al.*, 1999). For example, Sonnichsen *et al.* (2000) prepared fusion constructs of various Rab-GTPases and EGFP, ECFP, and EYFP and studied the localization of these fusion constructs in living cells by fluorescence video microscopy using a sensitive charge coupled device (CCD) camera and filters that allowed for simultaneous detection of either CFP and YFP (expressed as recombinant fusion proteins) or GFP and endocytosed rhodamine-conjugated probes.

Simultaneous expression of mutant GFP-fusion proteins of different colors also has the potential of allowing a determination of whether the two proteins interact on a scale of 1 – 10 nm by using the technique of fluorescence resonance energy transfer (FRET; Kenworthy, 2001). If two different fluors, such as CFP and YFP, are within 10 nm, then excitation of the CFP will lead to energy transfer to the YFP, quenching of the CFP fluorescence, and emission of photons at the (longer) YFP emission wavelength. If the two fluors do not interact at a distance of <10 nm, then this phenomenon will not occur. Whereas immunogold staining might be able to demonstrate molecular interactions of this sort at the electron microscope level, in practice, steric interactions may prevent dual immunogold staining of molecules that are this close.

Pitfalls and Limitations of the Use of GFP-tagged Proteins

GFP has a mass of 27 kDa and therefore could readily interfere with the biological function of the tagged protein. Thus, it is important to verify that the GFP-tagged protein faithfully reflects the distribution of the native protein and to verify that the expression of the GFP-tagged protein does not disturb the physiology of the host cell. It is not essential that the GFP-tagged protein retain biological activity, as long as expression of the fusion protein does not alter the host cell physiology. For example, Zerial *et al.* demonstrated that a non-functional truncated syntaxin 13-GFP fusion protein (containing the transmembrane portion of syntaxin 13) reflected faithfully the distribution of the native protein and did not interfere with host cell membrane trafficking. On the other hand, whereas low level stable expression of a TGN38-GFP fusion protein in a homologous cell line allowed correct functional localization of the tagged protein, nonhomologous high level expression caused a disruption of intracellular morphology and mislocalization of the tagged protein (Girotti and Banting, 1996). The position of the GFP-tag can also determine whether fusion protein is functional and whether it localizes correctly.

An intriguing alternative to GFP-tagged fusion proteins is the use of a tetracysteine fusion-tag. The cell permeable fluor, FLASH-EDT$_2$, specifically and covalently binds to fusion proteins bearing the tetra-cysteine tag (Griffin *et al.*, 1998). The FLASH-EDT$_2$ only becomes fluorescent when bound to the tetracysteine moiety, thereby increasing the specificity and decreasing the background fluorescence.

While not useful for fluorescence visualization of living cells, non-fluorescent tags can be made much smaller than GFP and can obviate the need for developing new antibodies for every protein of interest in immunofluorescence and immunoelectron microscopy studies. Many such fusion-tags can be localized by commercially available antibodies, e.g. influenza hemagglutinin, myc, and the FLAG epitope. However, myc should not be used in human cell lines such as THP-1, U937, or HeLa, because of the high level of endogenous expression of the myc antigen by these tumor cell lines.

Immunofluorescence and Immunoelectron Microscopy of Cells Overexpressing Recombinant Proteins Involved in the Endocytic and Phagocytic Pathways

Expression of recombinant proteins involved in the endocytic pathway serves two important purposes: 1) overexpression of the molecules enables the visualization of molecules that are otherwise below the detection limit (Clemens *et al.*, 2000a, 2000b); and 2) expression of dominant positive or dominant negative mutants allows studies of the function of these proteins in the membrane trafficking of the phagosome. With regard to the first purpose, it is important to verify that the overexpression of the wild-type protein does not alter the phenotype of the phagosome and the overexpressed protein is specific in its association with the phagosome. Internal and external positive and negative controls should be employed to verify that the localization of the molecule is specific. For example, overexpressed wild-type Rab5 and Rab7 should not be observed on mitochondria or on nuclear membranes (Clemens *et al.*, 2000a, 2000b).

Immunofluorescence, immunoelectron microscopy, and analysis of isolated phagosomes can readily indicate the presence of selected molecules on the *M. tuberculosis* phagosome. However, it is possible that some molecules that are present on the phagosome at levels below their detection limit are nonetheless functionally important in mediating membrane trafficking events. Conversely, some proteins that are present on the phagosome may have no functional role on the phagosome. Therefore the studies of distribution of various molecules must be complemented by functional studies of these molecules. In the case of the Rab-GTPases, dominant positive and dominant negative mutants have been identified. The functional role of selected molecules can also be addressed by microinjection or electroporation of antibodies to the protein, by the use of antisense oligonucleotides, and by genetic knock-out mutants. In the case of the Rab-GTPases, for example, although we observe that *M. tuberculosis* phagosomes in HeLa cells acquire Rab7, we found no change in the phenotype of the *M. tuberculosis* phagosome in cells expressing the dominant positive Rab7 (Clemens *et al.*, 2000b), suggesting that the Rab7 is not functional on the phagosome. Thus, although the phagosome has receptors that allow recruitment of Rab7, it lacks functional effectors for this Rab-GTPase. In contrast, expression of the dominant positive Rab5 in HeLa cells infected with *M. tuberculosis* causes dramatic changes in the phenotype of the *M. tuberculosis* phagosome, confirming that the phagosome has both receptors that allow the recruitment of the Rab5 and functional effectors for this Rab-GTPase (Clemens *et al.*, 2000a).

Analysis of Phagosomal pH

Fluorescence Methods

Acidification is an important process of phagosomal maturation, and there is a consensus that *M. tuberculosis* and other mycobacterial species prevent the acidification of their phagosomes. The pH of the phagosomal compartment has been evaluated both by immunoelectron microscopy and by a variety of fluorescence microscopy techniques.

Fluorescent Lysosomotropic Probes

Numerous cell permeant lysosomotropic fluorescent dyes are available for visualization of acidified intracellular compartments in living cells. These are weak bases that accumulate within acidified compartments and include acridine orange and neutral red. The probe is added to the culture medium and incubated for 30 seconds to a few minutes prior to viewing by fluorescence microscopy. Presence or absence of the dye from a compartment can be used to judge whether or not the compartment is acidified. For example, acridine orange has been shown to be excluded from the *M. tuberculosis* phagosome (Gordon *et al.*, 1980). Newer lysosomotropic dyes, such as the LysoTracker and LysoSensor series of fluorescent probes developed by Molecular Probes (Haugland, 1996) may have greater specificity as well as being available in a variety of different colors and in forms that can be fixed by aldehydes, allowing the fluorescent staining to be viewed in live or in fixed cells. For example, Deretic and co-workers (Via *et al.*, 1998) have demonstrated that the LysoTracker Red fluorescent dye is excluded from GFP-expressing *M. bovis* BCG phagosomes in unactivated mouse macrophages, but does colocalize with the BCG in murine macrophage activated with IFN-gamma and LPS.

Evaluation of pH using lysosomotropic probes tends to be all or none – i.e., acidified or not acidified, and exact pH measurements are not possible. However, it has been noted that the pH of acridine orange shifts to a longer wavelength when it stacks in high local concentration. Thus, red or orange vesicles are more acidic than yellow ones. The LysoSensor fluorescent probes have pH-dependent spectral emission properties and thus may also provide a qualitative assessment of the degree of acidification (Haugland, 1996).

Measurement of pH by Fluorescence Ratio Imaging

A more quantitative assessment of the pH of the phagosomal compartment can be obtained by covalently coupling fluorescein to the bacteria and using the technique of fluorescence ratio imaging. This technique has been applied to *M. avium* (Sturgill-Koszycki *et al.*, 1994; Oh and Straubinger, 1996) and to other bacterial pathogens (Horwitz and Maxfield, 1984), but has not yet been applied to *M. tuberculosis*, possibly because of biohazard problems. The technique relies upon the fact that the fluorescence emission spectrum of fluorescein is sensitive to pH. If fluorescein is excited with light at 450 nm, the emission intensity is independent of pH, but if the excitation is at 495 nm, the emission intensity of fluorescein is pH sensitive. Therefore, the pH of the environment of a fluorescein-tagged particle can be determined by measuring the ratio of the fluorescein emission intensity from the particle at 450 nm and 495 nm (provided that an appropriate calibration curve has been constructed). Using the ratio of emission intensities at the two excitation wavelengths automatically corrects for the concentration of fluorescein and can provide accurate assessment of the pH in the microenvironment of the fluorescein probe (Ohkuma and Poole, 1978; Schlesinger, 1994). An improvement on this method is dual labeling of the bacteria with a pH sensitive fluor (fluorescein) and a pH insensitive fluor (rhodamine) (Oh and Staubinger, 1996). This overcomes the problem with the single ratio method that fluorescein's emission intensity is not very strong at the pHs at which it is pH insensitive.

Measurements of the pH of phagosomal compartments by fluorescence ratio imaging have been made both by microfluorimetry on individual phagosomes, recording and quantitating by video fluorescence microscopy (Horwitz and Maxfield, 1984; Oh and Staubinger, 1996), and by spectrofluorimetry on populations of infected cells (Sturgill-Koszycki *et al.*, 1994; Schlesinger, 1994). Valuable information has been obtained using both techniques. Spectrofluorimetry has the advantage of being relatively inexpensive and utilizes equipment that is widely available, but it also has some potential disadvantages. First, it analyzes a population of bacteria rather than individual bacteria. Thus, bacteria that are extracellular or fragments of bacteria will be averaged into the signal. Second, the level of heterogeneity of the sample cannot be assessed. In other words, while the average pH may be found to be 6.0, it is possible that this is due to 50% of the bacteria being at pH 7 and the remaining 50% being at pH 5. In this case, reporting the phagosomal pH as 6 does not provide an accurate picture. Third, in order to maximize signal in spectrofluorimetry, it may be necessary to use a higher multiplicity of infection (MOI) than is required with microfluorimetry. The higher MOI may cause greater toxicity to the host cells, even killing some host cells. A fourth problem that is more likely to complicate measurement of pH by spectrofluorimetry than microfluorimetry is dissociation of the fluorescent dye from the bacterium and subsequent traffic of the dye into vesicles separate from the microbe. Dissociated dye will contribute to the fluorescence signal measurements with spectrofluorimetry, but this problem can be ignored by microfluorimetry by not measuring the pH of small vesicles and compartments that do not appear to contain bacteria. The disadvantage of microfluorimetry, apart from the greater cost and specialization of the equipment, is that one must examine a large number of bacteria in order to assure that the data are representative. Nevertheless, many studies have obtained sufficient valuable data from microfluorimetry, indicating that this task, though laborious, is not insurmountable.

There is danger that the function or viability of a microbe may be altered by the fluorescent labeling reactions. Fortunately, commercially available fluorescein and rhodamine derivatives are available that react at neutral or slightly alkaline pH with the amino groups on the surface of the bacteria and in many studies have not appreciably damaged the bacteria. Nevertheless, it is important to verify that the fluorescently labeled bacteria retain viability and virulence.

Whereas the vital staining techniques can be used to study the acidification of phagosomes at any time after infection, the use of the fluorescence ratio technique has usually been restricted to times within a few hours of infection, because of the problem of fluorescent dye dissociating from the bacteria. A problem with both the ratio fluorimetric method and vital fluorescent stains that are not reliably fixed by aldehyde fixation is that these methods can only be used with living cells. With *M. tuberculosis*, this can introduce logistical biohazard problems if the fluorescence microscopy apparatus is not in a biohazard room.

Measurement of Intraphagosomal pH by DAMP and Immunoelectron Microscopy

Anderson and Orci developed DAMP (3-[2,4-dinitroanilino]3' amino-N-methyldipropylamine) for use in immunoelectron microscopy as a lysosomotropic agent that could be used in postembedding immunoelectron microscopy to assess the pH of intracellular compartments (Anderson *et al.*, 1984; Orci *et al.*, 1986; Anderson and Orci, 1988). DAMP is a weak base that accumulates in acidic compartments, can be fixed by aldehydes, and contains a highly antigenic dinitrophenol group that can be immunolocalized by monoclonal or polyclonal antibodies. Crowle *et al..* employed this technique to demonstrate that *M. tuberculosis* and *M. avium* phagosomes in human macrophages are not acidified (Crowle *et al.*, 1991). Although these authors used the technique qualitatively, the DAMP technique can be used to calculate pH quantitatively by using a neutral pH compartment (e.g. the nucleus) as a reference and counting the number of gold particles per unit cross sectional area in the phagosome and in the neutral pH compartment (Orci *et al.*, 1986). As with vital fluorescent stains, the combination of DAMP and immunoelectron microscopy can be used to measure the pH of the phagosome at any time after infection. The technique allows the correlation of phagosomal pH with the ultrastructural morphology of the mycobacterium and the morphology of the phagosome. The use of dual-label immunogold staining also allows the pH of the phagosome to be correlated with the levels of other immunogold stained antigens of interest.

Controls in Evaluation of Phagosomal pH

With each of these methods of measurement of phagosomal pH, it is important to employ positive and negative controls. Positive controls include inert particles that are targeted to acidified phagolysosomes, e.g. FITC-zymosan (Schlesinger, 1994), or latex beads, or lysosomes pre-labeled with BSA-gold. An important "negative" control with all of the above techniques is to demonstrate that labeling of acidified compartments is disrupted with agents such as monensin, chloroquine, or ammonium chloride (Schlesinger, 1994; Anderson *et al.*, 1984, Orci *et al.*, 1986).

Analysis of Isolated Phagosomes

Immunocytochemical techniques are best suited for detection of relatively abundant antigens and these techniques may not be able to determine whether certain relatively scarce regulatory molecules are present or absent on the *M. tuberculosis* phagosome. While this problem can sometimes be solved by overexpressing the molecule of interest, there is always the danger that overexpression may alter the properties of the phagosome. In addition, immunofluorescence and immuoelectron microscopy have a very limited capacity to determine whether or not any post-translational modification are present on proteins present on the *M. tuberculosis* phagosome. Biochemical analysis of isolated phagosomes is a powerful technique that can complement the immunocytochemical and ultrastructural analysis of the *M. tuberculosis* phagosome and, in many cases, can detect molecules that are too scarce to be detected by the preceding immunomicroscopic methods. For example, Rab-GTPases are present in relatively scarce amounts in cells and often must be overexpressed in order to be detected reliably. However, very sensitive radiolabelling with γ-P^{32} GTP overlays of 2-dimensional Western blots has allowed their detection in isolated *M. bovis* BCG phagosomes (Via *et al.*, 1997; Deretic *et al.*, 1997; Scianimanico *et al.*, 1997). In addition, biochemical analysis of isolated phagosomes allows a determination of the biophysical properties of the molecules on the phagosome and can readily detect post-translational modifications of the proteins associated with the phagosome. For example, in their SDS-PAGE/Western blot analysis of isolated *M. avium* vacuoles, Ullrich *et al.* (1999) demonstrated that the vacuoles contained cathepsin D in its immature procathepsin D form and suggest that the vacuole may have acquired the enzyme by direct interaction with the biosynthetic pathway. [However, the presence of procathepsin D in the phagosome may indicate the maturational arrest of the phagosome at a minimally acidified stage that interacts with early endosomes, rather than direct interaction with biosynthetic vesicles from the TGN. It has been reported that cathepsin D is transiently present in early endosomes and that it traffics rapidly through early endosomes before it accumulates in late endosomes and lysosomes (Ludwig *et al.*, 1991; Press *et al.*, 1998). Consistent with this, 2-dimensional proteomic analysis of isolated latex bead phagosomes revealed that they acquire procathepsin D before they acquire cathepsin D (Garin *et al.*, 2001).] More recently, Fratti *et al.* (2002) have made the intriguing observation that cellubrevin (VAMP-3) exists in a truncated form in isolated *M. bovis* BCG phagosomes.

Techniques for Isolation of Phagosomes

Several different techniques have been published for the isolation of phagosomes of mycobacteria and of other bacterial pathogens. To date, none of these techniques have been applied to the isolation of phagosomes of fully virulent *M. tuberculosis,* possibly because of biohazard considerations. Russell and co-workers (Chakraborty *et al.*, 1994; Sturgill-Koszycki *et al.*, 1994) developed a technique for isolation of *M. avium* phagosomes from mouse bone marrow macrophages employing a low speed centrifugation to remove debris, a sucrose density centrifugation to pellet the vacuoles, and filtration through a 3 μm filter. Luhrmann and Haas (2001) have also recently published a detailed procedure for discontinuous sucrose density gradient isolation of a variety of bacterial phagosomes. Ramachandra *et al.* (2001) isolated *M. tuberculosis* H37Ra phagosomes from IFN-γ activated mouse bone marrow macrophages by a combination of differential centrifugation, percoll density gradient centrifugation, and flouresecence activated organelle sorting. Hasan *et al.* (1997) have reported the use of an electrophoretic technique to separate *M. bovis* BCG phagosomes from other subcellular organelles. Biochemical analysis of the purified mycobacterial phagosomes revealed the absence of the endosomal/lysosomal markers LAMP-1 and beta-hexosaminidase on phagosomes of live, but not killed *M. bovis* BCG. A very promising new technique is the purification of bacterial phagosomes by flow cytometry and fluorimetric sorting. Technical advances have led to high speed sorters that now allow sufficient quantities of phagosomes to be sorted based on their fluorescence. Thus, phagosomes containing GFP expressing bacteria can be sorted and subjected to analysis by 2-D gel electrophoresis and mass spectrometry (Meresse *et al.*, 1999; Ashcroft and Lopez, 2000). Unfortunately, fluorescence activated cell sorters are relatively expensive ($250,000) and may not be available in biohazard facilities. Therefore some logistical problems may arise in applying this technique to *M. tuberculosis* phagosomes. However, closed systems for FACS are being developed and could help alleviate this problem.

Latex Bead Phagosomes as Model Phagosomes

Whichever method is employed to isolate mycobacterial phagosomes, it may be useful to have control phagolysosomes as a reference. Phagosomes containing latex beads have been used extensively to study the composition and membrane trafficking events of phagosomes of inert particles. Indeed, biochemical analysis of isolated latex bead phagosomes has led to many advances in our understanding of the membrane trafficking interactions of maturing phagosomes (Desjardins *et al.*, 1994a., 1994b; Desjardins, 1995; Garin *et al.*, 2001). As control particles, latex beads have many theoretical and practical advantages: they are of uniform physical properties, they are physically stable and are not degraded by the macrophage as the phagosome matures, they contribute no extraneous proteins to the phagosome: any proteins found in the phagosomal preparation are attributable to the macrophage (or to proteins present in the culture medium that adsorbed to the latex bead). Isolation of magnetic latex bead control particles by the use of magnets has been described (Sturgill-Koszycki *et al.*, 1994; Chakraborty *et al.*, 1994). Isolation of non-magnetic latex beads in pure form by sucrose density gradient centrifugation has been described by Desjardins and co-workers (Desjardins *et al.*, 1994a, 1994b). By virtue of their very low density, latex beads and the phagosomes that contain them float above all other subcellular organelles on sucrose density step gradients. (Desjardins *et al.*, 1994a, 1994b).

Analysis of Proteins of Isolated Phagosomes

To examine the host proteins present on isolated bacterial phagosomes, investigators have labeled host proteins by metabolic incorporation of [^{35}S]-methionine; the proteins present on isolated bacterial phagosomes have been resolved by 2-dimensional gel electrophoresis and visualized by autoradiography, as described by Russell and co-workers (Chakraborty *et al.*, 1994; Sturgill-Koszycki *et al.*, 1994, 1997; Scianimanico *et al.*, 1997). Selected host proteins have also been detected on isolated phagosomes after resolution of the proteins by 1 or 2-dimensional gel electrophoresis and Western immunoblotting. Rab-GTPases have been identified in blots of phagosomal proteins separated by 2-D gel electrophoresis by sensitive radiolabeled-GTP overlay and autoradiography (Via *et al.*, 1997; Deretic *et al.*, 1997; Scianimanico *et al.*, 1997).

The total protein profile of isolated phagosomes can be evaluated by silver staining, but this will inevitably reveal a mixture of host and bacterial proteins. Russell's laboratory (Chakraborty *et al.*, 1994) has described a method for selectively solubilizing the macrophage constituents from *M. avium* phagosomes with 0.03% Nonidet P-40 for 15 min on ice. The bacilli are then removed from the extract by centrifugation at 6000 *g* for 10 min (Chakraborty *et al.*, 1994). This procedure clearly has some limitations: while the extraction does not lyse mycobacteria and thus will not extract cytoplasmic mycobacterial proteins, it will extract mycobacterial secreted proteins and possibly some cell wall associated proteins. Gentle detergent extractions also may not extract all host proteins from the phagosomal membrane. A brute force approach is now feasible with proteomic techniques (Garin *et al.*, 2001): i.e. the complete protein profile of a 2-D gel can be visualized by silver staining and the proteins on replicate 2-D gels can be analysed and sequenced by mass spectroscopy and their identity determined by comparison with the *M. tuberculosis* and human genome data bases.

Lipids present in the isolated *M. avium* phagosomes have been analysed by thin layer chromatography (TLC), as described by Russell *et al.* (1996). To date, similar analysis of *M. tuberculosis* phagosomal lipids has not been reported. In addition to analysis by TLC, a more detailed analysis of lipids present in isolated phagosomes should be possible using techniques such as gas chromatography/mass spectroscopy and high pressure liquid chromatography (HPLC) with electrospray ionization tandem mass spectroscopy.

Analysis of Isolated Phagosomes by Organelle Flow Cytometry and Immune Lymphocyte Response

Ramachandran *et al.* (1999, 2001) have pioneered an intriguing technique for analysis of isolated phagosomes by a combination of organelle flow cytometry and immune lymphocyte response. These authors covalently labeled *M. tuberculosis* H37Ra with a fluorescent protein modification reagent, fed the bacteria to interferon-gamma activated mouse macrophages, and isolated phagosomes bearing these fluorescent bacteria by differential centrifugation and Percoll density gradient centrifugation. The phagosomes were fixed, permeabilized with saponin, and immunostained for major histocompatability molecules and other molecules of interest with fluorescent antibodies of a different color. The isolated, stained phagosomes were then analyzed by dual-fluorescence flow organellometry. With this technique, Ramachandran *et al.*

(2001) demonstrated that the isolated phagosomes have all the molecules required for formation of MHC II – antigen peptide complexes. In addition, they demonstrated that such complexes are indeed formed within the *M. tuberculosis* phagosomes by probing these phagosomes with HLA-restricted immune lymphocytes that responded to mycobacterial antigens.

The flow organellometry technique has a clear advantage over standard protein analysis of isolated phagosomes in that the entire spectrum of individual phagosomes can be appreciated. In addition, dual-color flow organellometry data may be influenced much less by contamination of the phagosomal preparation by other extraneous organelles. However, some false negatives may be obtained if the isolated phagosomal preparation that is analyzed by flow organellometry contains naked mycobacteria (free of any phagosome). Western blotting techniques and analysis of total protein of isolated phagosomes yields an assessment of the average phagosome; this "average phagosome" will include any contaminating organelles and extreme/unusual cases of mycobacterial phagosomes that are present in the preparation. The "extreme/ unusual phagosome" category can include phagosomes that have been internalized into autophagosomes and phagosomes from *M. tuberculosis* infected cells that have been phagocytosed by other host cells in the monolayer. Such cases are readily excluded (or analysed as a separate category) in immunoelectron microscopy, but will be included in the analysis of the average *M. tuberculosis* phagosome in isolated phagosomal preparations. *M. tuberculosis* phagosomes are heterogeneous, just as infecting inocula of *M. tuberculosis* generally has some heterogeneity. It is not credible that LAMP's or acid hydrolases or vacuolar proton pumps would be completely excluded from the phagosomal preparation, since there are always some phagosomes that mature to phagolysosomes, and there are always some phagosomes that are acidified.

TEM Analysis of Isolated Phagosomes

Isolated phagosomes should be examined by standard TEM in order to demonstrate that they are indeed mycobacterial phagosomes and in order to estimate the degree of contamination of the preparation. In addition, electron microscopic techniques have the potential to provide additional valuable information about the isolated phagosomes. For example, freeze etch microscopy has been applied to the study of isolated *M. avium* phagosomes to demonstrate that the surface of *M. avium* phagosomes is smooth and devoid of the peg-like structures that represent the vacuolar proton pump (Sturgill-Koszycki *et al.*, 1994). To date, similar studies have not been reported with the *M. tuberculosis* phagosome.

The arrangement of antigens on a phagosome is information that is lost in thin sections through the phagosome, which convert the 3-dimensional membrane into a thin line. Thus, whether or not particular antigens are arranged together in domains, or patches is information that is lost in immunogold staining of thin sections. It should be possible to stain the surface of isolated phagosomes (whole mounted and fixed on EM grids) by immunogold staining to determine whether domains (patches) of Rab-GTPases and their effectors are present on the phagosomal membrane in a fashion similar to what has been described with isolated preparations of early endosomes (Sonnichsen *et al.*, 2000).

Freeze fracture microscopy is ideal for examining the structure of membranes and this technique might provide information regarding the presence or absence of pores in the *M. tuberculosis* phagosomal membrane. It is likely that images provided by this technique, if applied to infected host cells, would be exceedingly difficult to interpret. Freeze fracture images of isolated phagosomal preparations, on the other hand, would be much more likely to yield information about the membrane structure.

Quality Control of Isolated Phagosome Preparations

It is essential that the purity of the isolated phagosomal preparation be assessed. As noted above, the quality of the preparation should be evaluated by transmission electron microscopy. Russell and co-workers (Chakraborty *et al.*, 1994) have employed electron microscopy and have also described a method of quantifying and characterizing the contamination of isolated phagosomal preparations. Radiolabeled infected macrophages are mixed with uninfected cold macrophages and non-radiolabeled infected macrophages are mixed with radiolabeled uninfected macrophages. The two mixed cell populations are processed in parallel and the presence of radiolabel from uninfected macrophages in the mycobacterial phagosome preparation is an indicator of contamination. The severity of the contamination can be assessed by comparing the amount of contaminating radioactivity with the total radioactivity recovered in the phagosomal preparation that included radiolabeled infected cells. The identity of the contaminants can be determined by 2-dimensional gel electrophoresis and autoradiography (Chakraborty *et al.*, 1994).

Advantages, Problems, and Limitations of the Different Approaches

Resolution

Standard TEM can readily resolve structures separated by 1 nm, and this degree of resolution is more than adequate to follow membrane trafficking events. Under ideal circumstances, the resolution in fluorescence microscopy is determined by the wavelength of the emitted light, the numerical aperture (NA) of the objective lens, and the refractive indices (RI) of the mounting and immersion media (Inoué and Spring, 1997). For fluorescein (emitting at 520 nm), an NA 1.4 oil immersion lens, and a mounting/immersion medium with an RI of 1.5, the microscope can resolve structures separated by 0.23 μm in the x-y dimensions and 0.8 μm in the axial (z) dimension. [The minimum resolvable distance in the z dimension is greater than that in the x and y dimensions because of the shape of the 3-dimensional diffraction pattern (or point spread function, PSF) of an infinitely small point of light by a perfect optical lens (Inoué and Spring, 1997).] Resolution is compromised further both by the microscope and by the specimen. Spherical aberration of the lens, vibration of the equipment, instability of the illuminating beam are features of the microscope that can degrade resolution. Differences in refractive index of the specimen will scatter light and degrade the resolution. Differences in refractive index occur at the following interfaces: lens/immersion medium, immersion medium/coverslip, coverslip/mounting medium, mounting medium/ plasma membrane, and plasma membrane/cytosol. Light is also scattered and resolution degraded by structures with different RIs within cells, such as granules and lipid droplets. In the examination of living cells, the movement of the organelles themselves within the cell can cause blurring of the image and loss of resolution. In addition, observations of living cells may require more rapid scanning and binning of data in order to obtain a greater speed of data collection and reduced phototoxicity. The resolution achieved in practice with CLSM of living cells is reported to be on the order of 1.2 μm in the x, y, and z dimensions (Fricker and Meyer, 2001). Whereas the electron microscope has resolving power to waste in following membrane trafficking interactions, fluorescence microscopy methods operate at the limit of their resolving power to obtain meaningful data.

Both immunoelectron microscopy and fluorescence microscopy allow the examination of individual phagosomes and allow an assessment of phagosomal heterogeneity. However, with immunoelectron microscopy, it can be determined readily whether a phagosome contains a morphologically intact mycobacterium or a partially degraded bacterium. With immunofluorescence microscopy, on the other hand, this distinction cannot readily be made. Likewise, other ultrastructural features of the phagosome – such as the presence of internal membranes or tubular extension – can be made at the EM level but not at the fluorescence microscope level.

Sensitivity

Among the TEM immunostaining techniques, cryosection immunogold staining usually provides the most intense immunostaining because of the excellent preservation of antigenicity and the great accessibility of antigen to antibody. Post embedding and pre-embedding immunostaining of resin embedded samples usually have lower levels of immunostaining because of problems with antigen accessibility and, in the case of post-embedding immunostaining, there may also be more loss of the antigenicity due to the dehydration and embedding. As long as permeabilization is adequate and does not cause extraction of the antigen of interest, immunofluorescence tends to be much more sensitive than immunoelectron microscopy. Antigen signals that are undetectable or barely detectable by immunogold staining are often easily observed by immunofluorescence. In these cases, it is possible that the only way to observe the signal at the EM level is to use recombinant DNA techniques to overexpress the antigen of interest.

Fluorescence microscopy can also have an advantage over cryosection immunogold staining in visualization of subdomains, or patches of antigens. Such subdomains may be missed entirely in a thin section through a phagosome, but would be appreciated in fluorescence microscopy which captures the entire cell. However, ultimately such ultrastructural subdomains may be best visualized and documented by immunogold staining and electron microscopy of whole mounts of isolated phagosomal preparations.

Biochemical and proteomic analysis of isolated phagosomes has even greater sensitivity to detect particular molecules on phagosomes than either immuno-EM or immunofluorescence. First, whereas immuno-EM can only examine thin sections through phagosomes and immunofluorescence can only examine individual phagosomes, biochemical analysis of isolated phagosomes can benefit from scaling-up of the phagosomal preparation to whatever number is required to obtain the desired signal. Second, extremely sensitive immunochemical and autoradiographic methods, as well as tandem mass-spectroscopy, are available to analyze electrophoretically separated phagosomal proteins.

Sample Size, Sampling Error, and Appreciation of Sample Heterogeneity

With TEM, there is a danger of analyzing too few cells, or even of analyzing the same cells multiple times. (To avoid this, it is important to obtain sections from different cell blocks or to do additional trimming of a cell block to ensure that different cell populations are being sampled.) It is usually easier to examine a larger number of infected cells by immunofluorescence than by electron microscopy. Both immuno-EM and immunofluorescence can provide information about the heterogeneity of phagosomes and the distribution of antigen staining in the phagosomal population. For example, both of these methods can determine whether the phagosomes consist of two populations: one with no LAMP and one with extremely rich LAMP staining. Biochemical analysis of isolated phagosomes, on the other hand, will provide a picture of the average phagosome with no information about how many, if any, phagosomes actually look like the average phagosome. Likewise, both immunofluorescence and immuno-EM can evaluate phagosomes by dual-labeling and can determine to what extent particular antigens do or do not co-localize. For example, whereas immunofluorescence and immuno-EM can determine that the phagosomes with transferrin receptor are not the phagosomes with LAMP staining, biochemical analysis of isolated phagosomes can reveal the presence of both antigens but cannot identify that these two antigens are present on separate populations of phagosomes. It should be noted however, that the dual-color flow organellometry of isolated phagosomes (Ramachandra *et al.*, 2001) does have the capacity to evaluate the heterogeneity of phagosomal populations and to assess the extent to which markers do or do not colocalize.

One potential problem with the study of isolated phagosomes is the possibility of sample alteration during the isolation procedure. For example, soluble Rab-GTPase effectors, such as EEA1, may dissociate from early endosomes during the preparation and processing of the endosomes if the Rab proteins are not maintained in a GTP-bound state. In addition, Rab proteins themselves could be extracted from the membrane by chaperon proteins if they are in their GDP-bound form, as opposed to their GTP-bound form. Inclusion of GTP and a nucleotide triphosphate regenerating system in the buffer solutions during processing may help to preserve the distribution of Rab-proteins and their effectors on endosomes (Sonnichsen *et al.*, 2000). There is also the danger of proteolytic processing of the phagosomal proteins by acid hydrolases liberated from broken lysosomes and phagolysosomes. This process can be minimized by reduced temperature and the inclusion of protease inhibitor cocktails, but it is difficult to eliminate proteolysis completely, and some proteins are much more susceptible to proteolysis than others.

Summary

Each of the techniques described - standard TEM, immuno-electron microscopy, immunofluorescence microscopy, and the analysis of isolated phagosomes - has advantages and disadvantages when compared with the other techniques. Electron microscopy and immunoelectron microscopy offers excellent ultrastructural resolution but cannot analyze living cells. Immunofluorescence microscopy is ideal for examining living cells and visualizing 3-dimensional interactions within the cells, but has a limited capacity for assessing ultrastructure. Biochemical analysis of isolated phagosomes has far greater sensitivity than immunofluorescence or immuno-EM in evaluating the composition of the phagosome and offers the capacity of evaluating post-translational modification of proteins, but has the disadvantage of averaging the phagosomes together rather than evaluating distinct populations of phagosomes. In addition, analysis of isolated phagosomes can have problems with contamination from other organelles and modification of phagosomal content due to loss of molecules during sample preparation, problems that are not significant concerns with immuno-fluorescence and immunoelectron microscopy.

Each of the techniques has the capability of complementing the shortcomings of other techniques. Thus, it is important that findings made by one technique be verified independently with other techniques in order to arrive at a complete picture of the membrane trafficking events involving the *M. tuberculosis* phagosome.

Acknowledgments

This work was supported by research grant AI-35275 from the National Institutes of Health.

References

Anderson, R.G., and Orci, L. 1988. A view of acidic intracellular compartments. J. Cell Biol. 106: 539-543.

Anderson, R.G., Falck, J.R., Goldstein, J.L., and Brown, M.S. 1984. Visualization of acidic organelles in intact cells by electron microscopy. Proc. Natl. Acad. Sci. USA 81: 4838-4842.

Antoine, J.-C., Prina, E., Jouanne, C., and Bongrand, P. 1990. Parasitophorous vacuoles of *Leishmania amazonensis*-infected macrophages maintain an acidic pH. Infect. Immun. 58: 779-787.

Antonin, W., Holroyd, C., Fasshauer, D., Pabst, S., von Mollard, G.F., and Jahn, R. 2000a. A SNARE complex mediating fusion of late endosomes defines conserved properties of SNARE structure and function. EMBO (Eur. Mol. Biol. Organ.) J. 19: 6453-6464.

Antonin, W., Holroyd, C., Tikkanen, R., Honing, S., and Jahn, R. 2000b. The R-SNARE endobrevin/VAMP-8 mediates homotypic fusion of early and late endosomes. Mol. Biol. Cell 11: 3289-3298.

Arias, M., Zabaleta, J., Rodriguez, J.I., Rojas, M., Paris, S.C., and Garcia, L.F. 1997. Failure to induce nitric oxide production by human monocyte-derived macrophages. Manipulation of biochemical pathways. Allergol. Immunopathol. (Madr) 25: 280-288.

Armstrong, J.A., and Hart, P.D. 1971. Response of cultured macrophages to *Mycobacterium tuberculosis* with observations on fusion of lysosomes with phagosomes. J. Exp. Med. 134: 713-740.

Ashcroft, R.G., and Lopez, P.A. 2000. Commercial high speed machines open new oppurtunities in high throughput flow cytometry. J. Immunol. Methods 243: 13-24.

Barbieri, M., Li, G., Colombo, M., and Stahl, P. 1994. Rab5, an early acting endosomal GTPase, supports in vitro endosome fusion without GTP hydrolysis. J. Biol. Chem. 269: 18720-18722.

Bennett, B., Check, I.J., Olsen, M.R., and Hunter, R.L. 1992. A comparison of commercially available adjuvants for use in research. J. Immunol. Methods. 153: 31-40.

Bottger, G., Nagelkerken, B., and van der Sluijs, P. 1996. Rab4 and Rab7 define distinct nonoverlapping endosomal compartments. J. Biol. Chem. 271: 29191-29197.

Bucci, C., Lutcke, A., Steele-Mortimer, O., Olkkonen, V., Dupree, P., Chiariello, M., Bruni, C., Simons, K., and Zerial, M. 1995. Cooperative regulation of endocytosis by three Rab5 isoforms. FEBS Lett. 366: 65-71.

Bucci, C., Parton, R.G., Mather, I.H., Stunnenberg, H., Simons, K., Hoflack, B., and Zerial, M. 1992. The small GTPase rab5 functions as a regulatory factor in the early endocytic pathway. Cell 70: 715-728.

Burton, P.R., Kordova, N., and Paretsky, D. 1971. Electron microscopic studies of the rickettsia *Coxiella burnetii*: entry, lysosomal response, and fate of rickettsial DNA in L-cells. Can. J. Micro. 17: 143-150.

Chakraborty, P., Sturgill-Koszycki, S., and Russell, D.G. 1994. Isolation and characterization of pathogen-containing phagosomes. Methods in Cell Biology 45: 261-276.

Chan, J., Zin, R., Magliozzo, S., and Bloom, B.R. 1992. Killing of virulent *Mycobacterium tuberculosis* by reactive nitrogen intermediates produced by activated murine macrophages. J. Exp. Med. 175: 1111-1122.

Chavrier, P., Parton, R.G., Hauri, H.P., Simons, K., and Zerial, M. 1990. Localization of low molecular weight GTP binding proteins to exocytic and endocytic compartments. Cell 62: 317-329.

Christoforidis, S., Miaczynska, M., Ashman, K., Wilm, M., Zhoa, L., Yip, S., Waterfield, M., Backer, J., and Zerial, M. 1999a. Phosphoinositide-3-OH kinases are Rab5 effectors. Nat. Cell Biol. 4: 249-252.

Christoforidis, S., McBride, H.M., Burgoyne, R.D., and Zerial, M. 1999b. The Rab5 effector EEA1 is a core component of endosome docking and fusion. Nature 397: 621-625.

Clemens, D.L. 1996. Characterization of the *Mycobacterium tuberculosis* phagosome. Trends Microbiol. 4: 113-118.

Clemens, D.L., and Horwitz, M.A. 1995. Characterization of the *M. tuberculosis* phagosome and evidence that phagosomal maturation is inhibited. J. Exp. Med. 181: 257-270.

Clemens, D.L., and Horwitz, M.A. 1996. The *Mycobacterium tuberculosis* phagosome interacts with early endosomes and is accessible to exogenously administered transferrin. J. Exp. Med. 184: 1-7.

Clemens, D.L., Lee, B.-Y., and Horwitz, M.A. 2000a. Deviant expression of Rab5 on phagosomes containing the intracellular pathogens *Mycobacterium tuberculosis* and *Legionella pneumophila* is associated with altered phagosomal fate. Infect. Immun. 68: 2671-2684.

Clemens, D.L., Lee, B.-Y., and Horwitz, M.A. 2000b. *Mycobacterium tuberculosis* and *Legionella pneumophila* phagosomes exhibit arrested maturation despite acquisition of Rab7. Infect. Immun. 68: 5154-5166.

Clerc, P.L., Ryter, A., Mounier, J., and Sansonetti, P.J. 1987. Plasmid-mediated killing of eucaryotic cells by *Shigella flexneri* as studied by infection of J774 macrophages. Infect. Immun. 55: 521-527.

Crowle, A., Dahl, R., Ross, E., and May, M. 1991. Evidence that vesicles containing living virulent *Mycobacterium tuberculosis* or *Mycobacterium avium* in cultured human macrophages are not acidic. Infect. Immun. 59: 1823 - 1831.

de Chastellier, C., and Thilo, L. 1997. Phagosome maturation and fusion with lysosomes in relation to surface property and size of the phagocytic particle. Eur. J. Cell Biol. 74: 49-62.

de Duve, C. 1963. The lysosome concept. In: Lysosomes. Ciba Foundation Symposium. A.V.S. de Reuck, and M.P. Cameron, eds. Churchill Press, London. p. 1-35.

Deretic, V., Via, L.E., and Deretic, D. 1997. Mycobacterial phagosome maturation, rab proteins, and intracellular trafficking. Electrophoresis 18: 2542-2547.

Desjardins, M. 1995. Biogenesis of phagolysosomes: the "kiss and run" hypothesis. Trends Cell Biol. 5: 183-186.

Desjardins, M., Celis, J.E., van Meer, G., Dieplinger, H., Jahraus, A., Griffiths, G., and Huber, L.A. 1994a. Molecular characterization of phagosomes. J. Biol. Chem. 269: 32194-32200.

Desjardins, M., Huber, L.A., Parton, R.G., and Griffiths, G. 1994b. Biogenesis of phagolysosomes proceeds through a sequential series of interactions with the endocytic apparatus. J. Cell Biol. 124: 677-688.

Ellenberg, J., Lippincott,-Schwartz, J., and Presley, J.F. 1999. Dual-color imaging with GFP variants. Trends Cell Biol. 9: 52-56.

Ferro-Novick, S., and Jahn, R. 1994. Vesicle fusion from yeast to man. Nature 370: 191-193.

Fratti, R.A., Chua, J., and Deretic, V. 2002. Cellubrevin alterations and *Mycobacterium tuberculosis* phagosome maturation arrest. J. Biol. Chem. 277: 17320-17326.

Fratti, R.A., Backer, J.M., Gruenberg, J., Corvera, S., and Deretic, V. 2001. Role of phophatidylinositol 3-kinase and Rab5 effectors in phagosomal biogenesis and mycobacterial phagosome arrest. J. Cell Biol. 154: 631-644.

Frehel, C., de Chastellier, C., Lang, T., and Rastogi, N. 1986. Evidence for inhibition of fusion of lysosomal and prelysosomal compartments with phagosomes in macrophages infected with pathogenic *Mycobacterium avium*. Infect. Immun. 52: 252-262.

Fricker, A.J., and Meyer., M.D. 2001. Confocal imaging of metabolism *in vivo*: Pitfalls and possibilities. J. Exp. Bot. 52: 631-640.

Gaillard, J.-L., Berche, P., Mounier, J., Richard, S., and Sansonetti, P. 1987. *In vitro* model of penetration and intracellular growth of *Listeria monocytogenes* in the human enterocyte-like cell line Caco-2. Infect. Immun. 55: 2822-2829.

Garin, J., Diez, R., Kieffer, S., Dermine, J., Duclos, S., Gagnon, E., Sadoul, R., Rondeau, C., and Desjardins, M. 2001. The phagosome proteome: insight into phagosome functions. J. Cell Biol. 152: 165-180.

Gatfield, J., and Pieters, J. 2000. Essential role for cholesterol in entry of mycobacteria into macrophages. Science 288: 1647-1650.

Geuze, H., Slot, J., van der Ley, P., and Scheffer, R. 1981. Use of colloidal gold particles in double-labeling immunoelectron microscopy of ultrathin frozen tissue sections. J. Cell Biol. 89: 653-665.

Girotti, M., and Banting, G. 1996. TGN38-green fluorescent protein hybrid proteins expressed in stably transfected eukaryotic cells provide a tool for the real-time, *in vivo* study of membrane traffic pathways and suggest a possible role for rat TGN38. J. Cell Sci. 109: 2915-2926.

Gordon, A.H., Hart, P. D'Arcy, and Young, M. R. 1980. Ammonia inhibits phagosome-lysosome fusion in macrophages. Nature 286: 79-80.

Gorvel, J.P., Chavrier, P., Zerial, M., and Gruenberg, J. 1991. Rab5 controls early endosome fusion *in vitro*. Cell 64: 915-925.

Griffin, A.B., Adams, S.R., and Tsien, R.Y. 1998. Specific covalent labeling of recombinant protein molecules inside cells. Science 281: 269-272.

Griffiths, G., Simons, K., Warren, G., and Tokuyasu, K.T. 1983. Immunoelectron microscopy using thin, frozen sections: application to studies of the intracellular transport of Semiliki Forest virus spike glycoproteins. In: Methods in Enzymology. S. Fleischer, and B. Fleischer, eds. Academic Press, New York. p. 199-216.

Griffiths, G. 1993. Fine structure immunocytochemistry. Springer-Verlag, Heidelberg.

Hall, A. 1990. The cellular functions of small GTP-binding proteins. Science 249: 635-640.

Hanson, P.I., Heuser, J.E., and Jahn, R. 1997. Neurotransmitter release - four years of SNARE complexes. Curr. Opin. Neurobiol. 7: 310-315.

Hasan, Z., Schlax, C., Kuhn, L., Lefkovits, I., Young, D., Thole, J., and Pieters, J. 1997. Isolation and characterization of the mycobacterial phagosome: segregation from the endosomal/lysosomal pathway. Mol. Microbiol. 24: 545-553.

Haugland, R.P. 1996. Handbook of fluorescent probes and research chemicals, Sixth edition. Molecular Probes, Inc., Eugene, OR.

Hirsch, C.S., Ellner, J.J., Russell, D.G., and Rich, E.A. 1993. Complement receptor mediated uptake and TNF-alpha-mediated growth inhibition of *M. tuberculosis* by human alveolar macrophages. J. Immunol. 152: 743-753.

Horwitz, M.A. 1983. Formation of a novel phagosome by the Legionnaires' disease bacterium (*Legionella pneumophila*) in human monocytes. J. Exp. Med. 158: 1319-1331.

Horwitz, M.A. 1988. Intracellular Parasitism. Curr. Opin. Immunol. 1: 41-46.

Horwitz, M.A., and Maxfield, F.R. 1984. *Legionella pneumophila* inhibits acidification of its phagosome in human monocytes. J. Cell. Biol. 99: 1936-1943.

Inoué, S. 1995. Foundations of confocal scanned imaging in light microscopy. In: Handbook of Biological Confocal Microscopy, 2nd ed. J. Pawley, ed. Plenum Press, New York.

Inoué, S., and Spring, K.R. 1997. Video Microscopy, 2nd ed. Plenum Press, New York.

Kenworthy, A. 2001. Imaging protein-protein interactions using fluorescence resonance energy transfer microscopy. Methods 24: 289-296.

Lee, B.-Y., and Horwitz, M.A. 1995. Identification of marcrophage and stress-induced proteins of *Mycobacterium tuberculosis*. J. Clin. Invest. 96: 245-249.

Liou, W., and Slot, J.W. 1994. Improved fine structure in immunolabeled cryosections after modifying the sectioning and pick-up conditions. In: 13th International Congress on Electron Microscopy, Paris. p. 253-254.

Liou, W., Geuze, H.J., Slot, J.W. 1996. Improving structural integrity of cryosections for immunogold labeling. Histochem. Cell Biol. 106: 41-58.

Lombardi, D., Soldati, T., Riederer, M., Goda, Y., Zerial, M., and Pfeffer, S. 1993. Rab9 functions in transport between late endosomes and the trans Golgi network. EMBO J. 12: 677-682.

Ludwig, T., Griffiths, G., and Hoflack, B. 1991. Distribution of newly synthesized lysosomal enzymes in the endocytic pathway of normal rat kidney cells. J. Cell Biol. 115: 1561-1572.

Luhrmann, A., and Haas, A. 2001. A method to purify bacteria-containing phagosomes from infected macrophages. Methods in Cell Science 22: 329-341.

Malik, Z.A., Denning, G.M., and Kusner, D.J. 2000. Inhibition of Ca(2+) signaling by *Mycobacterium tuberculosis* is associated with reduced phagosome-lysosome fusion and increased survival within human macrophages. J. Exp. Med. 287-302.

Malik, Z.A., Iyer, S.S., and Kusner, D.J. 2001. *Mycobacterium tuberculosis* phagosomes exhibit altered calmodulin-dependent signal transduction: contribution to inhibition of phagosome-lysosome fusion and intracellular survival in human macrophages. J. Immunol. 166: 3392-3401.

McBride, H.M., Rybin, V., Murphy, C., Giner, A., Teasdale, R., and Zerial, M. 1999. Oligomeric complexes link Rab5 effectors with NSF and drive membrane fusion via interactions between EEA1 and syntaxin13. Cell 98: 377-386.

McCafferey, J.M., and Farquhar, M.G. 1995. Localization of GTPases by indirect immunofluorescence and immunoelectron microscopy. In: Small

GTPases and Their Regulators, Part C. W. E. Balch, C. J. Der, and A. Hall, eds. Academic Press, New York. p. 259-279

McDonough, K.A., and Kress, Y. 1995. Cytotoxicity for lung epithelial cells is a virulence-associated phenotype of *Mycobacterium tuberculosis*. Infect. Immun. 63: 4802-4811.

McDonough, K.A., Kress, Y., and Bloom, B.R. 1993. Pathogenesis of tuberculosis: Interaction of *Mycobacterium tuberculosis* with macrophages. Infect. Immun. 61: 2763-2773.

McLean, I.W., and Nakane, P.K. 1974. Periodate-lysine-paraformaldehyde fixative for immunoelectron microscopy. J. Histochem. Cytochem. 22: 1077-1083.

McNally, J.G., Karpova, T., Cooper, J., and Conchello, J.A. 1999. Three-dimensional imaging by deconvolution microscopy. Methods 19: 373-385.

Meresse, S., Gorvel, J., and Chavrier, P. 1995. The Rab7 GTPase resides on a vesicular compartment connected to lysosomes. J. Cell Sci. 108: 3349-3358.

Meresse, S., Steele-Mortimer, O, Moreno, E., Desjardins, M., Finlay, B., and Gorvel, J.P. 1999. Controlling the maturation of pathogen-containing vacuoles: a matter of life and death. Nat. Cell Biol. 1: E183-E188.

Myrvik, Q.N., Leake, E.S., and Wright, M.J. 1984. Disruption of phagosomal membranes of normal alveolar macrophages by the H37Rv strain of *Mycobacterium tuberculosis*. American Review of Respiratory Diseases 129: 322-328.

Noguiera, N., and Cohn, Z. 1976. *Trypanosoma cruzi*: mechanism of entry and intracellular fate in mammalian cells. J. Exp. Med. 143: 1402-1420.

Novikoff, A. 1963. Lysosomes in the physiology and pathology of cells: contributions of staining methods. In: Lysosomes. Ciba Foundation Symposium. A. V. S. de Reuck, and M.P. Cameron, eds. Churchill Press, London. p. 36-77.

Nuoffer, C., and Balch, W. 1994. GTPases: Multifunctional molecular switches regulating vesicular traffic. Annu. Rev. Biochem. 63: 949-990.

Oh, Y.K., and Straubinger, R.M. 1996. Intracellular fate of *Mycobacterium avium*: use of dual-label spectrofluorometry to investigate the influence of bacterial viability and opsonization on phagosomal pH and phagosome-lysosome interaction. Infect. Immun. 64: 319-325.

Ohkuma, S., and Poole, B. 1978. Fluorescence probe measurement of the intralysosomal pH in living cells and the perturbation of pH by various agents. Proc. Natl. Acad. Sci. USA. 75: 3327-3331.

Orci, L., Ravazzola, M., Amherdt, M., Madsen, O., Perrelet, A., Vassali, J. and Anderson, R. 1986. Conversion of proinsulin to insulin occurs coordinately with acidification of maturing secretory vesicles. J. Cell Biol. 103: 2273-2281.

Pfeffer, S. 1994. Rab-GTPases: Master regulators of membrane trafficking. Curr. Biol. 6: 522-526.

Pitt, A., Mayorga, L.S., Schwartz, A.L., and Stahl, P.D. 1992a. Transport of phagosomal components to an endosomal compartment. J. Biol. Chem. 267: 126-132.

Pitt, A., Mayorga, L.S., Stahl, P.D., and Schwartz, A.L. 1992b. Alterations in the protein composition of maturing phagosomes. J. Clin. Invest. 90: 1978-1983.

Press, B., Feng, Y., Hoflack, B., and Wandinger-Ness, A. 1998. Mutant Rab7 causes the accumulation of cathepsin D and cation-independent mannose 6-phosphate receptor in an early endocytic compartment. J. Cell Biol. 140: 1075-1089.

Racoosin, E.L., and Swanson, J.A. 1993. Macropinosome maturation and fusion with tubular lysosomes in macrophages. J. Exp. Med. 121: 1011-1020.

Ramachandra, L., Noss, E., Boom, W.H., and Harding, C.V. 1999. Phagocytic processing of antigens for presentation by class II major histocompatability complex molecules. Cell Microbiol. 1: 205-214.

Ramachandra, L., Noss, E., Boom, W.H., and Harding, C.V. 2001. Processing of *Mycobacterium tuberculosis* antigen 85B involves intraphagosomal formation of peptide-major histocompatability complex II complexes and is inhibited by live bacilli that decrease phagosome maturation. J. Exp. Med. 194: 1421-1432.

Riederer, M., Soldati, T., Shapiro, A., Lin, J., and Pfeffer, S. 1994. Lysosome biogenesis requires Rab9 function and receptor recycling from endosomes to the trans-Golgi network. J. Cell Biol. 125: 573-582.

Roth, J. 1983. The colloidal gold marker system for light and electron microscopic cytochemistry. In: Techniques in Immunocytochemistry. Academic Press, London. p. 218-284.

Rothman, J.E. 1994. Mechanisms of intracellular protein transport. Nature 372: 55-63.

Roy, C., Berger, K., and Isberg, R. 1998. *Legionella pneumophila* DotA protein is required for early phagosome trafficking decisions that occur within minutes of bacterial uptake. Mol. Microbiol. 28: 663-674.

Russell, D.G. 1994. Immunoelectron microscopy of endosomal trafficking in macrophages infected with microbial pathogens. Methods in Cell Biology 45: 277-288.

Russell, D.G., Dant, J., and Sturgill-Koszycki, S. 1996. *Mycobacterium avium* and *M. tuberculosis* -containing vacuoles are dynamic, fusion-competent vesicles that are accessible to glycosphingolipids from the host cell plasmalemma. J. Immunol. 156: 4764-4773.

Sansonetti, P.J., Ryter, A., Clerc, P., Maurelli, A.T., and Mounier, J. 1986. Multiplication of *Shigella flexneri* within HeLa cells: lysis of the phagocytic vacuole and plasmid mediated contact hemolysis. Infect. Immun. 51: 461-469.

Schlesinger, P.H. 1994. Measuring the pH of pathogen-containing phagosomes. Methods in Cell Biology 45: 289-311.

Scianimanico, S., Pasquali, C., Lavoie, J., Huber, L.A., Gorvel, J.-P., and Desjardins, M. 1997. Two-dimensional gel electrophoresis analysis of endovacuolar organelles. Electrophoresis 18: 2566-2572.

Shapiro, A., Riederer, M., and Pfeffer, S.R. 1993. Biochemical analysis of Rab9, a *ras*-like GTPase involved in protein transport from late endosomes to the trans Golgi network. J. Biol. Chem. 268: 6925-6931.

Sibley, D.L., Franzblau, S.G., and Krahenbuhl, J.L. 1987. Intracellular fate of *Mycobacterium leprae* in normal and activated mouse macrophages. Infect. Immun. 55: 680-685.

Sinai, A.P., and Joiner, K.A. 1997. Safe Haven: The cell biology of nonfusogenic pathogen vacuoles. Annu. Rev. Microbiol. 51: 415-462.

Slot, J.W., and Geuze, H. J. 1985. A new method for preparing gold probes for multiple-labeling cytochemistry. Eur. J. Cell Biol. 38: 87-93.

Slot, J.W., Geuze, H.J., and Weerkamp, J. 1988. In: Methods Microbiol. F. Mayer, ed. Academic Press.

Sollner, T., Bennet, M., Whiteheart, S., Scheller, R., and Rothman, J.E. 1993. A protein assembly-disassembly pathway *in vitro* that may correspond to sequential steps of synaptic vesicle docking, activation, and fusion. Cell 75: 409-418.

Sonnichsen, B., de Renzis, S., Nielsen, E., Rietdorf, J., and Zerial, M. 2000. Distinct membrane domains on endosomes in the recycling pathway visualized by multicolor imaging of Rab4, Rab5, and Rab11. J. Cell Biol. 149: 901-913.

Steinhoff, U., Golecki, J.R., Kazda, J., and Kaufmann, S.H.E. 1989. Evidence for phagosome-lysosome fusion in *Mycobacterium leprae* infected murine schwann cells. Infect. Immun. 57: 1008-1010.

Stenmark, H., Parton, R., Steele-Mortimer, O., Lucte, A., Gruenberg, J., and Zerial, M. 1995. Inhibition of Rab5 GTPase activity stimulates membrane fusion in endocytosis. EMBO J. 13: 1287-1296.

Straub, M., Lodemann, P., Holroyd, P., Jahn, R., and Hell, S.W. 2000. Live cell imaging by multifocal multiphoton microscopy. Eur. J. Cell Biol. 79: 726-734.

Sturgill-Koszycki, S., Haddix, P.L., and Russell, D.G. 1997. The interaction between *Mycobacterium* and the macrophage analyzed by two-dimensional polyacrylamide gel electrophoresis. Electrophoresis 18: 2558-2665.

Sturgill-Koszycki, S., Schlesinger, P., Chakraborty, P. Haddix, P., Collins, H., Fok, A., Allen, R., Gluck, S., Heuser, J., and Russell, D. 1994. Lack of acidification in *Mycobacterium phagosomes* produced by exclusion of the vesicular proton-ATPase. Science 263: 678-681.

Tanowitz, H., Wittner, M., Kress, Y., and Bloom, B. 1975. Studies of *in vitro* infection by *Trypanosoma cruzi*. I. Ultrastructural studies on the invasion of macrophages and L-cells. Am. J. Trop. Med. Hyg. 24: 25-33.

Tokuyasu, K.T. 1986. Application of cryoultramicrotomy to immunocytochemistry. J. Microscocopy 143(2): 139-149.

Ullrich, H.J., Beatty, W., and Russell, D.G. 1999. Direct delivery of procathepsin D to phagosomes: implications for phagosome biogenesis and parasitism by *Mycobacterium*. Eur. J. Cell Biol. 78: 739-748.

Ullrich, O., Reinsch, S., Urbe, S., Zerial, M., and Parton, R.G. 1996. Rab11 regulates recycling through the pericentriolar recycling endosome. J. Cell Biol. 135: 913-924.

Van der Sluijs, P., Hull, M., Webster, P., Male, P., Goud, B., and Mellman, I. 1992. The small GTP-binding protein rab4 controls an early sorting event on the endocytic pathway. Cell 70: 729-740.

Via, L.E., Deretic, D., Ulmer, R.J., Hibler, N.S., Huber, L.A., and Deretic, V. 1997. Arrest of mycobacterial phagosome maturation is caused by a block in vesicle fusion between stages controlled by rab5 and rab7. J. Biol. Chem. 272: 13326-13331.

Via, L.E., Fratti, R.A., McFalone, M., Pagan-Ramos, E., Deretic, D., and Deretic, V. 1998. Effects of cytokines on mycobacterial phagosome maturation. J. Cell Sci. 111: 897-905.

Wallace, W., Schaefer, L.H., and Swedlow, J. R. 2001. A working persons's guide to deconvolution in light microscopy. Biotechniques 31: 1076-1082.

Ward, D., Pevsner, J., Scullion, M., Vaughn, M., and Kaplan, J. 2000. Syntaxin 7 and VAMP-7 are soluble N-ethylmaleimide-sensitive factor attachment protein receptors required for late endosome-lysosome and homotypic lysosome fusion in alveolar macrophages. Mol. Biol. Cell 11: 2327-2333.

White, J.G., Squirrell, J.M., and Eliceri, K.W. 2001. Applying multiphoton imaging to the study of membrane dynamics in living cells. Traffic 2: 775-780.

Winkler, H.H. 1990. Rickettsia species (as organisms). Annu. Rev. Microbiol. 44: 131-153.

Wyrick, P.B., and Brownridge, E.A. 1978. Growth of *Chlamydia psittaci* in macrophages. Infect. Immun. 19: 1054-1060.

Xu, S., Cooper, A., Sturgill-Koszycki, S., van Heyningen, T., Chatterjee, D., Orme, I., Allen, P., and Russell, D. 1994. Intracellular trafficking in *Mycobacterium tuberculosis* and *Mycobacterium avium* infected macrophages. J. Immunol. 153: 2568-2578.

From: Tuberculosis: The Microbe Host Interface
Edited by: Larry S. Schlesinger and Lucy E. DesJardin

Chapter 3

Analysis of Macrophage Signaling Following *M. tuberculosis* Infection

David Kusner

Abstract

The interactions of *Mycobacterium tuberculosis* with human macrophages are central to all aspects of the pathogenesis of tuberculosis, from initial infection to reactivation disease. Characterization of the biochemical mechanisms by which macrophages respond to *M. tuberculosis* has provided exciting insights into pathophysiology and will foster unique approaches for therapy and prevention. This chapter will outline the macrophage signal transduction pathways that regulate two critical phases of the host-pathogen interaction in tuberculosis; phagocytosis and phagosome maturation. In addition to reviewing our current understandings and gaps in knowledge, emphasis will be placed on the experimental approaches that are currently employed, as well as those emerging technologies that are beginning to inform our studies of mycobacterial pathogenesis.

Introduction

The pathogenesis of tuberculosis is complex and manifold, as befits a disease characterized by a lifetime of dynamic interactions between microbial virulence factors and the human immune system. This pathophysiologic complexity, diversity, and longevity have presented daunting obstacles to experimental modeling and analysis. However, recent efforts to integrate genetic, microbiologic, immunologic, and cell biologic approaches upon a common biochemical foundation have begun to yield penetrating insights into the mechanisms of tuberculous pathogenesis. Thus, all modern efforts at understanding the fundamental characteristics and mechanistic principles of this disease, including the analysis of the regulatory signaling pathways, must be formulated and critiqued within an experimental framework of relevance to this uniquely human pathogen and disease.

The molecular and cellular interactions of *Mycobacterium tuberculosis* with human macrophages are central to all aspects of the pathogenesis of tuberculosis (Schlesinger 1996; Ernst 1998). From the initial infection of alveolar macrophages via aerosolized droplets, to the lysis of these naive macrophages and the resultant lympho-hematogenous dissemination, to the bacilli's containment by the adaptive immune system, and finally to the reactivation of latent infection via breakdown of cell-mediated immune defenses, the dynamic interactions between tubercle bacilli and human macrophages are the focus of the pathophysiology. Characterization of the biochemical mechanisms by which macrophages detect and respond to *M. tuberculosis* has provided both a unique view of pathogenesis and a foundation for advances in therapy and prevention (Russell 2001; Dannenberg and Collins 2001). In addition, it has shed light on fundamental processes of the human immune response from the initiation of innate immunity to the multivariate linkages to adaptive mechanisms. By deepening our understanding of the ways in which *M. tuberculosis* inhibits and evades innate immunity and thus impairs the elicitation of adaptive immune responses, studies of macrophage signal transduction during tuberculous infection have enriched our knowledge of phagocytosis, maturational processes that regulate the transformation of nascent phagosomes to phagolysosomes, antigen processing and presentation, and the effector mechanisms exerted by activated macrophages acting as agents of adaptive immunity (Bloom *et al.*, 1999; Ferguson and Schlesinger 2000; Kusner *et al.*, 1996; Russell 2001; Clemens and Horwitz 1995; Master *et al.*, 2001; Malik *et al.*, 2000; Ramachandra *et al.*, 2001; Iwasaki *et al.*, 1999; Gumperz and Brenner 2001; Stenger *et al.*, 1997).

On a broader level, characterization of signal transduction in host cells has become an invaluable component of the study of bacterial pathogenesis, yielding the dynamic new field of cellular microbiology (Cossart *et al.*, 1996). This focus on definition of the biochemical mechanisms that underlie physiologic and pathologic aspects of microbial-host interactions has yielded crucial insights into diseases caused by *Salmonella, Shigella, Listeria, Legionella, Toxoplasma, Yersinia,* as well as *Mycobacteria* (Detweiler *et al.,* 2001; Navarre and Zychlinsky 2000; Bierne *et al.,* 2001; Roy and Tilney 2002; Pelletier *et al.,* 2002; Cornelis 2002; Clemens *et al.,* 2000b; McKinney *et al.,* 2000). In fact, one could argue that there is a special applicability of cellular microbiology to tuberculosis, as the unique complexity, diversity, and chronicity of mycobacterial infection presents both a challenge to the definition of the fundamental regulatory mechanisms as well as a valuable opportunity to provide a unifying causal molecular framework to understand the multiple steps in pathogenesis.

My focus in this chapter will be to describe the major macrophage signaling pathways that regulate two of the critical steps in the host-pathogen interaction in tuberculosis: phagocytosis and phagosome maturation. The approach will be to review our current understanding of the biochemical mechanisms that regulate these responses, their modification by mycobacterial components, and the current gaps in knowledge of these processes. An emphasis will be placed on the experimental approaches that are currently employed, as well as those emerging technologies that are beginning to inform our studies of mycobacterial pathogenesis. Since *M. tuberculosis* is a pathogen unique to humans, my primary focus will be on experimental models utilizing human macrophages.

Macrophage Activation During Phagocytosis of *M. tuberculosis*

The initial interaction of *M. tuberculosis* with the human immune system involves the binding of the organism to specific receptors on the surface of alveolar macrophages (Schlesinger 1996; Ernst 1998). This primary event initiates a series of biochemical responses in the macrophage that mediates phagocytosis of tubercle bacilli and sets in motion subsequent innate immune defense responses. In this latter respect, signal transduction events that are initiated prior to and during ingestion are critically important to future stages in the host-pathogen interaction. As detailed in Chapter 1, numerous macrophage surface receptors are involved in the phagocytosis of *M. tuberculosis*. Complement receptors, especially the β_2 integrins, CR3 and CR4, play a primary role, since specific blockade with anti-receptor antibodies markedly inhibits phagocytosis (Schlesinger *et al.,* 1990; Hirsch *et al.,* 1994; Schlesinger 1993). The macrophage mannose receptor functions in phagocytosis of virulent, but not attenuated, strains of *M. tuberculosis*, and is probably most important in the non-opsonic ingestion of tubercle bacilli (Schlesinger 1993; Schlesinger *et al.,* 1994; Kang and Schlesinger 1998). In addition to the direct roles of these phagocytic receptors, *M. tuberculosis* interacts with numerous macrophage surface receptors that modulate the host-pathogen interaction via activation of host signaling pathways, including Toll-like receptors (TLR) 2 and 4, as well as CD14 and CD43 (Means *et al.*, 1999; Jones *et al.*, 2001; Peterson *et al.*, 1995; Fratazzi *et al.*, 2000). Given the physical and biochemical complexity of the mycobacterial surface and ensuing multiplicity of receptor-ligand binding events initiated by bacterial adherence, assessment of macrophage signal transduction events and their relation to specific immune effector responses is a challenging task. Progress in this area has been greatly aided by the development of *in vitro* models utilizing primary human macrophage that accurately reproduce phagocytosis and intracellular survival of *M. tuberculosis* following infection at physiologically-relevant multiplicities of infection (Schlesinger *et al.*, 1990; Schlesinger 1993; Clemens and Horwitz 1995; Clemens and Horwitz 1996). In addition, comparison of the biochemical signaling events initiated by live virulent *M. tuberculosis* with those stimulated by killed tubercle bacilli or other complement-opsonized particles have provided insights regarding those signals required for phagocytosis vs. those relevant to subsequent mycobacterial-macrophage interactions.

Tyrosine Kinases

The earliest biochemical signals that are detectable in human macrophages following exposure to *M. tuberculosis* are the phosphorylation of multiple proteins on tyrosine residues (Kusner *et al.*, 1996). Within 30 sec of addition of *M. tuberculosis*, increased tyrosine phosphorylation is detectable in proteins with molecular masses of 53-63 kDa. By 1 min, increases in the tyrosine phosphorylation of multiple other proteins are detectable, most prominently in those of 150, 120, 95, 72, and 42 kDa. The global extent of tyrosine phosphorylation of macrophage proteins increases progressively over 120 min following incubation with *M. tuberculosis*. Current evidence indicates that these tyrosine phosphorylation events are necessary for phagocytosis of *M. tuberculosis*. Several different classes of tyrosine kinase inhibitors, with distinct mechanisms of action, produce concentration- and time-dependent inhibition of both the phagocytosis of tubercle bacilli as well as the accompanying accumulation of tyrosine-phosphorylated proteins (Kusner *et al.*, 1996). In addition, both live and killed *M. tuberculosis*, as well as complement-opsonized zymosan (COZ), induce tyrosine phosphorylation of macrophage proteins. It is likely that ligation of complement receptors (CRs) by *M. tuberculosis* plays a prominent role in stimulation of tyrosine phosphorylation. Antibodies to CD11b and CD11c, the α-chains of the β_2 integrins CR3 and CR4, as well as to their common β-chain, CD18, stimulate a similar pattern of increased tyrosine phosphorylation of multiple macrophage proteins. In support of the role of CRs in the initial signaling events, other β_2 integrin-dependent functions, such as cellular adhesion, are also accompanied by significant increases in tyrosine phosphorylation (Fuortes *et al.,* 1999, 1994, 1993; Berton *et al.,* 1994). Phagocytosis of multiple other pathogens by macrophages, monocytes, and neutrophils, as well microbial invasion of epithelial cells, is similarly accompanied by increases in tyrosine phosphorylation of host proteins (Sanguedolce *et al.,* 1993; Zaffran *et al.,* 1995; Galan and Zhou 2000). This commonality of mechanism, supports the hypothesis that protein tyrosine phosphorylation is causally related to phagocytosis of *M. tuberculosis.*

Alterations in protein tyrosine phosphorylation reflect a complex summation of changes in the activity of both kinases and phosphatases (Madhani 2001; Ostman and Bohmer 2001). The extent of this complexity is increased by the fact that a given enzymatic activity (e.g., phosphatase-mediated removal of inorganic phosphate from tyrosine residues) can be coupled to both cell activation or deactivation, depending on the substrate, and even, on the specific tyrosine within a given substrate. Identification of the key tyrosine kinases, phosphatases, and substrates involved in phagocytosis of *M. tuberculosis* is still at an early stage of investigation. However, important insights regarding likely candidates have been provided by research on other immune recognition receptors. Stimulation of protein tyrosine phosphorylation is a critical proximal step following the activation of multiple immune recognition receptors, including the phagocytic Fcγ receptors, as well as the antigen receptors on T and B lymphocytes. In these latter cases, receptor ligation increases tyrosine phosphorylation primarily via activation of protein tyrosine kinases of the Src family. This large family of non-receptor tyrosine kinases (m.w. range 53-62 kDa) includes nine members, three of which (Src, Yes, and Fyn) are ubiquitously expressed, whereas the remaining six are differentially distributed among hematopoetic cells (Korade and Corey 2000). Of the leukocyte-specific isoforms, human macrophages express Lyn and Hck (Kusner, *et al.*, unpublished data). *M. tuberculosis* stimulates the rapid activation of both Lyn and Hck, which are the first two macrophage proteins in which tyrosine phosphorylation is detectable following exposure to tubercle bacilli.

Determination of the specific roles of individual Src kinase in physiologic responses, eg., phagocytosis of *M. tuberculosis*, is a difficult experimental problem. Genetic disruptions of individual Src kinases, i.e., generation of "knock-out" mice, have yielded conflicting results, and several technical and conceptual problems in interpretation. Most importantly, disruption of a single Src-family gene is accompanied by significant increases in the expression of one or more other members of this large gene family (Martin 2001; Bjorge *et al.,* 2000; Korade and Corey 2000). Increased levels of other Src family kinases often provide compensatory or redundant functions for the missing gene product, seriously hampering the definition of the "normal" function of the absent kinase. Furthermore, complimentary approaches, such as anti-sense mediated inhibition or the use of dominant-negative mutants, have yielded data which conflict with those derived from genetic disruption (Martin 2001;

Bjorge *et al.,* 2000; Korade and Corey 2000). It is likely that improved technologies, including use of targeted RNA interference (RNAi), will be required to determine specific functional roles of individual Src family kinases. In summary, current evidence indicates that tyrosine phosphorylation is required for macrophage phagocytosis of *M. tuberculosis,* and that Src family tyrosine kinases likely play a primary role in signal initiation. Areas for future study include; (1) identification of other tyrosine phosphorylated proteins, especially those of 150, 120, 95, 72, and 42 kDa (2) The effects of *M. tuberculosis-*induced ligation of other macrophage surface receptors, e.g., mannose receptor, TLR2/4, CD14 on protein tyrosine phosphorylation, (3) definition of the pathways by which tyrosine phosphorylation is coupled to other processes required for mycobacterial phagocytosis, including dynamic rearrangements of the actin cytoskeleton, and (4) determination of whether live *M. tuberculosis* specifically modifies tyrosine phosphorylation pathways to promote its intracellular survival. In reference to this latter point, Reiner and colleagues have demonstrated that the cell wall glycolipid of *M. tuberculosis,* lipoarabinomannan (LAM), activates the protein tyrosine phosphatase, SHP-1, in human macrophages (Knutson *et al.,* 1998). Evidence from experiments with other intracellular pathogens of macrophages, including *Leishmania* and *Salmonella* species, indicate that stimulation of tyrosine phosphatases results in reductions in host antimicrobial functions (Nandan *et al.,* 2000, 1999; Murli *et al.,* 2001).

Phospholipase D

Phospholipases function as critical proximal mediators of signal transduction cascades initiated by a wide range of inflammatory stimuli, including complement components, cytokines and activators of the immune recognition receptors, including FcγR, TCR, and BCR (Exton 2002b; Melendez and Allen 2002; Ott and Cambier 2002; Fruman and Cantley 2002). By catalyzing the hydrolysis of membrane phospholipids, phospholipases release a series of bioactive lipid signaling molecules that trigger physiologic and pathologic processes in immune cells and adjacent tissues. In macrophages, phospholipases regulate chemotaxis, phagocytosis, secretion, and the respiratory burst (Exton 2002b; Melendez and Allen 2002; Ott and Cambier 2002). Phospholipase D (PLD), which cleaves the predominant membrane phospholipid, phosphatidylcholine, to yield phosphatidic acid (PA) and choline, has a central role in the activation of macrophage antimicrobial properties. Stimulation of PLD activity is required for phagocytosis of both virulent and attenuated strains of *M. tuberculosis* (Kusner *et al.,* 1996). Complement-opsonized, as well as unopsonized tubercle bacilli, rapidly stimulate a significant increase in the activity of macrophage PLD. This increase in PLD activity is most marked within the first 30 min, when approximately 90% of total phagocytosis occurs, and PLD activity remains elevated for at least 2 hr following exposure to *M.*

tuberculosis (Kusner *et al.*, 1996). Addition of a several different PLD inhibitors with distinct mechanisms of action results in dose-dependent decreases in phagocytosis of *M. tuberculosis*. Furthermore, addition of purified PLD enzymes restores control levels of phagocytosis to macrophages in which endogenous PLD has been inhibited (Kusner *et al.*, 1996). The importance of PLD to phagocytosis is not restricted to ingestion of *M. tuberculosis*. Macrophage or neutrophil phagocytosis of a range of complement- or antibody-opsonized particles exhibits a similar dependence on stimulation of PLD (Kusner *et al.*, 1996, 1999; Fallman *et al.*, 1992, 1993; Serrander *et al.*, 1996). Two mammalian PLD isoforms, PLD1 and PLD2, have been identified that exhibit distinct tissue and cellular distributions, sub-cellular localizations, modes of activation, and coupling to physiologic responses (Exton 2002b). Macrophages express both PLD isoforms. Recent evidence indicates that PLD1 and PLD2 regulate phagocytosis of both complement- and Ab opsonized microbes and other particles (Kusner, D.J., *et al.*, unpublished data). Catalytically-inactive PLD1 and PLD2 mutants suppress phagocytosis upon transfection into human THP-1 promonocytes that have been differentiated to a macrophage-like phenotype by cytokine treatment. These data are consistent with a dominant negative phenotype conferred by the catalytically-inactive PLD mutants (Denmat-Ouisse *et al.*, 2001; Humeau *et al.*, 2001).

The critical importance of PLD to phagocytosis likely derives from the direct effects of the product, PA, as well as via the conversion of PA to other lipid signaling molecules. PA stimulates several protein kinases, lipid kinases, and phospholipases (Exton 2002b) that are potential regulators of phagocytosis. In addition, PA serves as the major source of the protein kinase C activator, diacylglycerol (DAG), via removal of its phosphate group by phosphatidate phosphohydrolase (Fallman *et al.*, 1992; Billah *et al.*, 1989; Exton 2002a; Palicz *et al.*, 2001; Regier *et al.*, 2000). Like PA, DAG, serves both to stimulate phagocytosis as well as to promote subsequent macrophage antimicrobial effector responses, including generation of the respiratory burst and secretion of cytokines and other inflammatory mediators (Fallman *et al.*, 1992; Kanaho *et al.*, 1993; Kusner *et al.*, 1996, 1999; Serrander *et al.*, 1996; Kanaho *et al.*, 1991; Xie *et al.*, 1991; Agwu *et al.*, 1991; Waite *et al.*, 1997). Thus, similar to the stimulation of PTKs discussed above, activation of PLD is important both to regulation of the phagocytic event itself, as well as to stimulation of subsequent macrophage microbicidal and inflammatory activities. In this regard, it is worth noting that PLD1 is also the isoform that is coupled to activation of the superoxide-generating NADPH oxidase, based on antisense-mediated specific inhibition (Melendez *et al.*, 2001).

The mechanism by which complement receptors stimulate this requisite activation of PLD during phagocytosis of *M. tuberculosis* has not been fully elucidated, though strong evidence for the involvement of protein tyrosine kinases (PTKs) has been presented (Kusner *et al.*, 1996). Thus, dose-dependent

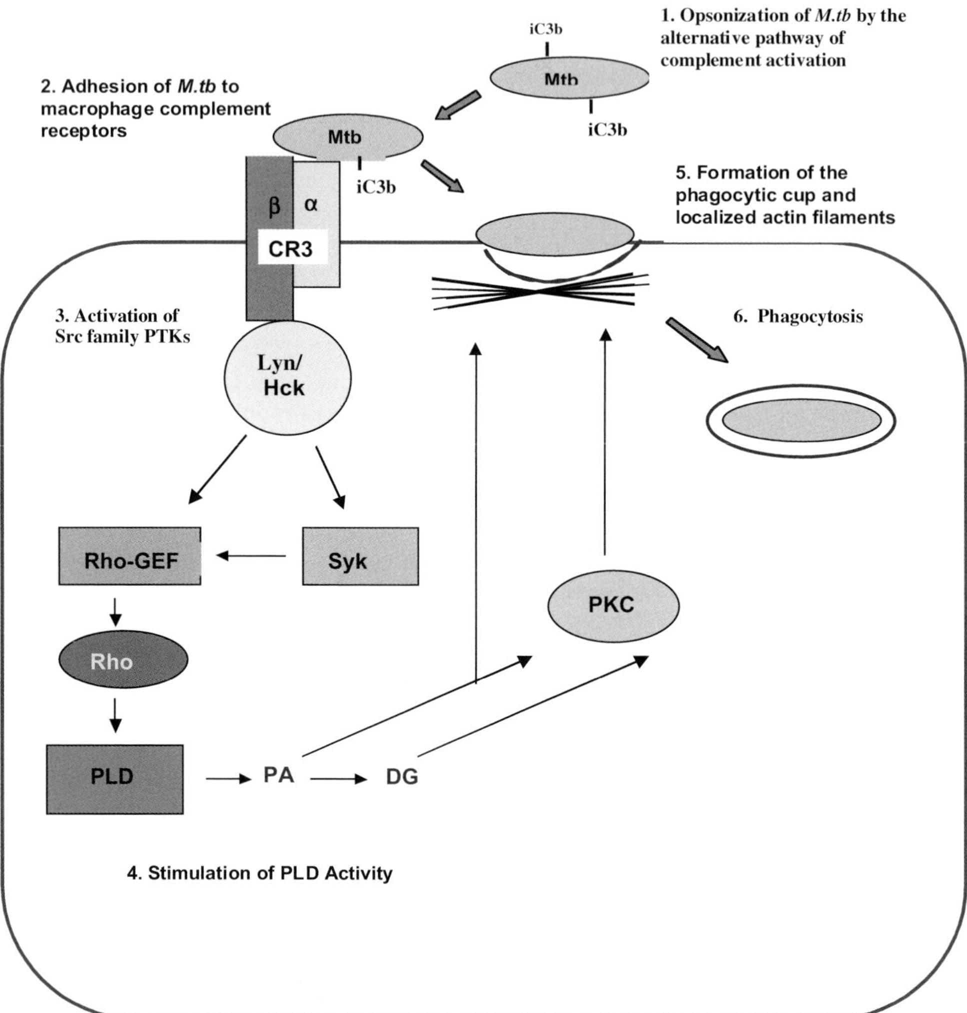

Figure 1. Signal transduction pathways that regulate macrophage phagocytosis of *M. tuberculosis*. Following its opsonization by the complement component, iC3b, phagocytosis of *M. tuberculosis* is initiated by adhesion to macrophage complement receptors, including the heterodimeric β_2 integrin, CR3. Stimulation of the Src family PTKs, Lyn and Hck, leads to Rho-dependent activation of PLD, via stimulation of Rho-GEF by pathways that are dependent, as well as independent of the tyrosine kinase, Syk. PLD-mediated generation of the lipid second messengers PA and DG stimulates PKC and localized rearrangements of the actin cytoskeleton at the forming phagocytic cup. Actin-dependent ingestion results in membrane invagination and ultimately fission of the nascent phagosome from the plasma membrane. Stimulation of actin polymerization by Rho GTPases is not shown in the diagram.

inhibition of CR-stimulated PLD activity by three mechanistically-distinct classes of PTK inhibitors is accompanied by parallel reductions in phagocytosis (Kusner *et al.*, 1996). Based on work in other cellular systems, PTK-mediated activation of PLD during macrophage phagocytosis of *M. tuberculosis* is likely to proceed via activation of a Rho GTP-binding protein by its specific Rho-Guanine nucleotide Exchange Factors (Rho-GEFs) (Caron and Hall 1998).

In addition to PTKs and PLD, it is likely that several other signal transduction pathways, that have been demonstrated to participate in phagocytosis of other particles, also have a similar role during macrophage ingestion of *M. tuberculosis*, though no direct evidence has yet been presented. These additional components include the tyrosine kinase Syk, the serine-threonine kinase, protein kinase C (PKC), and the actin cytoskeleton. An integrated model for the regulation of macrophage phagocytosis of *M. tuberculosis* is presented in Figure 1.

Phagosome Maturation in Macrophage Antimicrobial Activity and its Inhibition in Tuberculosis

Macrophage bactericidal activity is comprised of a myriad of mechanisms. Although characteristically classified as oxidative or non-oxidative systems of host defense, there is substantial evidence of synergy between biochemically-distinct pathways (Weiss 1989; Kusner and King 1989; Elsbach and Weiss 1999; Weiss *et al.*, 2003). The NADPH oxidase-dependent generation of superoxide anion and its subsequent conversion to hydrogen peroxide and other reactive oxygen species (ROS) is crucial to microbicidal defense, as evidenced by the severe morbidity and mortality associated with its dysfunction in Chronic Granulomatous Disease (Weiss *et al.*, 2003). Classically-defined "non-oxidative" defense mechanisms include peptides, proteases, and other enzymes (Elsbach and Weiss 1999; Weiss *et al.*, 2003). Despite a formidable fund of knowledge on phagocyte bactericidal activity, much remains unanswered regarding the microbicidal mechanisms of human macrophages, including: (1) What are the kinetic and spatial determinants of macrophage microbicidal activity? (2) Do reactive nitrogen intermediates produced by Inducible Nitric Oxide Synthase (iNOS) function in microbial killing? (3) What is the relative importance of these diverse bactericidal mechanisms vs. specific pathogens, such as *M. tuberculosis*?

The maturation of phagosomes to microbicidal phagolysosomes is a key mechanism of innate immune defense in macrophages and other phagocytic leukocytes. Phagolysosomes serve as a locus for focusing the diverse antimicrobial responses of activated macrophages within a membrane-enclosed compartment, thus limiting access of cytotoxic products to host components.

The mechanisms that regulate phagosome maturation have been the subject of intensive investigation, especially since many intracellular pathogens, including *M. tuberculosis*, interfere with this process as a central feature of pathogenesis (Russell 2001; Meresse *et al.*, 1999; Clemens and Horwitz 1996). The characteristic alterations in intracellular trafficking that occur in *M. tuberculosis*-infected macrophages have been detailed in Chapter 2 of this volume. Herein, I will limit my focus to the regulatory mechanisms responsible for the normal sequence of phagosome maturation and its inhibition by *M. tuberculosis*.

The recognition that the transition of phagosomes to phagolysosomes occurs via multiple fission and fusion reactions with vesicles of the endosomal-lysosomal pathway has formed the framework for current work in this field (Desjardins *et al.*, 1994; Mayorga *et al.*, 1991). These dynamic interactions with endosomal trafficking pathways have provided numerous hypotheses regarding candidate regulatory components and mechanisms. However, interpretation of the literature is complicated by the use of murine macrophages (MPs) despite substantial evidence that these do not accurately model the interaction of human MPs with intracellular pathogens (Ellner, 1990, 1997a, 1997b; Nauseef, 2001; Weiss *et al.*, 2003; Redpath *et al.*, 2001; Mosser and Karp 1999; Haas, 1998; Weiss *et al.*, 2003; Nauseef, 2001). For example, murine and rodent species are relatively-to-highly resistant to diseases caused by intracellular pathogens, including legionellosis, listeriosis, plague, leishmaniaisis, tularemia, as well as tuberculosis; whereas humans are highly susceptible to all of these. Murine and rodent macrophages produce large amounts of nitric oxide via inducible nitric oxide synthase (iNOS) which kills Legionella, Listeria, Yersinia, Leishmania, Francisella and Mycobacteria species, whereas it has been difficult to detect iNOS or nitric oxide in human macrophages or to demonstrate microbicidal activity toward these pathogens (Nathan and Shiloh 2000; Thomassen and Kavuru 2001; Albina, 1995). Furthermore the relevance of studies utilizing non-tuberculous mycobacteria to our understanding of the pathogenesis of tuberculosis is unknown. *Mycobacterium bovis* BCG (BCG) is an attenuated vaccine strain and *Mycobacterium avium-intracellulare* (*MAI*) is an organism of very low virulence, neither of which cause disease in immunocompetent individuals. Thus, it is unclear to what extent the mechanisms regulating their intracellular trafficking resemble those responsible for inhibition of the maturation of phagosomes containing *M. tuberculosis*.

In evaluating the various signal transduction mechanisms that have been proposed to account for *M. tuberculosis*-induced inhibition of phagosome maturation, I will emphasize three questions: (1) Has the specific pathway been demonstrated to be required for phagosome maturation under normal conditions? (2) Has the inhibition of this pathway been demonstrated in human macrophages infected with virulent *M. tuberculosis*? (3) Has experimental reconstitution of the inhibited signaling pathway been demonstrated to remove the *M. tuberculosis*-induced inhibition of phagosome maturation, i.e., to lead to the maturation of mycobacterial phagosomes to phagolysosomes?

Inhibition of Macrophage Ca^{2+}-Mediated Signal Transduction

The only signaling pathway for which all three of the criteria noted above have been rigorously established is mycobacterial-induced inhibition of macrophage Ca^{2+}-mediated signal transduction. Thus, (1) complement-opsonized particles, including killed *M. tuberculosis* and COZ, stimulate a significant increase in levels of cytosolic Ca^{2+} ($[Ca^{2+}]_c$) in human macrophages, and this increase in $[Ca^{2+}]_c$ is required for phagosome maturation (Malik *et al.*, 2000, 2001). (2) In contrast, complement-opsonized live virulent *M. tuberculosis* inhibit this Ca^{2+}-signaling pathway in human macrophages (Malik *et al.*, 2000, 2001, 2003), and (3) Pharmacologic reversal of the block in Ca^{2+}-signaling with a Ca^{2+}-specific ionophore results in maturation of phagosomes containing live *M. tuberculosis* to phagolysosomes (Malik *et al.*, 2000). Although the mycobacterial component(s) responsible for inhibition of macrophage Ca^{2+}-mediated signal transduction remains unknown, a key host locus of this effect is the macrophage enzyme, sphingosine kinase (Malik *et al.*, 2003). Through catalyzing the conversion of sphingosine to sphingosine-1-phosphate (S1P), sphingosine kinase normally functions to increase levels of $[Ca^{2+}]_c$ by triggering its release form intracellular stores in the endoplasmic reticulum (Spiegel and Milstien 2002). Utilizing both intact human macrophages, *ex vivo*, as well as a cell-free *in vitro* system comprised of purified cytosol and membrane fractions, we have recently demonstrated that live *M. tuberculosis* inhibits macrophage sphingosine kinase whereas killed *M. tuberculosis* or COZ stimulate this enzyme (Malik *et al.*, 2003). Though the inhibitory effect of *M. tuberculosis* in the cell-free assay suggests that sphingosine kinase itself may be the direct target of mycobacterial-interference, this hypothesis will require evaluation utilizing the purified macrophage enzyme.

Increases in $[Ca^{2+}]_c$ trigger many aspects of cellular activation and, in phagocytes, are coupled to several key antimicrobial responses, including synthesis and secretion of cytokines, generation of reactive oxidants via the respiratory burst, as well as promotion of phagosome maturation (Korchak *et al.*, 1988; Malik *et al.*, 2001, 2000; Jaconi *et al.*, 1990). The contribution that each of these Ca^{2+}-dependent antimicrobial responses makes to the killing of ingested microbes remains to be determined, as does the means by which $[Ca^{2+}]_c$ regulates these complex physiologic functions. We have recently determined that phagocytosis-associated increases in $[Ca^{2+}]_c$ stimulate the translocation of the Ca^{2+} effector protein, calmodulin (CaM), from cytosol to the phagosome membrane (Malik *et al.*, 2001). Furthermore, this Ca^{2+}-dependent increase in phagosomal CaM is coupled to localized activation of

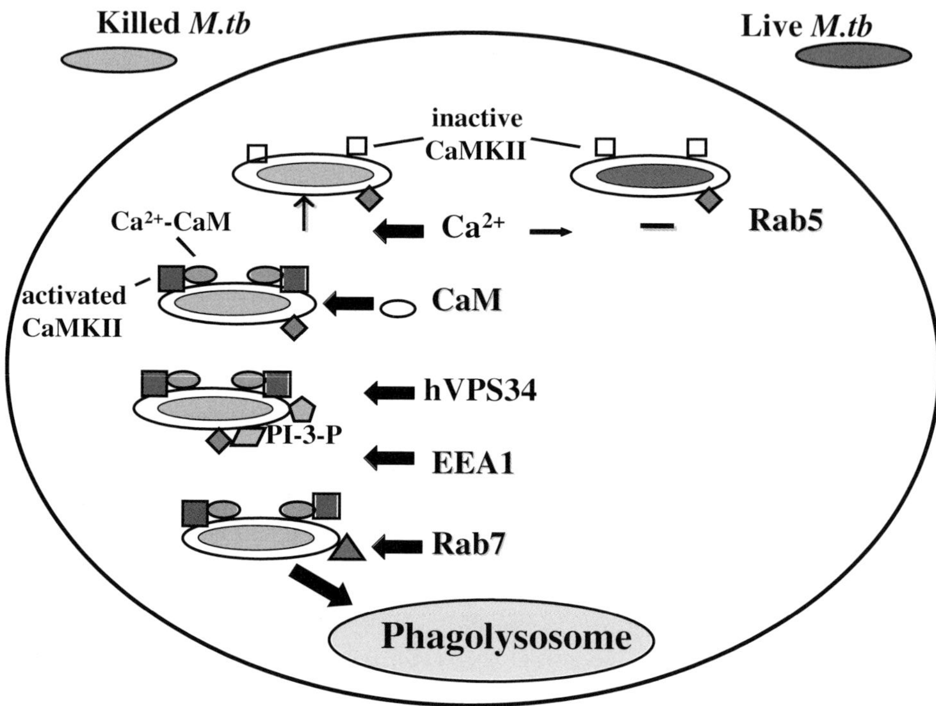

Figure 2. Regulate of phagosome maturation in macrophages and its inhibition by live *M. tuberculosis*. The signal transduction pathways that regulate the maturation of macrophage phagosomes to phagolysosomes are illustrated on the left side of the diagram, utilizing killed *M. tuberculosis* as an example. Elevation of cytosolic Ca^{2+} levels results in recruitment of CaM to the phagosome membrane and focal conversion of inactive CaMKII to its activated form . The PI-specific PI 3-kinase, hVPS34, is also recruited to the surface of the phagosome and catalyzes the formation of PI-3-P, which, in turn, recruits the Rab5 effector EEA-1. The loss of Rab5 and recruitment of Rab7 signals the terminal stages of phagosome maturation. The marked contrast with live *M. tuberculosis* is illustrated on the right side of the diagram. Mycobacterial-inhibition of Ca^{2+}-signaling results in the failure of subsequent maturation, and the phagosome is restricted to a Rab5-containing recycling endosome-like phenotype. The presence of the vacuolar H+-ATPase on maturing phagosomes, and its absence from those containing live *M. tuberculosis*, are not illustrated, but are described in the text.

Ca^{2+}/CaM-dependent protein kinase II (CaMKII) on the phagosome surface. Specific inhibition of either CaM or CaMKII results in decreased phagosome maturation, even in the setting of elevated levels of [Ca^{2+}]$_c$. In the case of live *M. tuberculosis*, the inhibition of MP Ca^{2+}-signaling results in a lack of CaM recruitment to the membrane of the nascent phagosome and a corresponding lack of activation of CaMKII (Malik *et al.*, 2001). The role of this Ca^{2+}/CaM/ CaMKII pathway in the regulation of phagosome maturation (Figure 2) parallels its function in the control of endosome-endosome fusion in MPs (Hodgkin *et al.*, 1997), further illustrating the tight coupling of phagosome maturation to endosomal-lysosomal trafficking. Of interest, CaM also functions in regulating the fusion of vesicles with the yeast vacuole, the acidic digestive organelle that is functionally analogous to mammalian lysosomes (Peters *et al.*, 2001a; Peters, 1999).

Alterations in Rab GTPases

Low molecular weight GTPases of the Rab family regulate several steps in vesicular trafficking (Rybin *et al.*, 1996; Mukhopadhyay *et al.*, 1997a). Of specific relevance to phagosome maturation, Rab5 regulates interactions with early endosomes, whereas Rab7 performs a similar function for the late endosomal compartment (Rybin *et al.*, 1996; Mukhopadhyay *et al.*, 1997a, 1997b). BCG-containing phagosomes in the J774 murine macrophage-like cell line acquire Rab5, but not Rab7 (Via *et al.*, 1997). Thus, it has been hypothesized that the failure to remove Rab5 and/or acquire Rab7 is causally related to mycobacterial inhibition of phagosome maturation. Currently, there is no experimental evidence directly testing this hypothesis via modulation of the levels of Rab5 and Rab7 on mycobacterial phagosomes, and determining the effect of this alteration on phagosome. Interestingly, the levels of Rab5 do not differ between BCG phagosomes and those containing latex beads (Via *et al.*, 1997; Fratti *et al.*, 2001). However, despite this similarity in Rab5 levels, the Rab5 effector, Early Endosomal Antigen 1 (EEA1), which is known to regulate trafficking of early endosomes, is absent from BCG phagosomes, suggesting that the Rab5 on these vesicles may not be functional (Fratti *et al.*, 2001). In addition to Rab5 and EEA1, a third signaling component in this pathway is the Class III phosphatidylinositol (PI) 3-kinase, hVPS34, which catalyzes the conversion of PI to PI-3-phosphate (PI-3-P) (Vieira *et al.*, 2001). The acquisition of EEA1 by control latex bead phagosomes is blocked by chemical inhibitors of PI-3-kinase or injection of Abs specific for hVPS34 (Fratti *et al.*, 2001). Furthermore, these maneuvers, as well as injection of anti-EEA1 Abs, reduced the maturation of latex bead phagosomes. Conversely, phagosomes containing BCG did not acquire detectable EEA1 and exhibited delayed accumulation of hVPS34. Coating of latex beads with purified LAM from *M. tuberculosis* inhibited phagosomal acquisition of EEA1 and blocked maturation (Fratti *et al.*, 2001), suggesting a direct inhibitory role for this surface glycolipid of tubercle bacilli. However, the activity of hVPS34 and the levels of its product PI-3-P were not determined in this study. Nevertheless, these results are consistent with a model in which defective signaling via the Rab5 effector EEA1 and hVPS34 contributes to mycobacterial inhibition of phagosome maturation (Figure 2). Important unresolved questions include: (1) Is the activity of hVPS34 and the levels of its product, PI-3-P, decreased on the surface of mycobacterial phagosomes? (2) Does the Rab5/PI 3-kinase/EEA1 pathway regulate phagosome maturation in human macrophages? (3) Is this pathway disrupted by virulent *M. tuberculosis*?, and (4) Would reversal of mycobacterial inhibition promote the maturation of phagosomes containing live tubercle bacilli?

Conflicting data have been presented from a human epithelial cell culture system in which infection of HeLa cells with virulent *M. tuberculosis* results in phagosomes that share several markers characteristic of those found in human macrophages (Clemens *et al.*, 2000a, 2000b). In this system, phagosomes containing live *M. tuberculosis* retain Rab5 for up to 72 hr, in marked contrast to the transient acquisition of Rab5 by phagosomes containing killed tubercle bacilli or latex beads (Clemens *et al.*, 2000a). Furthermore, in direct contrast to the results reviewed above, abundant levels of Rab 7 are detected on phagosomes containing live *M. tuberculosis* (Clemens *et al.*, 2000b). These differences in levels of Rab5 and Rab7 are difficult to reconcile considering the differences in host cells (J774 vs. HeLa) and organisms (BCG vs. *M. tuberculosis*). It is likely that studies with human macrophages infected with virulent *M. tuberculosis* will be required to settle these important questions concerning the role of Rab GTPases as regulators of phagosome maturation and as key targets of mycobacterial inhibition. Since Rab5 and Rab7 have been difficult to detect by immunofluorescence or immunoelectron microscopy in human macrophages (Clemens *et al.*, 2000a, 2000b), it is likely that purification of phagosomes, followed by Western blotting with Abs specific for Rab5 or Rab7 will be required.

The Vacuolar Proton-ATPase

A distinguishing feature of lysosomes that is essential to their degradative function is the maintenance of an acidic luminal environment. The vacuolar proton ATPase (V/H$^+$-ATPase) is primarily responsible for this acidification via the vectorial transport of protons across the lyosomal membrane, coupled to the hydrolysis of ATP (Nishi and Forgac 2002). As phagosomes containing inert particles or killed bacteria mature to phagolysosomes, they acquire the V/H$^+$-ATPase and undergo progressive acidification to a pH of approximately (5.0 - 5.5) (Hackam *et al.*, 1998). In contrast, phagosomes containing live virulent *M. tuberculosis* fail to acidify significantly (Crowle *et al.*, 1991; Xu *et al.*, 1994). In murine macrophages, these *M. tuberculosis*-containing phagosomes lack detectable V/H$^+$-ATPase (Xu *et al.*, 1994), which is likely the mechanism for defective acidification. In addition to its direct role in microbicidal activity, acidification of the phagosome and/or the presence of components of the V/H$^+$-ATPase may act as signaling elements to promote phagosome maturation. Evidence in support of this latter hypothesis includes, (a) specific inactivation of the V/H$^+$-ATPase by bafilomycin A1 blocks conversion of late endosomes to lysosomes (van Weert *et al.*, 1995), (b) fusion of vesicles with the yeast vacuoles requires the transmembrane V$_o$ component of the V/H$^+$-ATPase and luminal acidification (Peters *et al.*, 2001a) (Ungermann *et al.*, 1999), and (c) acidification of BCG-containing phagosomes in resistant mice (wild-type for the Natural Resistance-Associated Macrophage Protein 1 [N*ramp1*]) correlates with phagosome maturation, in contrast to susceptible mice (mutant N*ramp1*) in which both acidification and maturation of phagosomes are defective (Hackam *et al.*, 1998). Important unresolved questions include: (1) What is the pH and V/H$^+$-ATPasecontent of phagosomes containing live *M. tuberculosis* in human macrophages? (2) Are there alterations in other regulatory proteins known to regulate acidification of lysosomes (and presumably phagosomes), including the Na$^+$/H$^+$ exchanger and the Na$^+$/K+-ATPase (Hackam *et al.*, 1997), and (3) Are alterations in phagosomal V/H$^+$-ATPase and pH levels causally related to the block in maturation of *M. tuberculosis*-containing phagosomes, or rather an important physiologic consequence of inhibition of a more proximate step in the regulatory pathways. Of note, the diversity in luminal pH in BCG-containing phagosomes among different murine strains (Hackam *et al.*, 1998) strongly suggests that the evaluation of the pathophysiologic significance of V/H$^+$-ATPase and pH be evaluated in human macrophages utilizing virulent *M. tuberculosis*.

Emerging Technologies

Despite notable advances in our understanding of the pathogenesis of tuberculosis, including the biochemical mechanisms that regulate the interaction of *M. tuberculosis* with human macrophages, many critical questions remain. These issues have recently been clarified and shaped by concurrent developments in intersecting fields of cell biology, biochemistry, microbiology, and molecular physiology. Thus, there is a pressing need to characterize the spatial determinants of the signal transduction pathways at focal points of the bacterial-host interaction, including the forming phagocytic cup and the intersection of the nascent phagosome with distinct vesicular compartments and cytoskeletal elements. Likewise, increased temporal resolution is required to understand the sequence of the earliest biochemical signaling events, including those initiated by adherence and phagocytosis, since these are likely to be key determinants of subsequent host-pathogen interactions, including the extent of phagosome maturation, the mobilization of host antimicrobial components, and ultimately, intracellular mycobacterial viability. Advances in microscopy, including continuous video imaging, differential interference techniques, and activation-specific probes (Marshall *et al.,* 2001; McMahon *et al.,* 2002; Malik *et al.,* 2001) will undoubtedly contribute to more detailed characterization of these spatial and temporal determinants. Likewise, biophysical approaches, including Fluorescence Resonance Energy Transfer (FRET) and its derivatives, will be utilized for greater resolution of intermolecular relationships and definition of the macromolecular signaling complexes responsible for critical steps in the host-pathogen interaction (Ting *et al.,* 2001; Miyawaki and Tsien 2000). Purification of phagosomes and comprehensive characterization of proteomic and lipid profiles via mass spectrometry (Garin *et al.,* 2001; Gagnon *et al.,* 2002; Ivanova *et al.,* 2001) will revolutionize our ability to identify novel regulators of pathogenesis. Finally, the recent development of *in situ* technologies, including laser capture microdissection and whole animal imaging, for probing the interactions of specific immune cells and pathogens with the complexities of organismal physiology *in vivo* (Hooper *et al.,* 2001; Olasz *et al.,* 2002; Carlsen *et al.,* 2002) will enable more detailed examination of our experimental paradigms and the exploration of novel hypotheses regarding the pathogenesis of tuberculosis.

Conclusion

In this brief review, we have seen that the experimental approaches and conceptual frameworks of signal transduction research have contributed novel insights into the fundamental mechanisms of the molecular pathogenesis of tuberculosis. Though our focus has been primarily directed at two of the critical initial steps at the host-pathogen interface, phagocytosis and phagosome maturation, significant advances have also been made in other key aspects of the dynamic interplay between mycobacterial virulence factors and the human immune system, including modulation of macrophage signaling via stimulation of TLR2 and TLR4 (Jones *et al.,* 2001; Means *et al.,* 2001; Thoma *et al.,* 2001) and transcriptional regulation of inflammatory mediators including cytokines, chemokines, and macrophage receptors (Peters *et al.,* 2001b; DesJardin *et al.,* 2002). As our knowledge of each of these processes progresses in complexity, numerous points of intersection and counter-regulation emerge. Defining the key focal points or nodes in this biochemical signaling network is an important conceptual and experimental challenge for future research.

References

Agwu, D.E., McPhail, L.C., Sozzani, S., Bass, D.A., and McCall, C.E. 1991. Phosphatidic acid as a second messenger in human polymorphonuclear leukocytes. Effects on activation of NADPH oxidase. J. Clin. Invest. 88: 531-539.

Albina, J.E. 1995. On the expression of nitric oxide synthase by human macrophages. Why no NO? J. Leukoc. Biol. 58: 643-649.

Berton, G., Fumagalli, L., Laudanna, C., and Sorio, C. 1994. β2 integrin-dependent protein tyrosine phosphorylation and activation of the FGR protein tyrosine kinase in human neutrophils. J.Cell. Biol. 126: 1111-1121.

Bierne, H., Gouin, E., Roux, P., Caroni, P., Yin, H.L., and Cossart, P. 2001. A role for cofilin and LIM kinase in Listeria-induced phagocytosis. J. Cell. Biol. 155: 101-112.

Billah, M.M., Eckel, S., Mullmann, T.J., Egan, R.W., and Siegel, M. I. 1989. Phosphatidylcholine hydrolysis by phospholipase D determines phosphatidate and diglyceride levels in chemotactic peptide- stimulated human neutrophils. J. Biol.Chem. 264: 17069-17077.

Bjorge, J.D., Jakymiw, A., and Fujita, D J. 2000. Selected glimpses into the activation and function of Src kinase. Oncogene 19: 5620-5635.

Bloom, B.R., Mazzaccaro, R.J., Flynn, J.A., Chan, J., Sousa, A., Salgame, P., Stenger, S., Modlin, R.L., Krensky, A., Demant, P., and Kramni, I. 1999. Immunology of an infectious disease: Pathogenesis and protection in tuberculosis. Immunologist 7: 54-59.

Carlsen, H., Moskaug, J., Fromm, S.H., and Blomhoff, R. 2002. *In vivo* imaging of NF-kappa B activity. J. Immunol. 168: 1441-1446.

Caron, E. and Hall, A. 1998. Identification of two distinct mechanisms of phagocytosis controlled by different Rho GTPases. Science 282: 1717-1721.

Clemens, D.L. and Horwitz, M.A. 1995. Characterization of the *Mycobacterium tuberculosis* phagosome and evidence that phagosomal maturation is inhibited. J. Exp. Med. 181: 257-270.

Clemens, D.L. and Horwitz, M.A. 1996. The *Mycobacterium tuberculosis* phagosome interacts with early endosomes and is accessible to exogenously administered transferrin. J. Exp. Med. 184: 1349-1355.

Clemens, D.L., Lee, B.Y., and Horwitz, M.A. 2000a. Deviant expression of Rab5 on phagosomes containing the intracellular pathogens *Mycobacterium tuberculosis* and *Legionella pneumophila* is associated with altered phagosomal fate. Infect. Immun. 68: 2671.

Clemens, D.L., Lee, B.Y., and Horwitz, M.A. 2000b. *Mycobacterium tuberculosis* and *Legionella pneumophila* phagosomes exhibit arrested maturation despite acquisition of Rab7. Infect. Immun. 68: 5154.

Cornelis, G.R. 2002. The Yersinia Ysc-Yop 'Type III' weaponry. Nat. Rev. Mol. Cell. Biol. 3: 742-754.

Cossart, P., Boquet, P., Normark, S., and Rappuoli, R. 1996. Cellular microbiology emerging. Science 271: 315-316.

Crowle, A.J., Dahl, R., Ross, E., and May, M.H. 1991. Evidence that vesicles containing living, virulent *Mycobacterium tuberculosis* or *Mycobacterium avium* in cultured human macrophages are not acidic. Infect. Immun. 59: 1823-1831.

Dannenberg, A.M. and Collins, F.M. 2001. Progressive pulmonary tuberculosis is not due to increasing numbers of viable bacilli in rabbits, mice and guinea pigs, but is due to a continuous host response to mycobacterial products. Tuberculosis (Edinb) 81: 229-242.

Denmat-Ouisse, L.A., Phebidias, C., Honkavaara, P., Robin, P., Geny, B., Min, D.S., Bourgoin, S., Frohman, M.A., and Raymond, M.N. 2001. Regulation of constitutive protein transit by phospholipase D in HT29-cl19A cells. J. Biol. Chem. 276: 48840-48846.

DesJardin, L.E., Kaufman, T.M., Potts, B., Kutzbach, B., Yi, H., and Schlesinger, L.S. 2002. *Mycobacterium tuberculosis*-infected human macrophages exhibit enhanced cellular adhesion with increased expression of LFA-1 and ICAM-1 and reduced expression and/or function of complement receptors, FcgammaRII and the mannose receptor. Microbiol. 148: 3161-3171.

Desjardins, M., Huber, L.A., Parton, R.G., and Griffiths, G. 1994. Biogenesis of phagolysosomes proceeds through a sequential series of interactions with the endocytic apparatus. J.Cell Biol. 124: 677-688.

Detweiler, C.S., Cunanan, D.B., and Falkow, S. 2001. Host microarray analysis reveals a role for the Salmonella response regulator phoP in human macrophage cell death. Proc Natl Acad Sci USA 98: 5850-5855.

Ellner, J.J. 1990. Killing intracellular mycobacteria. Sources of variability in assays of the interaction of mycobacteria with mononuclear phagocytes: of mice and men. Res.Micro. 141: 237-240.

Ellner, J.J. 1997a. Regulation of the human immune response during tuberculosis. J.Lab. Clin. Medi. 130: 469-475.

Ellner, J.J. 1997b. Review: the immune response in human tuberculosis— implications for tuberculosis control. [Review] [54 refs]. J. Infect. Dis. 176: 1351-1359.

Elsbach, P. and Weiss, J. 1999. Oxygen-independent antimicrobial systems of phagocytes. In: Inflammation: Basic Principles and Clinical Correlates. Gallin, J. I., Snyderman, R., and Nathan, C. eds. New York, Lippincott-Raven. p. 801-817.

Ernst, J.D. 1998. Macrophage receptors for *Mycobacterium tuberculosis*. Infect. Immun. 66: 1277-1281.

Exton, J.H. 2002b. Phospholipase D-structure, regulation and function. Rev. Physiol. Biochem. Pharmacol. 144: 1-94.

Exton, J.H. 2002a. Phospholipase D-structure, regulation and function. Rev. Physiol. Biochem. Pharmacol. 144: 1-94.

Fallman, M., Andersson, R., and Andersson, T. 1993. Signaling properties of CR3 (CD11b/CD18) and CR1 (CD35) in relation to phagocytosis of complement-opsonized particles. J.Immunol. 151: 330-338.

Fallman, M., Gullberg, M., Hellberg, C., and Andersson, T. 1992. Complement receptor-mediated phagocytosis is associated with accumulation of phosphatidycholine-derived diglyceride in human neutrophils. J.Biol.Chem. 267: 2656-2663.

Ferguson, J.S. and Schlesinger, L.S. 2000. Pulmonary surfactant in innate immunity and the pathogenesis of tuberculosis. Tuber. Lung Dis. 80: 173-184.

Fratazzi, C., Manjunath, N., Arbeit, R.D., Carini, C., Gerken, T.A., Ardman, B., Remold, O., and Remold, H.G. 2000. A macrophage invasion mechanism for mycobacteria implicating the extracellular domain of CD43. J. Exp. Med. 192: 183-192.

Fratti, R.A., Backer, J.M., Gruenberg, J., Corvera, S., and Deretic, V. 2001. Role of phosphatidylinositol 3-kinase and Rab5 effectors in phagosomal biogenesis and mycobacterial phagosome maturation arrest. J. Cell. Biol. 154: 631-644.

Fruman, D.A. and Cantley, L.C. 2002. Phosphoinositide 3-kinase in immunological systems. Seminars in Immunol. 14: 7-18.

Fuortes, M., Jin, W., and Nathan, C. 1993. Adhesion-dependent protein tyrosine phosphorylation in neutrophils treated with tumor necrosis factor. J. Cell Biol. 120: 777-784.

Fuortes, M., Jin, W., and Nathan, C. 1994. $\beta2$ integrin-dependent tyrosine phosphorylation of paxillin in human neutrophils treated with tumor necrosis factor. J. Cell Biol. 127: 1477-1483.

Fuortes, M., Melchior, M., Han, H., Lyon, G.J., and Nathan, C. 1999. Role of the tyrosine kinase pyk2 in the integrin-dependent activation of human neutrophils by TNF. J. Clin. Invest. 104: 327-335.

Gagnon, E., Duclos, S., Rondeau, C., Chevet, E., Cameron, P.H., Steele-Mortimer, O., Paiement, J., Bergeron, J.J., and Desjardins, M. 2002. Endoplasmic reticulum-mediated phagocytosis is a mechanism of entry into macrophages. Cell 110: 119-131.

Galan, J.E. and Zhou, D. 2000. Striking a balance: modulation of the actin cytoskeleton by Salmonella. Proc. Natl. Acad. Sci. USA. 97: 8754-8761.

Garin, J., Diez, R., Kieffer, S., Dermine, J.F., Duclos, S., Gagnon, E., Sadoul, R., Rondeau, C., and Desjardins, M. 2001. The Phagosome proteome: Insight into phagosome functions. J. Cell Biol. 152: 165.

Gumperz, J.E. and Brenner, M.B. 2001. CD1-specific T cells in microbial immunity. Curr. Opin. Immunol. 13: 471-478.

Haas, A. 1998. Reprogramming the phagocytic pathway—intracellular pathogens and their vacuoles (review). Mol. Membr. Biol. 15: 103-121.

Hackam, D.J., Rotstein, O.D., Zhang, W.J., Demaurex, N., Woodside, M., Tsai, O., and Grinstein, S. 1997. Regulation of phagosomal acidification. Differential targeting of Na+/H+ exchangers, Na+/K+-ATPases, and vacuolar-type H+-atpases. J. Biol. Chem 272: 29810-29820.

Hackam, D.J., Rotstein, O.D., Zhang, W.J., Gruenheid, S., Gros, P., and Grinstein, S. 1998. Host resistance to intracellular infection: mutation of the natural resistance-associated macrophage protein 1 (Nramp1) impairs phagosomal acidification. J. Exp. Med. 188: 351-364.

Hirsch, C.S., Ellner, J.J., Russell, D.G., and Rich, E.A. 1994. Complement receptor-mediated uptake and tumor.necrosis factor-a- mediated growth inhibition of *Mycobacterium tuberculosis* by human alveolar macrophages. J. Immunol. 152: 743-753.

Hodgkin, M., Rose, S., Clark, J., and Wakelam, M. 1997. Arf-interacting proteins regulate phospholipase D in human leukaemic cells. Bioch. Soc. Trans. 25: S587.

Hooper, L.V., Wong, M.H., Thelin, A., Hansson, L., Falk, P.G., and Gordon, J.I. 2001. Molecular analysis of commensal host-microbial relationships in the intestine. Science 291: 881-884.

Humeau, Y., Vitale, N., Chasserot, G., Dupont, J.L., Du, G., Frohman, M.A., Bader, M.F., and Poulain, B. 2001. A role for phospholipase D1 in neurotransmitter release. Proc. Natl. Acad. Sci. USA. 98: 15300-15305.

Ivanova, P.T., Cerda, B.A., Horn, D.M., Cohen, J.S., McLafferty, F.W., and Brown, H.A. 2001. Electrospray ionization mass spectrometry analysis of changes in phospholipids in RBL-2H3 mastocytoma cells during degranulation. Proc. Natl. Acad. Sci. USA. 98: 7152-7157.

Iwasaki, Y., Horiike, S., Matsushima, K., and Yamane, T. 1999. Location of the catalytic nucleophile of phospholipase D of Streptomyces antibioticus in the C-terminal half domain. Eur. J. Biochem. 264: 577.

Jaconi, M.E., Lew, D.P., Carpentier, J.L., Magnusson, K.E., Sjogren, M., and Stendahl. 1990. Cytosolic free calcium elevation mediates the phagosome-lysosome fusion during phagocytosis in human neutrophils. J. Cell. Biol. 110: 1555-1564.

Jones, B.W., Means, T.K., Heldwein, K.A., Keen, M.A., Hill, P.J., Belisle, J.T., and Fenton, M.J. 2001. Different Toll-like receptor agonists induce distinct macrophage responses. J. Leukoc. Biol. 69: 1036-1044.

Kanaho, Y., Kanoh, H., Saitoh, K., and Nozawa, Y. 1991. Phospholipase D activation by platelet-activating factor, leukotriene B4, and formyl-methionyl-leucyl-phenylalanine in rabbit neutrophils. Phospholipase D activation is involved in enzyme release. J. Immunol. 146: 3536-3541.

Kanaho, Y., Nakai, Y., Katoh, M., and Nozawa, Y. 1993. The phosphatase inhibitor 2, 3-diphosphoglycerate interferes with phospholipase D activation in rabbit neutrophils. J. Biol. Chem. 268: 12492.

Kang, B.K. and Schlesinger, L.S. 1998. Characterization of mannose receptor-dependent phagocytosis mediated by *Mycobacterium tuberculosis* lipoarabinomannan. Infect. Immun. 66: 2769-2777.

Knutson, K.L., Hmama, Z., Herrera-Velit, P., Rochford, R., and Reiner, N.E. 1998. Lipoarabinomannan of *Mycobacterium tuberculosis* promotes protein tyrosine dephosphorylation and inhibition of mitogen-activated protein kinase in human mononuclear phagocytes. J.Biol.Chem. 273: 645-652.

Korade, M. and Corey, S.J. 2000. Src kinase-mediated signaling in leukocytes. J. Leukoc Biol. 68: 603-613.

Korchak, H.M., Vosshall, L.B., Zagon, G., Ljubich, P., Rich, A.M., and Weissmann, G. 1988. Activation of the neutrophil by calcium-mobilizing ligands. I. A chemotactic peptide and the lectin concanavalin A stimulate superoxide anion generation but elicit different calcium movements and phosphoinositide remodeling. J. Biol. Chem. 263: 11090-11097.

Kusner, D.J., Hall, C.F., and Jackson, S. 1999. Fc gamma receptor-mediated activation of phospholipase D regulates macrophage phagocytosis of IgG-opsonized particles. J. Immunol. 162: 2266-2274.

Kusner, D.J., Hall, C.F., and Schlesinger, L.S. 1996. Activation of phospholipase D is tightly coupled to the phagocytosis of *Mycobacterium tuberculosis* or opsonized zymosan by human macrophages. J. Exp. Med. 184: 585-595.

Kusner, D.J. and King, C.H. 1989. Protease-modulation of neutrophil superoxide response. J. Immunol. 143: 1696-1702.

Madhani, H.D. 2001. Accounting for specificity in receptor tyrosine kinase signaling. Cell 106: 9-11.

Malik, Z.A., Denning, G.M., and Kusner, D.J. 2000. Inhibition of Ca+2 signaling by *Mycobacterium tuberculosis is* associated with decreased phagosome-lysosome fusion and increased survival within human macrophages. J. Exp. Med. 191, 287-303.

Malik, Z.A., Iyer, S.S., and Kusner, D.J. 2001. *Mycobacterium tuberculosis* phagosomes exhibit altered calmodulin-dependent signal transduction: Contribution to inhibition of phagosome-lysosome fusion and intracellular survival in human macrophages. J. Immunol. 166: 3392-3401.

Malik, Z.A., Thompson, C.R., H.S., Porter, B., Iyer, S.S., and Kusner, D.J. 2003. *Mycobacterium tuberculosis* blocks Ca^{2+}-signaling and phagosome maturation in human macrophages via specific inhibition of sphingosine kinase. J. Immunol. 170: 2811-2815.

Marshall, J.G., Booth, J.W., Stambolic, V., Mak, T., Balla, T., Schreiber, A.D., Meyer, T., and Grinstein, S. 2001. Restricted accumulation of phosphatidylinositol 3-kinase products in a plasmalemmal subdomain during Fc gamma receptor-mediated phagocytosis. J. Cell Biol. 153: 1369-1380.

Martin, G.S. 2001. The hunting of the Src. Nat. Rev. Mol. Cell. Biol. 2: 467-475.

Master, S., Zahrt, T.C., Song, J., and Deretic, V. 2001. Mapping of *Mycobacterium tuberculosis katG* promoters and their differential expression in infected macrophages. J. Bacteriol. 183: 4033-4039.

Mayorga, L.S., Bertini, F., and Stahl, P.D. 1991. Fusion of newly formed phagosomes with endosomes in intact cells and in a cell-free system. J. Biol. Chem. 266: 6511-6517.

McKinney, J.D., Honer zu, B.K., Munoz-Elias, E.J., Miczak, A., Chen, B., Chan, W.T., Swenson, D., Sacchettini, J.C., Jacobs, W.R., Jr., and Russell,

D.G. 2000. Persistence of *Mycobacterium tuberculosis* in macrophages and mice requires the glyoxylate shunt enzyme isocitrate lyase. Nature 406: 735-738.

McMahon, P.J., Barone, N., Allman, B.E., and Nugent, K.A. 2002. Quantitative phase-amplitude microscopy II: differential interference contrast imaging for biological TEM. J. Microsc.206: 204-208.

Means, T.K., Jones, B.W., Schromm, A.B., Shurtleff, B.A., Smith, J.A., Keane, J., Golenbock, D.T., Vogel, S.N., and Fenton, M.J. 2001. Differential effects of a Toll-like receptor antagonist on *Mycobacterium tuberculosis*-induced macrophage responses. J. Immunol. 166: 4074-4082.

Means, T.K., Wang, S., Lien, E., Yoshimura, A., Golenbock, D.T., and Fenton, M.J. 1999. Human toll-like receptors mediate cellular activation by *Mycobacterium tuberculosis*. J. Immunol. 163: 3920-3927.

Melendez, A.J. and Allen, J.M. 2002. Phospholipase D and immune receptor signalling. Semin. Immunol. .14: 49-55.

Melendez, A.J., Bruetschy, L., Floto, R.A., Harnett, M.M., and Allen, J.M. 2001. Functional coupling of Fcgamma RI to nicotinamide adenine dinucleotide phosphate (reduced form) oxidative burst and immune complex trafficking requires the activation of phospholipase D1. Blood 98: 3421.

Meresse, S., Steele-Mortimer, O., Moreno, E., Desjardins, M., Finlay, B., and Gorvel, J. P. 1999. Controlling the maturation of pathogen-containing vacuoles: a matter of life and death. Nat. Cell. Biol. 1: E183-E188.

Miyawaki, A. and Tsien, R.Y. 2000. Monitoring protein conformations and interactions by fluorescence resonance energy transfer between mutants of green fluorescent protein. Methods Enzymol. 327: 472-500.

Mosser, D.M. and Karp, C.L. 1999. Receptor mediated subversion of macrophage cytokine production by intracellular pathogens. Curr. Opin. Immunol.11: 406-411.

Mukhopadhyay, A., Barbieri, A.M., Funato, K., Roberts, R., and Stahl, P.D. 1997a. Sequential actions of Rab5 and Rab7 regulate endocytosis in the Xenopus oocyte. J. Cell Biol. 136: 1227-1237

Mukhopadhyay, A., Funato, K., and Stahl, P.D. 1997b. Rab7 regulates transport from early to late endocytic compartments in Xenopus oocytes. J. Biol.Chem 272: 13055-13059.

Murli, S., Watson, R.O., and Galan, J.E. 2001. Role of tyrosine kinases and the tyrosine phosphatase SptP in the interaction of Salmonella with host cells. Cell Microbiol. 3: 795-810.

Nandan, D., Knutson, K.L., Lo, R., and Reiner, N.E. 2000. Exploitation of host cell signaling machinery: activation of macrophage phosphotyrosine phosphatases as a novel mechanism of molecular microbial pathogenesis. J. Leukoc. Biol. 67: 464-470.

Nandan, D., Lo, R., and Reiner, N.E. 1999. Activation of phosphotyrosine phosphatase activity attenuates mitogen-activated protein kinase signaling and inhibits c-FOS and nitric oxide synthase expression in macrophages infected with *Leishmania donovani*. Infect. Immun. 67: 4055-4063.

Nathan, C.F. and Shiloh, M.U. 2000. Reactive oxygen and nitrogen intermediates in the relationship between mammalian hosts and microbial pathogens. Proc. Natl. Acad. Sci. 97: 8841-8848

Nauseef, W.M. 2001. The proper study of mankind. J. Clin. Invest. 107: 401-403.

Navarre, W.W. and Zychlinsky, A. 2000. Pathogen-induced apoptosis of macrophages: a common end for different pathogenic strategies. Cell. Microbiol. 2: 265-273.

Nishi, T. and Forgac, M. 2002. The vacuolar (H+)-ATPases—nature's most versatile proton pumps. Nat. Rev. Mol. Cell. Biol. 3: 94-103.

Olasz, E.B., Lang, L., Seidel, J., Green, M.V., Eckelman, W.C., and Katz, S.I. 2002. Fluorine-18 labeled mouse bone marrow-derived dendritic cells can be detected *in vivo* by high resolution projection imaging. J. Immunol. Methods 260: 137-148.

Ostman, A. and Bohmer, F.D. 2001. Regulation of receptor tyrosine kinase signaling by protein tyrosine phosphatases. Trends Cell Biol. 11: 258-266.

Ott, V.L. and Cambier, J.C. 2002. Introduction: multifaceted roles of lipids and their catabolites in immune cell signaling. Seminars in Immunol. 14: 1-6.

Palicz, A., Foubert, T.R., Jesaitis, A.J., Marodi, L., and McPhail, L.C. 2001. Phosphatidic acid and diacylglycerol directly activate NADPH oxidase by interacting with enzyme components. J. Biol. Chem. 276: 3090-3097.

Pelletier, L., Stern, C.A., Pypaert, M., Sheff, D., Ngo, H.M., Roper, N., He, C.Y., Hu, K., Toomre, D., Coppens, I., Roos, D.S., Joiner, K.A., and Warren, G. 2002. Golgi biogenesis in *Toxoplasma gondii*. Nature 418: 548-552.

Peters, C. 1999. Ca 2+/calmodulin signals the completion of docking and triggers a late step of vacuole fusion. Mayer, A. Nature 396, 575-580.

Peters, C., Bayer, M.J., Buhler, S., Andersen, J.S., Mann, M., and Mayer, A. 2001a. Trans-complex formation by proteolipid channels in the terminal phase of membrane fusion. Nature 409: 581-588.

Peters, W., Scott, H.M., Chambers, H.F., Flynn, J.L., Charo, I.F., and Ernst, J.D. 2001b. Chemokine receptor 2 serves an early and essential role in resistance to *Mycobacterium tuberculosis*. Proc. Natl. Acad. Sci. USA. 98: 7958-7963.

Peterson, P.K., Gekker, G., Hu, S., Sheng, W.S., Anderson, W.R., Ulevitch, R.J., Tobias, P.S., Gustafson, K.V., Molitor, T.W., and Chao, C.C. 1995. CD14 receptor-mediated uptake of nonopsonized *Mycobacterium tuberculosis* by human microglia. Infect. Immun. 63: 1598-1602.

Ramachandra, L., Noss, E., Boom, W.H., and Harding, C.V. 2001. Processing of *Mycobacterium tuberculosis* antigen 85B involves intraphagosomal formation of peptide-major histocompatibility complex II complexes and is inhibited by live bacilli that decrease phagosome maturation. J. Exp. Med. 194: 1421-1432.

Redpath, S., Ghazal, P., and Gascoigne, N.R. 2001. Hijacking and exploitation of IL-10 by intracellular pathogens. Trends Microbiol. 9: 86-92.

Regier, D.S., Greene, D.G., Sergeant, S., Jesaitis, A.J., and McPhail, L.C. 2000. Phosphorylation of p22phox is mediated by phospholipase D-dependent and -independent mechanisms. Correlation of NADPH oxidase activity and p22phox phosphorylation. J. Biol. Chem. 275: 28406-28412.

Roy, C.R. and Tilney, L.G. 2002. The road less traveled: transport of Legionella to the endoplasmic reticulum. J. Cell. Biol. 158: 415-419.

Russell, D.G. 2001. *Mycobacterium tuberculosis*: here today, and here tomorrow. Nat. Rev. Mol. Cell. Biol. 2: 569-577.

Rybin, V., Ullrich, O., Rubino, M., Alexandrov, K., Simon, I., Seabra, M.C., Goody, R., and Zerial, M. 1996. GTPase activity of Rab5 acts as a timer for endocytic membrane fusion. Nature 383: 266-269.

Sanguedolce, M., Capo, C., Bouhamdan, M., Bongrand, P., Huang, C., and Mege, J. 1993. Zymosan-induced tyrosine phosphorylations in human monocytes. J. Immunol. 151: 405-414.

Schlesinger, L S. 1993. Macrophage phagocytosis of virulent but not attenuated strains of *Mycobacterium tuberculosis is* mediated by mannose receptors in addition to complement receptors. J.Immunol. 150: 2920-2930.

Schlesinger, L.S. 1996. Role of mononuclear phagocytes in *M. tuberculosis* pathogenesis. J. Invest. Med. 44: 312-323, 1996.

Schlesinger, L S., Bellinger-Kawahara, C.G., Payne, N.R., and Horwitz, M.A. 1990. Phagocytosis of *Mycobacterium tuberculosis* is mediated by human monocyte complement receptors and complement component C3. J. Immunol. 144: 2771-2780.

Schlesinger, L.S., Hull, S.R., and Kaufman, T.M. 1994. Binding of the terminal mannosyl units of lipoarabinomannan from a virulent strain of *Mycobacterium tuberculosis* to human macrophages. J.Immunol. 152: 4070-4079.

Serrander, L., Fallman, M., and Stendahl, O. 1996. Activation of phospholipase D is an early event in integrin-mediated signalling leading to phagocytosis in human neutrophils. Inflammation 20: 439-450.

Spiegel, S. and Milstien, S. 2002. Sphingosine 1-phosphate, a key cell signaling molecule. J. Biol. Chem. 277: 25851.

Stenger, S., Mazzaccaro, R.J., Uyemura, K., Cho, S., Barnes, P.F., Rosat, J.P., Sette, A., Brenner, M.B., Porcelli, S.A., Bloom, B.R., and Modlin, R.L. 1997. Differential effects of cytolytic T cell subsets on intracellular infection. Science 276: 1684-1687.

Thoma, U., Stenger, S., Takeuchi, O., Ochoa, M.T., Engele, M., Sieling, P.A., Barnes, P.F., Rollinghoff, M., Bolcskei, P.L., Wagner, M., Akira, S., Norgard, M.V., Belisle, J.T., Godowski, P.J., Bloom, B.R., and Modlin, R.L. 2001. Induction of direct antimicrobial activity through mammalian toll-like receptors. Science 291: 1544-1547.

Thomassen, M.J. and Kavuru, M.S. 2001. Human alveolar macrophages and monocytes as a source and target for nitric oxide. Int. Immunopharmacol. 1: 1479-1490.

Ting, A.Y., Kain, K.H., Klemke, R.L., and Tsien, R.Y. 2001. Genetically encoded fluorescent reporters of protein tyrosine kinase activities in living cells. Proc. Natl. Acad. Sci. USA. 98: 15003-15008.

Ungermann, C., Wickner, W., and Xu, Z. 1999. Vacuole acidification is required for trans-SNARE pairing, LMA1 release, and homotypic fusion. Proc. Natl. Acad. Sci. USA. 96: 11194-11199.

van Weert, A.W.M., Dunn, K.W., Geuze, H.J., Maxfield, F.R., and Stoorvogel, W. 1995. Transport from late endosomes to lysosomes, but not sorting of integral membrane proteins in endosomes, depends on the vacuolar proton pump. J. Cell Biol. 130: 821-834.

Via, L.E., Deretic, D., Ulmer, R.J., Hibler, N.S., Huber, L.A., and Deretic, V. 1997. Arrest of mycobacterial phagosome maturation is caused by a block in vesicle fusion between stages controlled by rab5 and rab7. J. Biol. Chem 272: 13326-13331.

Vieira, O.V., Botelho, R.J., Rameh, L., Brachmann, S.M., Matsuo, T., Davidson, H.W., Schreiber, A., Backer, J.M., Cantley, L.C., and Grinstein, S. 2001. Distinct roles of class I and class III phosphatidylinositol 3-kinases in phagosome formation and maturation. J. Cell Biol. 155: 19-25.

Waite, K.A., Wallin, R., Qualliotine-Mann, D., and McPhail, L.C. 1997. Phosphatidic acid-mediated phosphorylation of the NADPH oxidase component p47-phox. Evidence that phosphatidic acid may activate a novel protein kinase. J. Biol. Chem 272: 15569-15578.

Weiss, J., DeLeo, F.R., and Nauseef, W.M. 2003. Antimicrobial activity of host cells.Molecular Cellular Microbiology. London, Academic Press.

Weiss, S. J. 1989. Tissue destruction by neutrophils. N. Engl. J. Med. 320: 365-376.

Xie, M., Jacobs, L.S., and Dubyak, G.R. 1991. Regulation of phospholipase D and primary granule secretion by P2-purinergic- and chemotactic peptide-receptor agonists is induced during granulocytic differentiation of HL-60 cells. J. Clin. Invest. 88: 45-54.

Xu, S., Cooper, A., Sturgill-Koszycki, S., van Heyningen, T., Chatterjee, D., Orme, I., Allen, P., and Russell, D.G. 1994. Intracellular trafficking in *Mycobacterium tuberculosis* and *Mycobacterium avium*-infected macrophages. J. Immunol. 153: 2568-2578.

Zaffran, Y., Escallier, J.C., Ruta, S., Capo, C., and Mege, J.L. 1995. Zymosan-triggered association of tyrosine phosphoproteins and lyn kinase with cytoskeleton in human monocytes. J. Immunol. 154: 3488-3497.

From: Tuberculosis: The Microbe Host Interface
Edited by: Larry S. Schlesinger and Lucy E. DesJardin

Chapter 4

The Acquired Immune Response to *M. tuberculosis*

W. Henry Boom

Abstract

M. tuberculosis remains one of the most successful human pathogens. The ability of *M. tuberculosis* to elicit vigorous acquired immune responses and use of the macrophage as primary cell to infect, suggest that the organism has evolved multiple strategies to survive and persist in the face of innate and acquired immune responses. Persistence of *M. tuberculosis* in otherwise healthy persons is one of the hallmarks of this organism. Survival and persistence require not only resistance to microbicidal mechanisms of phagocytes but also avoidance of recognition by multiple T cell subsets. The acquired immune response to *M. tuberculosis* requires participation by multiple T cells subsets. These include not only a central role for MHC-II restricted CD4+ T cells, but also MHC-I restricted CD8+, gamma-delta and CD-1 restricted T cells. These diverse T cell populations recognize a wide range of mycobacterial antigens, but share overlapping functions such as secretion of IFN-γ and TNF-α, CTL function, and ability to provide cell-contact dependent help to macrophages. How T cell responses are regulated *in vivo* and their roles in different stages of

M. tuberculosis infection and in protective immunity remain to be determined. Answers to these questions will impact vaccine development as well as understanding the host-pathogen interaction in *M. tuberculosis* infection.

Natural History of *M. tuberculosis* Infection

Aerosolized *M. tuberculosis* bacilli are efficiently transmitted from person to person. Only small numbers of bacilli need enter the distal alveoli of human lungs to establish infection. In most persons, local innate immunity in the lung, mediated primarily by alveolar macrophages, fails to control slowly replicating tubercle bacilli. This exposes the immune system to increasing amounts of mycobacterial antigen and allows acquired immunity to develop. In most healthy adults, acquired immunity mediated by T cells controls but does not eradicate *M. tuberculosis* infection. One third of the world's population is thought to harbor persistent *M. tuberculosis* bacilli and ongoing immune surveillance by T cells is required to maintain control over them (Raviglione *et al.,* 1995). Acquired immunity likely also protects against re-infection, which is important in parts of the world with high levels of *M. tuberculosis* transmission.

During the last ten years substantial progress has been made in understanding protective immunity to *M. tuberculosis*, which is principally dependent on acquired cellular immune mechanisms. There is no experimental or clinical evidence to suggest that humoral immune mechanisms (B cells, antibodies) have a major protective role. *M. tuberculosis* is not dependent on opsonizing antibodies or Fc receptor (FcR) for uptake by macrophages. *M. tuberculosis* infection is certainly associated with development of antibodies to mycobacterial antigens and patients with active tuberculosis have substantially elevated levels of such antibodies but their significance is unknown. They do not appear to benefit the host and may in some circumstances be detrimental.

The interaction of T cells and infected macrophages is central to protective immunity to *M. tuberculosis*. CD4+ T cells have an essential role but are supported by other T cell subsets such as CD8+, γδ TCR+ T cells (γδ T cells), and CD1 restricted T cells(Boom 1996). Antigens for these T cells are being defined, and their function in the immune response is starting to be elucidated. Insight has been gained into mechanisms used by macrophages to control *M. tuberculosis*. How T cells help them perform this task, however, remains poorly understood. TNF-alpha, IL-12 and IFN-gamma are central cytokines in regulation and effector phase of immune responses to *M. tuberculosis*.

Macrophages are not only primary effector cells for control of *M. tuberculosis* but also essential for processing and presentation of antigens to T cells. To survive (and thrive) in macrophages, *M. tuberculosis* has not only evolved mechanisms to live through the initial encounter with macrophages but also to block acquired immunity. Modulation of phagosomes, neutralization of macrophage effector molecules, inciting the secretion of inhibitory cytokines, and interference with processing of antigens for T cells, all represent strategies for bacterial survival. The relative importance of different T cell subsets and mechanisms employed by *M. tuberculosis* to interfere with macrophage and T cell function likely depends on the stage of infection. During primary infection an acute acquired immune response develops in the lung as innate immune mechanisms fail to control dividing bacilli. As replicating bacteria are controlled by activated T cells and macrophages, immune responses are down regulated, and the infection enters a chronic phase, in which memory T cells help macrophages in granulomas control persistent bacilli and provide surveillance against re-infection. Failure of acquired immunity during the acute and/or chronic phases allows *M. tuberculosis* infection to become clinically apparent, most commonly with pulmonary manifestations but in some with extra-pulmonary or disseminated hematogenous disease. The balance of the interaction between T cells and infected macrophages determines the outcome of the host-pathogen interaction in *M. tuberculosis* infection. The ability to resist microbicidal functions and to modulate antigen processing and presentation allows *M. tuberculosis* to survive inside macrophages for many years. When host immunity fails, expectoration of reactivated bacilli allows the organism to seek a new host, perpetuating a cycle that allows *M. tuberculosis* to remain one of the most successful human pathogens.

Models and Methods Used for Analysis of Acquired Immunity to *M. tuberculosis*: "Of Mice and Men"

Current knowledge of the complex interactions between T cells and macrophages in *M. tuberculosis* infection has been obtained using *in vivo* and *in vitro* models. *M. tuberculosis* infection in mice remains the primary animal model for analysis of acquired immunity to this organism *in vivo*. It is particularly useful for analysis of acute primary infection, for determination of local immune responses in the lung and for testing of new vaccines. The murine model is inadequate for studies of mycobacterial persistence, the condition of most healthy tuberculin skin test positive persons, and for studies of reactivation disease, the most common form of tuberculosis in adults. Other animal models of *M. tuberculosis* infection such as the guinea pig and rabbit do not lend themselves to sophisticated immunological analysis because of a lack of reagents to measure cytokines and cell surface markers, and of animals with genes deleted or over-expressed. Primates have recently been re-introduced as animal model for tuberculosis. Immunologic reagents are available for primates and the disease appears to mimic different stages of human *M. tuberculosis* infection. Primates can be used also to the study the interaction of human immunodeficiency virus-1 (HIV-1) and *M. tuberculosis*. Research in the immunology of *M. tuberculosis* has benefited from the use of a range of *in vivo* and *in vitro* models in animals and *in vitro* studies with human cells from both ill and healthy infected persons. Integration of findings from different experimental approaches and recognition of their limitations has allowed studies of immunity to *M. tuberculosis* to make rapid progress. The mouse model has been invaluable and very informative. For example, the importance of TNF–α as cytokine in *M. tuberculosis* immunity was established in murine gene knock-out systems well before recent clinical studies revealed that prolonged treatment with anti-TNF–α antibodies for chronic arthritis could reactivate persistent *M. tuberculosis* (Keane *et al.*, 2001).

Studies of human immune responses depend on *ex vivo* analysis of cells obtained from peripheral blood or from lung from persons in different stages of *M. tuberculosis* infection . Such cells are used as primary populations or as source of T cell lines and clones. Studies of cells from persons at different stages of infection (primary, latent, active tuberculosis) or contacts of active TB cases (tuberculin skin test negative and positive) and appropriate controls provide insight into human immunity. Acute primary *M. tuberculosis* infection in humans is difficult to identify. However, careful epidemiologic studies, such as households of tuberculosis patients, can identify persons with recent close contact and exposure to persons with active tuberculosis. Such studies can identify persons with acute primary *M. tuberculosis* infection, defined as tuberculin skin test conversion from negative to positive. Analysis of immune responses during skin test conversion will provide insight into acquired protective immunity, since the vast majority of such persons successfully control the primary infection.

The availability of increasing numbers of well-defined antigens and immuno-modulatory molecules of *M. tuberculosis* has greatly increased the specificity of assays employed to probe immune responses. Basic assays include multi-color flow cytometry in which increasing numbers of surface and intracellular molecules are measured on the same cell. The immunospot or elispot assay allows quantification of precursor frequencies of antigen-reactive T cells. The immunospot assay is more sensitive and reproducible than flow cytometric measures of cytokine producing cells in *M. tuberculosis* infection. Increased sensitivity is necessary because the range of antigens and diversity of responses among *M. tuberculosis* infected persons is broad and, to date, lacking in clear-cut immuno-dominance. Cytokine responses are measured by ELISA in supernatants of T cell cultures stimulated with different antigens, and can be used with general populations of immune cells such as peripheral blood T cells or T cells present in alveoli of *M. tuberculosis* infected persons. Alternatively T cell lines and clones of defined phenotype (CD4+, CD8+, γδTCR+, CD-1 restricted) can be used to define antigens and epitopes, for functional assays, for cytokine responses, and to measure activation markers and gene expression. In addition to cytokine production, an important function, other T cell functions include cytotoxicity (CTL), proliferation and ability to control growth of intracellular mycobacteria ("killing"). Increasingly, gene expression measured either by quantitative PCR for specific genes or by gene-chip technology for expression of large families of genes is being used to probe human immune responses.

Current research on the acquired immunity to *M. tuberculosis* is focused on characterizing more fully the repertoire of antigens recognized by CD4+, CD8+, γδ T cells and CD1 restricted T cells. The goals are to identify not only immuno-dominant protective antigens but also T cell antigens that might be detrimental to the host. Additionally, definition of roles, functions and kinetics of expansion of different T cell subsets during different stages of *M. tuberculosis* infection needs to be determined. Understanding the mechanisms used by *M. tuberculosis* to obviate acquired immunity and means developed by T cells and macrophages to overcome inhibitory properties of tubercle bacilli is necessary for further progress in understanding protective immunity to *M. tuberculosis*.

T Cells Subsets and *M. tuberculosis* Antigens

Studies in humans and animal models demonstrate that acquired immunity to *M. tuberculosis* requires contributions by multiple T cell subsets, including a dominant role for CD4+ T cells and significant roles for CD8+ and γδ T cells. The reasons for these multiple T cell subsets are not known. Differences in mechanisms for antigen processing and molecules for antigen presentation increases the diversity of T cell receptors (TCR) and co-receptors used by T cells to recognize mycobacterial antigens. The ability of *M. tuberculosis* to infect, survive and persist in macrophages, a major antigen presenting cell population, contributes further to ready T cell activation. Slow growth and chronicity of *M. tuberculosis* infection results in prolonged exposure to a diversity of antigens. From the host's perspective, diverse T cell repertoire allows recognition of a wide range of mycobacterial antigens presented by different families of antigen presenting molecules on many different cell types, and thus a greater ability to detect *M. tuberculosis*.

CD4+ αβ T Cell Receptor (TCR) Bearing T Cells (CD4+ T Cells)

CD4+ αβTCR+ T cells (CD4+ T cells) are central to the human immune response to *M. tuberculosis*. The HIV pandemic provides direct evidence that loss of CD4+ T cell number and function results in progressive primary infection, reactivation of endogenous *M. tuberculosis* and enhanced susceptibility to re-infection (Murray 1991; Hopewell 1992; Raviglione *et al.*, 1995). Early cell transfer studies in mice established that CD4+ T cells could transfer protection against *M. tuberculosis* (Lefford 1975; Orme *et al.*, 1983). *In vivo* depletion with anti-CD4 monoclonal antibodies confirmed these findings (Muller *et al.*, 1987; Pedrazzini *et al.*, 1987). Mice with deleted genes for CD4 or MHC class II (MHC-II) molecules are markedly susceptible to *M. tuberculosis*, firmly establishing a central role for CD4+ T cells in protection (Ladel *et al.*, 1995; Caruso *et al.*, 1999).

CD4+ T cells are activated by both live bacilli and soluble protein antigens (Tsukaguchi *et al.*, 1995). CD4+ T cells recognize mycobacterial peptides presented to their TCR's by MHC-II molecules on professional antigen presenting cells such as macrophages and dendritic cells. Thus soluble antigens such as purified protein derivative (PPD), mycobacterial culture filtrate (CF) or defined protein antigens can be used to generate CD4+ T cell lines and clones. Such reagents have been used extensively to characterize the antigen repertoire for human *M. tuberculosis* specific CD4+ T cells. From these studies it is clear that CD4+ T cell recognize a large number of mycobacterial proteins (Oftung *et al.*, 1987; Barnes *et al.*, 1989; Boom *et al.*, 1991; Haanen *et al.*, 1991; Havlir *et al.*, 1991; Faith *et al.*, 1992; Schoel *et al.*, 1992). For CD4+ T cells no single immunodominant antigen has emerged to date. A number of antigens are recognized by a majority of healthy tuberculin skin test positive persons. Most tuberculin skin test persons have controlled primary *M. tuberculosis* infection and therefore can be considered to have mounted a successful protective immune response. Proteins recognized by such persons include the three 30-32kD 85 complex proteins, ESAT-6 and CFP-10, the 19 and 38 kDa lipoproteins and two recently described proteins of 32 (serine protease) and 39 kDa (Havlir *et al.*, 1991; Sorensen *et al.*, 1995; Belisle *et al.*, 1997; Dillon *et al.*, 1999; Skeiky *et al.*, 1999; Dillon *et al.*, 2000; Faith *et al.*, 1991). The availability of the complete *M. tuberculosis* genome will facilitate the identification and expression of additional antigens for CD4+ T cells (Cole *et al.*, 1998).

The 30-32 kDa 85A,B and C complex proteins are mycolyl transferases involved in mycobacterial cell wall synthesis, and are restricted to mycobacteria(Wiker *et al.*, 1992; Belisle *et al.*, 1997). 85B is the major 30kD protein of *M. tuberculosis,* is recognized readily by T cells from *M. tuberculosis* infected persons, and a number of MHC class II restricted epitopes have been mapped(Havlir *et al.*, 1991; Silver *et al.*, 1995). ESAT-6 and CFP-10 are proteins encoded within the RD-1 region of the *M. tuberculosis* genome (Sorensen *et al.*, 1995; Mahairas *et al.*, 1996; Dillon *et al.*, 2000). RD-1 has been deleted in all BCG sub-strains (Behr *et al.*, 1999). These proteins thus have potential as diagnostic reagents to distinguish persons with *M. tuberculosis* infection from BCG vaccinees. CD4+ T cells from healthy tuberculin positive persons respond readily to 19 and 38 kDa lipoproteins, however these proteins have been disappointing as vaccines in animal studies (Faith *et al.*, 1991; Vordemeier *et al.*, 1991; Young *et al.*, 1991; Vordermeier *et al.*, 1992; Abou-Zeid *et al.*, 1997). Mycobacterial lipoproteins such as the 19 kDa protein activate macrophages through toll-like receptor-2 (TLR-2) (Brightbill *et al.*, 1999; Means *et al.*, 1999; Noss *et al.*, 2001). The 32 kDa serine protease and a 39 kDa protein were identified by screening antigens of *M. tuberculosis* for reactivity by T cells from healthy tuberculin skin test positive persons (Dillon *et al.*, 1999; Skeiky *et al.*, 1999). The first *M. tuberculosis* antigens cloned and expressed in large amounts for testing in humans in the early 1980's were heat shock proteins of 71, 65 and 12-14 kDa and represented mycobacterial homologues of DnaK, Gro-EL and GroES (reviewed in Young *et al.*, 1991). Although CD4+ T cell reactivity was measured, their role in protective immunity to *M. tuberculosis* appears to be minimal.

Vaccine studies in animal models combining ESAT-6 with 85B or the 32 with the 39 kDa protein have been promising, warranting their development for testing as subunit vaccines in humans. Studies in households, where index cases and contacts differ in susceptibility to infection and risk for developing active disease should allow determination of antigens and functions of CD4+ T cells necessary for protective immunity to *M. tuberculosis*. Such studies can determine if additional immuno-dominant *M. tuberculosis* antigens will emerge. Studies of T cell responses of tuberculosis patients during active disease have demonstrated diminished responses to a range of mycobacterial antigens without selective loss of CD4+ T cell response to any of the specific mycobacterial proteins studied to date.

MHC Class I Restricted CD8+ αβTCR+ T cells (CD8+ T Cells)

Early cell transfer studies in mice suggested that during acute primary infection CD8+ αβTCR+ T cells (CD8+ T cells) added to the protection provided by CD4+ T cells (Orme *et al.*, 1983). CD8+ T cell populations also could be isolated that responded to mycobacteria (DeLibero *et al.*, 1988). However studies in mice deficient in MHC class I (MHC-I) expression by deletion of the β2-microgobulin gene established a role for MHC-I restricted CD8+ T cells in protection against acute *M. tuberculosis* and *M. bovis* infection (Flynn *et al.*, 1992). Studies in mice with deleted genes for MHC-I and TAP confirmed observations made in β2-microgobulin deleted mice (Behar *et al.*, 1999; Sousa *et al.*, 2000). CD8+ T cell expansion in lungs of infected mice parallels that of CD4+ T cells, suggesting a role during primary T cell responses to *M. tuberculosis* (Serbina *et al.*, 1999; Fulton *et al.*, 2000).

Whereas cytotoxic effector function (CTL) by human CD4+ T cells to mycobacterial antigens and macrophages was readily demonstrated (see below), the discovery of human CD8+ CTL activated and responsive to mycobacterial antigens took longer. Human MHC-I restricted CD8+ T cells were first found in alveolar and peripheral T cells of healthy tuberculin skin test positive persons (Tan *et al.,* 1997). Subsequent studies found that CD8+ T cells were activated by *M. bovis*-BCG and *M. tuberculosis* (Canaday *et al.,* 1999; Smith *et al.,* 1999). The majority of *M. tuberculosis* reactive CD8+ T cells recognize mycobacterial peptides in the context of MHC-I molecules, although some may use non-classical HLA molecules (i.e. non-HLA-A, -B, -C) (Lewinsohn *et al.,* 2000). Epitope mapping studies, usually guided by epitope prediction algorithms, have defined HLA-A2 restricted epitopes on already defined *M. tuberculosis* antigens such as ESAT-6, 85B, 85A, the 19kDa lipoprotein and CFP-10 among others in tuberculin skin test positive persons (Lalvani *et al.,* 1998; Mohagheghpour *et al.,* 1998; Geluk *et al.,* 2000; Lewinsohn *et al.,* 2000; Smith *et al.,* 2000; Klein *et al.,* 2001). Elution of peptides from human MHC class I molecules (HLA-A2) have identified novel *M. tuberculosis* antigens for CD8+ T cells (Flyer *et al.,* 2002).

MHC-I and-II molecules have distinct intracellular trafficking pathways and mechanisms for loading of microbial peptides. In addition, these molecules accommodate peptides of different lengths, 9-11 amino acids for MHC-I and 10-30 for MHC-II, for presentation to T cell receptors. Differences in antigen processing pathways likely play key roles in determining the repertoire of mycobacterial antigens and epitopes recognized by CD4+ and CD8+ T cells. For example, soluble proteins are poorly processed and presented by antigen presenting cells to CD8+ T cells, because of pinocytosis does not deliver proteins efficiently to the MHC-I antigen processing pathway. In contrast, phagocytosis of whole bacteria can deliver antigens for processing and presentation by MHC-I molecules by the alternate MHC-I processing pathway (Canaday *et al.,* 1999). To what extent the antigen repertoire of CD8+ and CD4+ T cells will overlap remains to be determined as more antigens and epitopes for these two T cell subsets are defined by proteinomic and other approaches.

γδ TCR+ T Cells (γδ T Cells)

γδ TCR expressing T cells (γδ T cells) are characterized by a unique T cell antigen receptor (TCR) comprised of γ and δ chains. The majority of circulating γδ T cells in adults express Vγ9 (aka Vγ2) and Vδ2 elements (Vδ2+ T cells) (reviewed in Rojas *et al.,* 2001). *M. tuberculosis* bacilli readily activate Vδ2+ T cells (Havlir *et al.,* 1991). Individuals sensitized to mycobacterial antigens have a greater ability to activate γδ T cells in response to *M. tuberculosis* than tuberculin negative persons (Barnes *et al.,* 1992). Vaccination of adults with BCG increases *in vitro* expansion of Vδ2+ T cells after stimulation with *M. tuberculosis* antigens (Hoft *et al.,* 1998). Tuberculosis patients have a diminished ability to activate γδ T cells in response to *M. tuberculosis* (Barnes *et al.,* 1992). γδ T cells are found in the lungs of tuberculosis patients, and there is a correlation between the absence or loss of Vδ2+ T cells in blood and lung, and extent of disease(Li *et al.*, 1996). In a primate model γδ T cell number and reactivity to phosphate antigens increase during primary mycobacterial infection and during challenge after BCG vaccination, suggesting a role for γδ T cells in protective immunity to *M. tuberculosis* (Shen *et al.*, 2002).

Vδ2+ T cells react to small phosphate containing molecules that can be divided into two groups. There are nucleotide-conjugated phosphate molecules, such as TUBAg3-4, isopentenyl-ATP, and pyrophosphate molecules such as TUBag1-2, isopentenyl pyrophosphate [IPP], MEPP and others (Constant *et al.*, 1994; Tanaka *et al.*, 1995; Poquet *et al.*, 1996). Recognition of phospho-molecules is TCR- dependent but not restricted or dependent on any known MHC or MHC-like molecules (Bukowski *et al.*, 1995; Morita *et al.*, 1995). Specificity of prenylphosphate recognition depends on particular CDR3 regions in the Vγ9 chain and the Vδ2 chain. Phosphoantigens are intracellular and not secreted by *M. tuberculosis*. Prenyl pyrophosphates are ubiquitous precursors for cholesterol and its derivatives, and for terpenoids. Discrimination between infected and non-infected cells may involve recognition of metabolic intermediates produced by bacteria (Sicard *et al.*, 2000). Thus, Vδ2+ T cells recognize entirely different mycobacterial molecules than CD4+ and CD8+ T cells, that restrict their responses to peptides derived from *M. tuberculosis* proteins.

Other T Cell Populations: CD1 Restricted αβTCR+ T Cells

One additional T cell population that responds to *M. tuberculosis* is the CD1 restricted αβ TCR+ T cell (Porcelli *et al.*, 1992). It is the least common T cell subset in human peripheral blood and lung. In humans, most of these T cells express neither CD4 nor CD8 and are referred to as double negative (DN) cells. A minority express CD8 (Rosat *et al.*, 1999). CD1 is an MHC class I like molecule associated with β2-microglobulin that can be induced by cytokines or constitutively expressed on antigen presenting cells such as dendritic cells. Human CD1 molecules exist in 5 isoforms (CD1a-e), encoded by separate genes, and demonstrate little polymorphism compared to MCH I and II molecules. The structures of CD1 and MHC I molecules differ markedly in their antigen binding grooves. CD1 molecules have much deeper pockets that can accommodate large hydrophobic molecules. These structural considerations explain the unique ability of human CD1b and CD1c to bind polar lipid molecules of *M. tuberculosis*, including mycolic acid and phosphatidyl-inositol-mannan (PIM)(Beckman *et al.*, 1994). *M. tuberculosis* reactive CD1 restricted T cells have the same functional capacities as CD4+ and CD8+ T cells such as IFN-γ secretion and CTL function.

Murine CD1 restricted T cells do not respond to mycobacterial lipid antigens, and CD-1d knock-out mice are not more susceptible to *M. tuberculosis* challenge (Behar *et al.,* 1999). Guinea pigs do have CD1 restricted T cell responses to mycobacterial lipids and may provide insight into the role of these T cells in mycobacterial immunity. The ability of human γδ and CD1 restricted T cells to respond to non-protein molecules of *M. tuberculosis* expands the host's ability to recognize infected cells beyond the peptides presented to MHC I and II restricted CD8+ and CD4+ T cells. Some have suggested that γδ and CD1 restricted T cells function as "innate" T cells, bridging innate and acquired immune responses. Studies of γδ T cells in mycobacterial infection in primates argue against this view, since γδ T cells did not become activated before αβTCR+ T cells and showed evidence for memory (Shen *et al.,* 2002).

T Cell Function in the Immune Response to *M. tuberculosis*

Upon activation, human T cells use three major mechanisms to affect evolving immune responses: secretion of cytokines, expression of cytotoxic effector function (CTL) and cell-contact dependent help. Human T cell responses to *M. tuberculosis* are overwhelmingly Th-1 like both in health and disease, and thus are characterized by abundant IFN-γ production. There is no evidence for a role for Th-2 like cytokines such as IL-4 and IL-5 in either protective immunity or as a Th-2 switch in causing disease. In murine models concurrent Th-2 responses can modulate responses to mycobacterial antigens (Pearlman *et al.,* 1993). Whether concurrent Th-2 responses in persons with active or repeated helminth infection modulate Th-1 responses to *M. tuberculosis* and whether this affects the risk for active tuberculosis has not been determined. CD4+ T cell secrete macrophage-activating cytokines such as IFN-γ and TNFα/β in response to mycobacterial antigens. This is true for both T cell clones and bulk populations (Boom *et al.,* 1991; Mutis *et al.,* 1993; Tsukaguchi *et al.,* 1999). In addition, CD4+ T cells are cytotoxic (CTL) for *M. tuberculosis* infected macrophages (Ottenhoff *et al.,* 1988; Hancock *et al.,* 1989; Kumararatne *et al.,* 1990; Boom *et al.,* 1991; Tsukaguchi *et al.,* 1995). CD4+ T cells also help macrophages control intracellular mycobacteria (Silver *et al.,* 1998; Worku *et al.,* 2000; Canaday *et al.,* 2001). CD4+ T cells, activated by *M. tuberculosis*, express granzymes, Fas-L (CD95L), granulysin, and perforin (Canaday *et al.,* 2001). Whether CTL function is necessary for control of *M. tuberculosis* in macrophages is unclear as is the role of CTL effector function *in vivo*.

CD4+ T cells are a major source of IL-2 and thus provide help for γδ and CD8+ T cells that do not produce a lot of IL-2 but need it for T cell expansion and differentiation (Pechhold *et al.,* 1994; Serbina *et al.,* 2001). The importance of cell-contact mediated "help" by CD4+ T cells in the immune response to *M. tuberculosis* remains unclear. The interaction of CD40L (CD154) on T cells with CD40 on macrophages is one "help" mechanism, but it is not clear that this interaction has direct effects on the survival of intracellular mycobacteria (Larkin *et al.*, 2002). CD154 expression by CD4+ T cells is necessary to enhance antigen presenting cell function and cytokine secretion of macrophages. CD4 T cell's central role in protective immunity is dependent on these three distinct functional capacities. The importance of each CD4+ T cell function may vary according to the stage of *M. tuberculosis* infection: macrophage activating cytokine secretion when organism burden is low such during persistence, CTL function when bacterial burden is high such as during acute primary or reactivation infection, "help" functions (IL-2 or contact-dependent) may be necessary during all phases of the infection.

M. tuberculosis activated CD8+ T cells secrete IFN-γ, but less on a per cell basis than CD4+ T cells (Canaday *et al.*, 1999). They express granzymes, Fas-L (CD95L), granulysin, and perforin, enabling them to lyse infected macrophages(Stenger *et al.*, 1998; Canaday *et al.*, 2001). CD8+ T cells can help macrophages control intracellular mycobacteria (Stenger *et al.*, 1997; Canaday *et al.*, 2001). To what extent protection by CD8+ T cells depends on CTL activity is un-clear, nor is it known when during human *M. tuberculosis* infection CD8+ T cells are most critical. They are present in peripheral blood and lungs of healthy tuberculin skin test positive persons, suggesting a role in ongoing immune surveillance. In murine studies, CD8+ T cells expand and are recruited to the lung at the same time as CD4+ T cells, indicating a role during containment of acute infection. Vδ2+ T cells also secrete IFN$-\gamma$, lyse infected macrophages and can help contain mycobacterial growth (Tsukaguchi *et al.*, 1995; Dieli *et al.*, 2001). On a per cell basis Vδ2+ T cells produce more IFN-γ than CD4+ T cells.

A number of studies have described assays to measure growth of *M. tuberculosis* inside monocytes/macropages in an effort to determine immune mechanisms to restrict and possibly eliminate intracellular mycobacteria ("killing assays") (Stenger *et al.*, 1997; Silver *et al.*, 1998; Worku *et al.*, 2000). The latter is a misnomer because *in vitro M. tuberculosis* readily survives and usually replicates within human macrophages, even at low multiplicities of infection (MOI of 1:1 or less). Efforts to enhance the ability of human macrophages to control intracellular *M. tuberculosis* with cytokines have been unsuccessful (Douvas *et al.*, 1985). However, a number of laboratories have been able to demonstrate that addition of T cells "help" macrophages control *M. tuberculosis* growth. In general these studies can demonstrate T cell inhibition and some killing of mycobacteria in short-term assays (12-24 hrs). All T cell subsets tested (CD4, CD8, $\gamma\delta$, CD1-restricted) appear to be able to perform this function. The mechanism(s) for T cell mediated control remain unclear. Granulysin found in most cytotoxic T cells and introduced through perforin mediated lysis of infected macrophages may have a role (Stenger *et al.*, 1998). Apoptosis of infected macrophages through interaction of Fas-FasL or ATP activated P2X7 receptors may play a role as well (Molloy *et al.*, 1994; Lammas *et al.*, 1997; Oddo *et al.*, 1998; Kusner *et al.*, 2000). The amount of ATP released by activated T cells or lysed macrophages however may not be sufficient for apoptosis of macrophages (Canaday *et al.*, 2002). Current *in vitro* assays used to measure *M. tuberculosis* "killing" are limited because MOI's required for reproducible experimentation are much higher than the ratio of macrophages to bacilli *in vivo*. Current assay systems do not allow reliable and reproducible measurements of *M. tuberculosis* growth at MOI's of 1 mycobacterium per 10-100 macrophages, thus hampering analysis of mechanism(s) for T cell mediated mycobacterial "killing".

Thus, human T cell responses to *M. tuberculosis* are characterized by participation of multiple T cell subsets with similar functions. They all secrete IFN–γ and TNF–α, lyse infected cells as CTL and help macrophages control *M. tuberculosis*. However, they differ in the efficiency with which they express different effector functions. For example, γδ and CD4+ T cells secrete more IFN-γ than CD8+ T cells, and γδ T cells may secrete more than CD4+ T cells. CD4+ T cells secrete more IL-2 than CD8+ and γδ T cells and thus are more efficient at providing help. There are likely differences in CTL efficiency as well. They differ markedly however in the range of mycobacterial antigens recognized and the antigen-processing mechanisms used to process and present these ligands to their different TCR's. They also may differ in *in vivo* kinetics, ability to enter sites of active *M. tuberculosis* infection such as the lung and stage of *M. tuberculosis* infection where they are most critical or activated. Some suggest that γδ T cells and CD1 restricted T cells provide a link between innate and adaptive phases of the immune response to *M. tuberculosis*. CD4+ T cells are known to have a key role through all stages of *M. tuberculosis* infection but may differ as to when cytokine secretion vs. CTL function are the most important. Whether CD8+, γδ and CD-1 restricted T cells have the same importance throughout *M. tuberculosis* infection is not known.

When T cell responses to *M. tuberculosis* between healthy tuberculin skin test positive persons are compared to those with active tuberculosis, T cell responses in active tuberculosis are generally diminished. Whether T cell responses contribute to the tissue destruction characteristic of active tuberculosis remains a controversial issue in the pathogenesis of cavitary tuberculosis.

Cytokines and Acquired Immunity to *M. tuberculosis*

Acquired immunity to *M. tuberculosis* develops in a complex cytokine environment of chemokines and pro-inflammatory cytokines. These cytokines are secreted by macrophages and inflammatory cells that reside in or are recruited to the site infection, most commonly the lung. Mycobacterial pathogen associated molecular patterns (PAMP's), such as lipoproteins, lipo-arabinomannan and others, are recognized by TLR receptors on macrophages, resulting in IL-1, IL-6, IL-8, TNF-α and IL-12 secretion (Medzhitov *et al.*, 1997; Brightbill *et al.*, 1999; Means *et al.*, 1999). TNF–α and IL-12 are necessary for acquired immunity to *M. tuberculosis*. TNF–α activates macrophages and is important for granuloma formation. IL-12, a key cytokine for developing of Th-1 responses to *M. tuberculosis*, stimulates secretion of IFN-γ by NK and T cells. IFN–γ further activates macrophages and enhances their antigen presenting cell function, allowing of memory T cells to *M. tuberculosis* antigens. Phagocytosis of *M. tuberculosis* or exposure of macrophages to mycobacterial PAMP's results in IL-12 and TNF-α secretion (Fulton *et al.*, 1996; Ladel *et al.*, 1997; Brightbill *et al.*, 1999). T cells are the major source of IFN-γ in the acquired immune response to *M. tuberculosis*. T cells also produce TNF–α and TNF–β, but macrophages are the major source of TNF–α. Excessive inflammation mediated by TNF–α, TNF–β and possibly IFN-γ likely contributes to tissue destruction in pulmonary tuberculosis. Clinical or experimental proof for this cytokine hypothesis of tissue injury has been difficult to obtain.

There is no evidence to date that progression from infection to active tuberculosis is due to skewing of a Th-1 response to a Th-2-like response (i.e IL-4, IL-5 production) as described for the polar forms of leprosy. However there is strong evidence that *M. tuberculosis* can render IFN-γ ineffective by blocking its effects on IFN-γ receptor (IFN-γR) expressing cells such as macrophages. In addition, defects in IFN-γ production by T cells have been observed in advanced tuberculosis. Aside from inhibiting IFN-γ mediated immune activation, *M. tuberculosis* infection *in vivo* and *in vitro* is associated with secretion of inhibitory cytokines such as IL-10 and TGF–β. While these cytokines are part of normal immuno-regulation, excess production may contribute to inhibition of T cells directly or indirectly by interfering with antigen recognition. Thus the balance of pro-inflammatory and inhibitory cytokines may determine the outcome of the host-pathogen interaction in *M. tuberculosis* infection.

IFN-γ and IL-12

Proof of the importance of IFN-γ in protective immunity to *M. tuberculosis* and *M. bovis*–BCG was demonstrated in mice with deleted genes for IFN-γR and IFN-γ (Cooper *et al.*, 1993; Flynn *et al.*, 1993; Kamijo *et al.*, 1993). These mice were highly susceptible to mycobacterial infection and succumbed rapidly to sublethal challenge. In humans direct proof of the critical importance of IFN-γ has depended on more indirect evidence. Adults with active tuberculosis generally produce less IFN-γ, when peripheral blood T cells are stimulated with mycobacterial antigens *in vitro*, than healthy *M. tuberculosis* infected persons. Whether that is true at sites of infection such as the lung is not known. Children born with genetic defects in IFN-γR are markedly susceptible to mycobacterial infection, presenting with disseminated BCG or recurrent/ invasive atypical mycobacterial infection. Although defects in IFN-γR function do not prove directly that IFN-γ and IFNγ-R are critical for protecting humans against tuberculosis, they do indicate the importance of IFN-γ for control of chronic intracellular bacterial infections with *M. tuberculosis*, atypical mycobacteria and *Salmonella* (Jouanguy *et al.*, 1996; Newport *et al.*, 1996; Jouanguy *et al.*, 1997). Genetic defects in IFN-γR cannot explain reactivation tuberculosis in adults, since the majority of these persons will have successfully controlled primary *M. tuberculosis* infection. However, functional defects in IFN-γR signaling (i.e. IFN-γ present but ineffective) could contribute to progression of *M. tuberculosis* infection. *In vitro*, *M. tuberculosis* can inhibit signaling through the IFN-γR in human macrophages by blocking the interaction between STAT-1 and the transcriptional co-activators CREB binding protein and p300 protein (Ting *et al.*, 1999). The *M. tuberculosis* molecule(s) responsible for this inhibition remain to be determined.

Whereas most would concur that IFN–γ has a key role in protective immunity to *M. tuberculosis*, controversy remains as to the mechanism(s) used by IFN-γ to mediate this protective effect. IFN–γ alone is inefficient in enhancing the ability of human macrophages to control growth of *M. tuberculosis* (Douvas *et al.*, 1985; Rook *et al.*, 1986; Carvalho de Sousa *et al.*, 1992). In mice, protective immunity by CD4+ T cells is not explained by IFN–γ alone, since MHC class II and CD4 knock-out mice produce large amounts of IFN-γ yet fail to control *M. tuberculosis* (Caruso *et al.*, 1999). Besides activating microbial "killing" mechanisms, IFN-γ also regulates antigen presenting cell function by increasing expression of MHC I/II and co-stimulatory molecules (B7.1, B7.2). This latter activity is essential for optimal T cell responses, and may represent a primary role for IFN-γ in adaptive immunity to *M. tuberculosis*. All T cell subsets activated by *M. tuberculosis* and its antigen secrete IFN-γ, although they may differ in amounts produced. CD4+ and γδ T cells produce more than CD8+ T cells on a per cell basis (Tsukaguchi *et al.*, 1995; Canaday *et al.*, 1999).

    M. tuberculosis infection of human macrophages results in production of IL-12 p40 and p70, that enhances IFN-γ production by NK and T cells, and thus has a role in both development and expression of Th-1 responses to mycobacterial antigens. Mice with disruption of the gene for IL-12 are markedly susceptible to pulmonary challenge with *M. tuberculosis* (Cooper *et al.*, 1997). Furthermore, children with genetic defects in IL-12R expression or function are at risk for severe infections with intracellular bacterial pathogens (Altare *et al.*, 1998; de Jong *et al.*, 1998).

TNF-α and IL-10

Whereas IL-12 is produced primarily by macrophages and IFN-γ primarily by T cells, TNF–α and IL-10 are secreted by both cell types. How important TNF–α and IL-10 production by T cells is in immunity to *M. tuberculosis* compared to production of these cytokines by macrophages is unknown. The protective role of TNF–α in the immune response to *M. tuberculosis* was demonstrated in mice with defects in genes for TNF–α and TNF–αRII (55 kDa subunit) (Flynn *et al.*, 1995; Bean *et al.*, 1999). These mice succumbed rapidly to overwhelming *M. tuberculosis* infection with high bacterial burdens and inadequate granuloma formation. How TNF–α exerts its protective effect is not known but is likely mediated through effects on macrophages. Its critical role for humans has been recently revealed by the occurrence of reactivation tuberculosis in rheumatoid arthritis patients who received long-term therapy with anti-TNF–α antibodies (Keane *et al.*, 2001). This clinical observation suggests that TNF–α not only has a role in acute mycobaterial infection but also in maintaining control over persistent tubercle bacilli. Some studies have indicated that excess TNF–α production in *M. tuberculosis* infection is responsible for the systemic symptoms of tuberculosis such as weight loss (cachexia), fever and possibly tissue damage. Excess TNF-α associated with active tuberculosis may also promote progression of HIV-1 disease in persons dually infected. Inhibitors of TNF–α (thalidomide, steroids) during acute tuberculosis have met with modest therapeutic success and that includes persons co-infected with HIV-1. TNF–α in tuberculosis may be similar to that observed in gram negative infections where little TNF–α elicits a beneficial acute phase response and activation of inflammatory cells, and excess TNF–α may cause sepsis.

    The role of IL-10 human *M. tuberculosis* infection is less clear-cut. IL-10 knock-out mice were not more resistant to *M. tuberculosis* but mice over-expressing IL-10 had difficulty clearing mycobacterial infection (Murray *et al.*, 1997; North 1998). IL-10 not only inhibits CD4+ and $\gamma\delta$ T cell responses directly but also can affect expression of MHC II and co-stimulatory molecules on infected macrophages and thus affect antigen presenting cell function (Rojas *et al.*, 1999). Neutralizing anti-IL-10 antibodies improve T cell responses (proliferation and IFN-γ secretion) of tuberculosis patients (Gong *et al.*, 1996; Hirsch *et al.*, 1996). T cell responses of tuberculosis patients also are restored by neutralizing TGF–β (Hirsch *et al.*, 1996). Thus IL-10, TGF–β or both can down-regulate acquired immunity to mycobacterial antigens. In health these cytokines down-regulate the inflammation associated with T cell and macrophage activation in response to *M. tuberculosis*, but in disease excess production of these cytokines impairs acquired immunity.

Antigen Processing of *M. tuberculosis*

Antigen processing cell (APC) function of mononuclear phagocytes for *M. tuberculosis* specific CD4+ and CD8+ T cells requires uptake/phagocytosis of bacilli, digestion of proteins to peptides of appropriate length, delivery of peptides to cellular compartments for loading onto MHC molecules, and presentation of these peptide-MHC complexes on the surface of APC. For optimal T cell activation, costimulatory molecules such as B7.1, B7.2 and CD40 also need to be expressed. CD4 and CD8+ T cells recognize antigens as peptide fragments bound to MHC-I and -II proteins respectively. MHC-I and-II molecules have distinct intracellular trafficking pathways and mechanisms for peptide loading. The context for Vδ2+ T cell recognition of molecules of *M. tuberculosis* is unknown. These diverse antigen processing mechanisms not only allow the host to recognize a wide range of mycobacterial antigens on many different cells but also to circumvent efforts by *M. tuberculosis* to modulate macrophage APC function to avoid detection.

Antigen Processing of *M. tuberculosis* for MHC-II

For MHC-II antigen processing, soluble proteins are internalized by endocytosis and particulates such as *M. tuberculosis* bacilli by phagocytosis. Soluble proteins are concentrated within endosomes. As endosomes fuse with lysosomes, protein antigens are broken down into peptides by lysosomal proteases. MHC-II molecules are concentrated in a late endocytic compartment, called the MHC-II compartment or MIIC. MHC-II -peptide complexes are formed in MIIC when peptides bind to MHC-II molecules. MHC-II molecules are targeted for endosomes by specific sequences on the invariant chain that is associated with MHC-II α and β chains as the hetero-dimer emerges from the endoplasmatic reticulum (ER). Invariant chain contains a CLIP region that blocks the peptide-binding site on MHC-II. CLIP is exchanged for antigen peptide with help from HLA-DM molecules and promoted by the acidic pH of the late endosome. A number of studies have demonstrated that certain mycobacterial proteins readily traffic within the endosomal compartment of macrophages infected with live bacilli (Clemens *et al.*, 1996; Holsti *et al.*, 1998; Neyrolles *et al.*, 2001).

Peptides also can be loaded directly on to MHC-II molecules in phagosomes, without trafficking to MIIC (reviewed in Ramachandra *et al.,* 1999). *M. tuberculosis* phagosomes from murine macrophages contain MHC-II molecules capable of processing and presenting mycobacterial peptides directly to T cells (Ramachandra *et al.,* 2001). Phagosomes from human macrophages also contain MHC-II, and can process mycobacterial antigen for MHC-II restricted T cells (Clemens *et al.,* 1996) (Torres, personal communication). Thus in macrophages there are at least two locations where mycobacterial peptides can be loaded onto MHC-II molecules for traffic to the cell surface and recognition by CD4+ T cells. The relative importance of these two sites (phagosome vs. MIIC) is likely determined by the ability of a given mycobacterial protein to circulate within endosomes of macrophages or to remain primarily within *M. tuberculosis* phagosomes. The viability of bacilli within phagosomes is likely an additional factor.

Antigen Processing for MHC-I

Traditionally, MHC-I molecules present microbial peptides derived from antigens present in cytosol either through *de novo* synthesis (viral antigens) or carried by microbial agents such as *Listeria monocytogenes* capable of penetrating into cytoplasm. Macrophages also can take up antigens by macro-pinocytosis or phagocytosis and deliver these antigens to cytosol where they can enter the traditional MHC-I processing pathway. Once in cytosol, protein antigens are cleaved by proteasomes (a cylindrical proteolytic complex made up of 24 subunits of 650kD). Peptide fragments then are delivered to the lumen of the endoplasmatic reticulum (ER) by TAP molecules present in the ER membrane. In the ER, peptides bind to newly synthesized class I MHC heavy chain molecules and transported through Golgi complex to plasma membrane for recognition by CD8+ T cells.

However, particulate bacterial antigens can be processed for MHC-I presentation from the phago-lysosome by an alternate vacuolar pathway without penetration of antigen into cytosol (Harding *et al.,* 1994; Harding 1996). Thus, there may be multiple mechanisms for mycobacterial peptide antigens to become associated with MHC-I for presentation to CD8+ T cells. In human macrophages, *M. tuberculosis* antigens can be shunted into the alternate MHC-I processing pathway as well as through the cytosolic pathway (Lewinsohn *et al.,* 1998; Canaday *et al.,* 1999). The role of phagosomes in the alternate pathway for processing of *M. tuberculosis* proteins and whether proteasomes are required for generating mycobacterial peptides for MHC-I, remain to be determined. A better understanding of MHC-I antigen processing of *M. tuberculosis* will help with design of vaccines that will stimulate both CD4+ and CD8+ T cell responses.

Recognition of Phosphate Containing Molecules by Vδ2+ T Cells

Synthetic and native phosphate containing molecules (phosphate molecules) for Vδ2+ T cells do not require intracellular processing, but cell surface presentation and possibly extra-cellular processing may play a role in optimal recognition of these ligands by Vδ2+ T cells. How phosphate molecules traffic from *M. tuberculosis* phagosomes to the surface of macrophages is not understood. Vδ2+ T cell responses to *M. tuberculosis* phosphate molecules are not restricted by MHC-I, -II, or CD1a, 1b, or 1c (Morita *et al.,* 1995). Vδ2+ T cell activation by phosphate molecules is dependent on γδTCR, as is cell to cell contact, but evidence for direct binding of phosphate molecules to γδTCR is lacking(Bukowski *et al.,* 1995; Lang *et al.,* 1995). Lack of direct γδTCR binding of phosphate molecules indirectly supports the presence of either to be defined "presenting" molecules or further antigen-processing on the cell surface. Reactivity to phosphate molecules was lost by chemical substitution with non-hydrolyzable groups in synthetic analogs, indicating that the phosphate moiety is a potential substrate for enzymatic processing. Even if antigen processing or presentation of phosphate molecules is not required, macrophages remain important accessory cells for Vδ2+ T cell activation because they represent a concentrated source of phosphate molecules when infected with *M. tuberculosis* and they provide co-stimulation for cytokine production and proliferation by Vδ2+ T cells (Rojas *et al.,* 2002).

Mechanisms for Interference by *M. tuberculosis* With Acquired Immunity

Inhibition of T Cell Function

The ability of *M. tuberculosis* to survive in the face of a complex and diverse acquired immune response implies that the organism developed mechanisms to resist host immunity. *M. tuberculosis* evolved to survive and persist in a key immune effector cell, the macrophage. Macrophages take up microbial pathogens, are microbicidal, process and present antigens to T cells, and express the costimulatory molecules required for T cell activation.

M. tuberculosis has evolved at least two mechanisms to interfere with innate immune defenses of macrophages: resisting phago-lysosomal fusion and microbicidal mechanisms. Macrophages use CR-1, CR3, CR4 or the mannose receptor for receptor mediated phagocytosis of *M. tuberculosis* (Schlesinger *et al.,* 1990; Schlesinger 1993; Hirsch *et al.,* 1994). Macrophage type, alveolar vs. monocyte, and differences among strains of *M. tuberculosis* in activating complement and lipoarabinomannan have a role in determining the dominant receptor used for uptake (Schlesinger *et al.*, 1996). *M. tuberculosis* does not penetrate into cytoplasm and modulates its phagosome by preventing fusion with acidic lysosomal compartments and actively excludes vesicular proton ATP-ases, resulting in an elevated pH of 6.3 to 6.5 (compared to the normal lysosomal pH of 4.5) (Sturgill-Koszycki *et al.*, 1994). IFN–γ may partially overcome *M. tuberculosis* mediated inhibition of phagosomal acidification.

M. tuberculosis resist killing by oxygen radicals, through production of superoxide dismutases (Zhang *et al.*, 1991). The importance of this microbial defense is reflected in mice deficient in superoxide production, which are only slightly more susceptible than wild-type mice to *M. tuberculosis* (Cooper *et al.*, 2000). *M. tuberculosis* is sensitive to nitric oxide (NO) made by inducible nitric oxide synthase (iNOS) and mice deficient in iNOS are highly susceptible (Chan *et al.*, 1992; MacMicking *et al.*, 1997). However, whether human macrophages express enough iNOS to produce enough NO to kill *M. tuberculosis* has not been definitely established(Aston *et al.*, 1998). Thus resistance to oxygen radicals may be a major defense against human macrophages.

M. tuberculosis also has evolved means to interfere with the host's acquired immune mechanisms. First, molecules of *M. tuberculosis* induce macrophages to produce inhibitory cytokines such as IL-10 and TGF–β . Excess production of IL-10 and TGF–β observed in active tuberculosis inhibits the effects of pro-inflammatory cytokines such as IFN-γ and TNF-α and also directly inhibit T cell function (Hirsch *et al.*, 1996; Rojas *et al.*, 1999; Boussiotis *et al.*, 2000). Recognition of pathogen-associated molecular patterns (PAMP's) by Toll-like receptors (TLR) may be a primary signal for cytokine release by macrophages. There are 10 known TLR's of which many are involved in PAMP recognition and a number are expressed on macrophages (Janeway *et al.*, 2002). TLR-4 is a primary receptor for LPS, and TLR-2 recognizes a number of molecules found in gram-positive bacteria (Medzhitov *et al.*, 1997; Lien *et al.*, 1999). TLR-2 and TLR-4 recognize *M. tuberculosis* associated PAMP's. TLR-2 recognizes the 19kDa mycobacterial lipoprotein and the specific mycobacterial ligand(s) for TLR-4 have not been characterized (Brightbill *et al.*, 1999; Means *et al.*, 1999; Noss *et al.*, 2001). Most studies on signaling through TLR's focus on secretion of pro-inflammatory cytokines, and thus involvement of mycobacterial PAMP's in inhibitory cytokine secretion (IL-10) remains unknown.

Second, active *M. tuberculosis* infection is associated with increased apoptosis of mycobacterial antigen-specific T cells. In contrast to macrophage apoptosis which is though to favor the host and may represent a mechanism for controlling mycobacterial growth, T cell apoptosis during active tuberculosis is detrimental and may have a prolonged effect on *M. tuberculosis*-specific T cell responses (Li *et al.*, 1998; Hirsch *et al.*, 1999; Hirsch *et al.*, 2001). Up to 12 months after initiation of TB treatment of tuberculosis and control of bacterial replication, CD4+ T cell responses to proteins of *M. tuberculosis* remain markedly diminished compared to responses to control antigens in some tuberculosis patients. T cell apoptosis during acute infection may cause prolonged defects in *M. tuberculosis*-specific T cell repertoire and function. The molecules of *M. tuberculosis* and mechanism(s) responsible for T cell apoptosis have not been defined, although infection can upregulate FasL (CD95L) on human macrophages(Dockrell *et al.*, 1998).

Modulation of Macrophage Antigen Presenting Cell Function

A third means for *M. tuberculosis* to evade acquired immune responses is by interfering with the antigen processing function of macrophages. Processing of antigens is essential for $\alpha\beta$TCR+ T cell recognition of infected cells and requires that peptides are generated for presentation by two distinct molecules: MHC II for CD4+ and MHC I for CD8+ $\alpha\beta$TCR+ T cells. Little is known about mycobacterial inhibitors of antigen processing for MHC molecules. Recent studies are starting to shed light on mechanisms for inhibition of MHC-II antigen processing and the molecule(s) of *M. tuberculosis* responsible for this inhibition.

M. tuberculosis infected monocytes do not present tetanus toxoid as well as uninfected cells (Gercken *et al.*, 1994). In murine macrophages, IL-6 secretion has been shown under some circumstances to inhibit T cell function, and *M. tuberculosis* can interfere with maturation of MHC class II and IFN$-\gamma$ signaling in human macrophages (VanHeyningen *et al.*, 1997; Hmama *et al.*, 1998; Hussain *et al.*, 1999; Ting *et al.*, 1999). In murine macrophages, mycobacterial infection decreased CIITA (class II transactivator) expression, resulting in decreased MHC-II levels (Wojciechowski *et al.*, 1999). Mycobacterial molecules responsible for these inhibitory activities had not been defined until recently, when the 19 kDa lipoprotein of *M. tuberculosis* was found to inhibit MHC-II expression and antigen processing in murine macrophages by prolonged signaling through TLR-2 (Noss *et al.*, 2000; Noss *et al.*, 2001). *M. tuberculosis* and the 19 kDa lipoprotein also inhibit IFN$-\gamma$ mediated regulation of human HLA-DR in human macrophages (Gehring *et al.*, in press, 2003). The ability to interfere with CD4+ T cell activation by "hiding" from the immune responses may represent a major mechanism for *M. tuberculosis* to avoid detection and elimination during the persistent phase of infection. Using carefully designed assays to measure specific stages of antigen processing for MHC-II and -I molecules, it is likely that additional mycobacterial molecules will be found that block antigen processing.

Thus, *M. tuberculosis* has evolved a number of mechanisms to interfere with activation of both innate and adaptive phases of the host's immune response. The importance of these different mechanisms likely will differ for the different stages of *M. tuberculosis* infection. Suppressive cytokines and apoptosis may be most important during phases when *M. tuberculosis* is replicating rapidly such as during primary infection and upon reactivation. Inhibition of antigen processing may be an effective strategy during the persistent phase of infection that affects the majority of *M. tuberculosis* infected persons. The balance between the host's ability to activate T cell subsets capable of recognizing a wide range of mycobacterial molecules on a diversity of antigen presenting molecules and express a range of effector mechanism, and *M. tuberculosis*' ability to interfere with immune recognition and effector phases of adaptive immunity defines the relationship between host and pathogen in *M. tuberculosis* infection. Applying modern biology to in creasing our understanding of the molecular intricacies, that make up the delicate host-pathogen balance in *M. tuberculosis* infection, will result in better vaccines, diagnostic tools and therapies for this age-old pathogen.

References

Abou-Zeid, C., M.P. Gares, J. Inwald, R. Janssen, Y. Zhang, D.B. Young, C. Hetzel, J.R. Lamb, S.L. Baldwin, I.M. Orme, V. Yeremeev, B.V. Nikonenko and A.S. Apt. 1997. Induction of a type 1 immune response to a recombinant antigen from *Mycobacterium tuberculosis* expressed in *Mycobacterium vaccae*. Infect. Immun. 65(5): 1856-1862.

Altare, F., A. Durandy, D. Lammas, J.F. Emile, S. Lamhamedi, F. Le Deist, P. Drysdale, E. Jouanguy, R. Doffinger, F. Bernaudin, O. Jeppsson, J.A. Gollob, E. Meinl, A.W. Segal, A. Fischer, D. Kumararatne and J.L. Casanova. 1998. Impairment of mycobacterial immunity in human interleukin-12 receptor deficiency. Science 280(5368): 1432-1435.

Aston, C., W.N. Rom, A.T. Talbot and J. Reibman. 1998. Early inhibition of mycobacterial growth by human alveolar macrophages is not due to nitric oxide. Am. J. Respir. Crit. Care Med. 157(6 Pt 1): 1943-1950.

Barnes, P.F., C.L. Grisso, J.S. Abrams, H. Band, T.H. Rea and R.L. Modlin. 1992. Gamma delta T lymphocytes in human tuberculosis. J. Infect. Dis. 165(3): 506-512.

Barnes, P.F., V. Mehra, G.R. Hirschfield, S.J. Fong, Z.C. Abou, G.A. Rook, S.W. Hunter, P.J. Brennan and R.L. Modlin. 1989. Characterization of T cell antigens associated with the cell wall protein-peptidoglycan complex of *Mycobacterium tuberculosis*. J. Immunol. 143(8): 2656-2662.

Bean, A.G., D.R. Roach, H. Briscoe, M.P. France, H. Korner, J.D. Sedgwick and W.J. Britton. 1999. Structural deficiencies in granuloma formation in TNF gene-targeted mice underlie the heightened susceptibility to aerosol *Mycobacterium tuberculosis* infection, which is not compensated for by lymphotoxin. J. Immunol. 162(6): 3504-3511.

Beckman, E.M., S.A. Porcelli, C.T. Morita, S.M. Behar, S.T. Furlong and M.B. Brenner. 1994. Recognition of a lipid antigen by CD1-restricted alpha beta+ T cells [see comments]. Nature 372(6507): 691-694.

Behar, S.M., C.C. Dascher, M.J. Grusby, C.R. Wang and M.B. Brenner. 1999. Susceptibility of mice deficient in CD1D or TAP1 to infection with *Mycobacterium tuberculosis*. J. Exp. Med. 189(12): 1973-1980.

Behr, M.A., M.A. Wilson, W.P. Gill, H. Salamon, G.K. Schoolnik, S. Rane and P.M. Small. 1999. Comparative genomics of BCG vaccines by whole-genome DNA microarray. Science 284(5419): 1520-1523.

Belisle, J.T., V.D. Vissa, T. Sievert, K. Takayama, P.J. Brennan and G.S. Besra. 1997. Role of the major antigen of *Mycobacterium tuberculosis* in cell wall biogenesis. Science 276(5317): 1420-1422.

Boom, W.H.. 1996. The role of T-cell subsets in *Mycobacterium tuberculosis* infection. Infect. Agents Dis. 5(2): 73-81.

Boom, W.H., R.S. Wallis and K.A. Chervenak. 1991. Human *Mycobacterium tuberculosis*-reactive CD4+ T-cell clones: heterogeneity in antigen recognition, cytokine production, and cytotoxicity for mononuclear phagocytes. Infect. Immun. 59(8): 2737-2743.

Boussiotis, V.A., E.Y. Tsai, E.J. Yunis, S. Thim, J.C. Delgado, C.C. Dascher, A. Berezovskaya, D. Rousset, J.M. Reynes and A.E. Goldfeld. 2000. IL-10-producing T cells suppress immune responses in anergic tuberculosis patients. J. Clin. Invest. 105(9): 1317-1325.

Brightbill, H.D., D.H. Libraty, S.R. Krutzik, R.B. Yang, J.T. Belisle, J.R. Bleharski, M. Maitland, M.V. Norgard, S.E. Plevy, S.T. Smale, P.J. Brennan, B.R. Bloom, P.J. Godowski and R.L. Modlin. 1999. Host defense mechanisms triggered by microbial lipoproteins through toll-like receptors. Science 285(5428): 732-736.

Bukowski, J.F., C.T. Morita, Y. Tanaka, B.R. Bloom, M.B. Brenner and H. Band. 1995. V gamma 2V delta 2 TCR-dependent recognition of non-peptide antigens and Daudi cells analyzed by TCR gene transfer. J. Immunol. 154(3): 998-1006.

Canaday, D.H., R. Beigi, R.F. Silver, C.V. Harding, W.H. Boom and G.R. Dubyak. 2002. ATP and control of intracellular growth of mycobacteria by T cells. Infect. Immun. 70(11): 6456-6459.

Canaday, D.H., R.J. Wilkinson, Q. Li, C.V. Harding, R.F. Silver and W.H. Boom. 2001. CD4(+) and CD8(+) T cells kill intracellular *Mycobacterium tuberculosis* by a perforin and Fas/Fas ligand-independent mechanism. J. Immunol. 167(5): 2734-2742.

Canaday, D.H., C. Ziebold, E.H. Noss, K.A. Chervenak, C.V. Harding and W.H. Boom. 1999. Activation of human CD8+ alpha beta TCR+ cells by *Mycobacterium tuberculosis* via an alternate class I MHC antigen-processing pathway. J. Immunol. 162(1): 372-379.

Caruso, A.M., N. Serbina, E. Klein, K. Triebold, B.R. Bloom and J.L. Flynn. 1999. Mice deficient in CD4 T cells have only transiently diminished levels of IFN-gamma, yet succumb to tuberculosis. J. Immunol. 162(9): 5407-5416.

Carvalho de Sousa, J.P. and N. Rastogi. 1992. Comparative ability of human monocytes and macrophages to control the intracellular growth of *Mycobacterium avium* and *Mycobacterium tuberculosis*: effect of interferon-gamma and indomethacin. FEMS Microbiol. Immunol. 4(6): 329-334.

Chan, J., Y. Xing, R.S. Magliozzo and B.R. Bloom. 1992. Killing of virulent *Mycobacterium tuberculosis* by reactive nitrogen intermediates produced by activated murine macrophages. J. Exp. Med. 175(4): 1111-1122.

Clemens, D.L. and M.A. Horwitz. 1996. The *Mycobacterium tuberculosis* phagosome interacts with early endosomes and is accessible to exogenously administered transferrin. J. Exp. Med. 184(4): 1349-1355.

Cole, S.T., R. Brosch, J. Parkhill, T. Garnier, C. Churcher, D. Harris, S.V. Gordon, K. Eiglmeier, S. Gas, C.E. Barry, 3rd, F. Tekaia, K. Badcock, D. Basham, D. Brown, T. Chillingworth, R. Connor, R. Davies, K. Devlin, T. Feltwell, S. Gentles, N. Hamlin, S. Holroyd, T. Hornsby, K. Jagels, B.G. Barrell and et al.,. 1998. Deciphering the biology of *Mycobacterium tuberculosis* from the complete genome sequence. Nature 393(6685): 537-544.

Constant, P., F. Davodeau, M.A. Peyrat, Y. Poquet, G. Puzo, M. Bonneville and J.J. Fournie. 1994. Stimulation of human gamma delta T cells by nonpeptidic mycobacterial ligands. Science 264(5156): 267-270.

Cooper, A.M., D.K. Dalton, T.A. Stewart, J.P. Griffin, D.G. Russell and I.M. Orme. 1993. Disseminated tuberculosis in interferon-gamma gene-disrupted mice. J. Exp. Med. 178: 2243-2247.

Cooper, A.M., J. Magram, J. Ferrante and I.M. Orme. 1997. Interleukin 12 (IL-12) is crucial to the development of protective immunity in mice intravenously infected with *Mycobacterium tuberculosis*. J. Exp. Med. 186(1): 39-45.

Cooper, A.M., B.H. Segal, A.A. Frank, S.M. Holland and I.M. Orme. 2000. Transient loss of resistance to pulmonary tuberculosis in p47(phox-/-) mice. Infect. Immun. 68(3): 1231-1234.

de Jong, R., F. Altare, I.A. Haagen, D.G. Elferink, T. Boer, P.J. van Breda Vriesman, P.J. Kabel, J.M. Draaisma, J.T. van Dissel, F.P. Kroon, J.L. Casanova and T.H. Ottenhoff. 1998. Severe mycobacterial and Salmonella infections in interleukin-12 receptor-deficient patients. Science 280(5368): 1435-1438.

DeLibero, G., I. Flesch and S.H.E. Kaufmann. 1988. Mycobacteria reactive Lyt2+ T cell lines. Eur. J. Immunol. 18: 59-66.

Dieli, F., M. Troye-Blomberg, J. Ivanyi, J.J. Fournie, A.M. Krensky, M. Bonneville, M.A. Peyrat, N. Caccamo, G. Sireci and A. Salerno. 2001. Granulysin-dependent killing of intracellular and extracellular *Mycobacterium tuberculosis* by Vgamma9/Vdelta2 T lymphocytes. J. Infect. Dis. 184(8): 1082-1085.

Dillon, D.C., M.R. Alderson, C.H. Day, T. Bement, A. Campos-Neto, Y.A. Skeiky, T. Vedvick, R. Badaro, S.G. Reed and R. Houghton. 2000.

Molecular and immunological characterization of *Mycobacterium tuberculosis* CFP-10, an immunodiagnostic antigen missing in *Mycobacterium bovis* BCG. J. Clin. Microbiol. 38(9): 3285-3290.

Dillon, D.C., M.R. Alderson, C.H. Day, D.M. Lewinsohn, R. Coler, T. Bement, A. Campos-Neto, Y.A. Skeiky, I.M. Orme, A. Roberts, S. Steen, W. Dalemans, R. Badaro and S.G. Reed. 1999. Molecular characterization and human T-cell responses to a member of a novel *Mycobacterium tuberculosis* mtb39 gene family. Infect. Immun. 67(6): 2941-2950.

Dockrell, D.H., A.D. Badley, J.S. Villacian, C.J. Heppelmann, A. Algeciras, S. Ziesmer, H. Yagita, D.H. Lynch, P.C. Roche, P.J. Leibson and C.V. Paya. 1998. The expression of Fas Ligand by macrophages and its upregulation by human immunodeficiency virus infection. J. Clin. Invest. 101(11): 2394-2405.

Douvas, G.S., D.L. Looker, A.E. Vatter and A.J. Crowle. 1985. Gamma interferon activates human macrophages to become tumoricidal and leishmanicidal but enhances replication of macrophage-associated mycobacteria. Infect. Immun. 50(1): 1-8.

Faith, A., C. Moreno, R. Lathigra, E. Roman, M. Fernandez, S. Brett, D.M. Mitchell, J. Ivanyi and A.D. Rees. 1991. Analysis of human T-cell epitopes in the 19,000 MW antigen of *Mycobacterium tuberculosis*: influence of HLA-DR. Immunology 74(1): 1-7.

Faith, A., D.M. Schellenberg, A.D. Rees and D.M. Mitchell. 1992. Antigenic specificity and subset analysis of T cells isolated from the bronchoalveolar lavage and pleural effusion of patients with lung disease. Clin. Exp. Immunol. 87(2): 272-278.

Flyer, D.C., V. Ramakrishna, C. Miller, H. Myers, M. McDaniel, K. Root, C. Flournoy, V.H. Engelhard, D.H. Canaday, J.A. Marto, M.M. Ross, D.F. Hunt, J. Shabanowitz and F.M. White. 2002. Identification by mass spectrometry of CD8(+)-T-cell *Mycobacterium tuberculosis* epitopes within the Rv0341 gene product. Infect. Immun. 70(6): 2926-2932.

Flynn, J.L., J. Chan, K.J. Triebold, D.K. Dalton, T.A. Stewart and B.R. Bloom. 1993. An essential role for interferon gamma in resistance to *Mycobacterium tuberculosis* infection. J. Exp. Med. 178(6): 2249-2254.

Flynn, J.L., M.M. Goldstein, J. Chan, K.J. Triebold, K. Pfeffer, C.J. Lowenstein, R. Schreiber, T.W. Mak and B.R. Bloom. 1995. Tumor necrosis factor-alpha is required in the protective immune response against *Mycobacterium tuberculosis* in mice. Immunity 2(6): 561-572.

Flynn, J.L., M.M. Goldstein, K.J. Triebold, B. Koller and B.R. Bloom. 1992. Major histocompatibility complex class I-restricted T cells are required for resistance to *Mycobacterium tuberculosis* infection. Proc. Natl. Acad. Sci. USA. 89(24): 12013-12017.

Fulton, S.A., J.M. Johnsen, S.F. Wolf, D.S. Sieburth and W.H. Boom. 1996. Interleukin-12 production by human monocytes infected with *Mycobacterium tuberculosis*: role of phagocytosis. Infect. Immun. 64(7): 2523-2531.

Fulton, S.A., T.D. Martin, R.W. Redline and W. Henry Boom. 2000. Pulmonary immune responses during primary *Mycobacterium bovis-* Calmette- Guerin bacillus infection in C57Bl/6 mice. Am. J. Respir. Cell. Mol. Biol. 22(3): 333-343.

Gehring, A.J., Rojas, R.E., Canaday, D.H., Lakey, D.L., Harding, C.V., and Boom, W.H. 2003. The 19-kDa lipoprotein of *Mycobacterium tuberculosis* inhibits IFN-γ regulated HLA-DR and FcgammaRI on human macrophages through TLR-2. Infect. Immun. In press.

Geluk, A., K.E. van Meijgaarden, K.L. Franken, J.W. Drijfhout, S. D'Souza, A. Necker, K. Huygen and T.H. Ottenhoff. 2000. Identification of major epitopes of *Mycobacterium tuberculosis* AG85B that are recognized by HLA-A*0201-restricted CD8(+) T cells in HLA- transgenic mice and humans. J. Immunol. 165(11): 6463-6471.

Gercken, J., J. Pryjma, M. Ernst and H.D. Flad. 1994. Defective antigen presentation by *Mycobacterium tuberculosis*-infected monocytes. Infect. Immun. 62(8): 3472-3478.

Gong, J.H., M. Zhang, R.L. Modlin, P.S. Linsley, D. Iyer, Y. Lin and P.F. Barnes. 1996. Interleukin-10 downregulates *Mycobacterium tuberculosis*-induced Th1 responses and CTLA-4 expression. Infect. Immun. 64(3): 913-918.

Haanen, J.B., R. de-Waal-Malefijt, P.C. Res, E.M. Kraakman, T.H. Ottenhoff, R.R. de-Vries and H. Spits. 1991. Selection of a human T helper type 1-like T cell subset by mycobacteria. J. Exp. Med. 174(3): 583-592.

Hancock, G.E., Z.A. Cohn and G. Kaplan. 1989. The generation of antigen-specific, major histocompatability complex-restricted cytotoxic T lymphocytes of the CD4+ phenotype. J. Exp. Med. 169: 909-919.

Harding, C., and R. Song. 1994. Phagocytic processing of exogenous particulate antigens by macrophages for presentation by class I MHC molecules. J. Immunol. 153: 4925-4935.

Harding, C.V.. 1996. Class I MHC presentation of exogenous antigens. J. Clin. Immunol. 16(2): 90-96.

Havlir, D.V., J.J. Ellner, K.A. Chervenak and W.H. Boom. 1991. Selective expansion of human gamma delta T cells by monocytes infected with live *Mycobacterium tuberculosis*. J. Clin. Invest. 87(2): 729-733.

Havlir, D.V., R.S. Wallis, W.H. Boom, T.M. Daniel, K. Chervenak and J.J. Ellner. 1991. Human immune response to *Mycobacterium tuberculosis* antigens. Infect. Immun. 59(2): 665-670.

Hirsch, C.S., J.J. Ellner, D.G. Russell and E.A. Rich. 1994. Complement receptor-mediated uptake and tumor necrosis factor-alpha-mediated growth inhibition of *M. tuberculosis* by human alveolar macrophages. J. Immunol. 152(2): 743-753.

Hirsch, C.S., R. Hussain, Z. Toossi, G. Dawood, F. Shahid and J.J. Ellner. 1996. Cross-modulation by transforming growth factor beta in human tuberculosis: suppression of antigen-driven blastogenesis and interferon gamma production. Proc. Natl. Acad. Sci. USA. 93(8): 3193-3198.

Hirsch, C.S., Z. Toossi, J.L. Johnson, H. Luzze, L. Ntambi, P. Peters, M. McHugh, A. Okwera, M. Joloba, P. Mugyenyi, R.D. Mugerwa, P. Terebuh and J.J. Ellner. 2001. Augmentation of apoptosis and interferon-gamma production at sites of active *Mycobacterium tuberculosis* infection in human tuberculosis. J. Infect. Dis. 183(5): 779-788.

Hirsch, C.S., Z. Toossi, G. Vanham, J.L. Johnson, P. Peters, A. Okwera, R. Mugerwa, P. Mugyenyi and J.J. Ellner. 1999. Apoptosis and T cell hyporesponsiveness in pulmonary tuberculosis. J. Infect. Dis. 179(4): 945-953.

Hmama, Z., R. Gabathuler, W.A. Jefferies, G. de Jong and N.E. Reiner. 1998. Attenuation of HLA-DR expression by mononuclear phagocytes infected with *Mycobacterium tuberculosis* is related to intracellular sequestration of immature class II heterodimers. J. Immunol. 161(9): 4882-4893.

Hoft, D.F., R.M. Brown and S.T. Roodman. 1998. Bacille Calmette-Guerin vaccination enhances human gamma delta T cell responsiveness to mycobacteria suggestive of a memory-like phenotype. J. Immunol. 161(2): 1045-1054.

Holsti, M.A., J.S. Schorey, E.J. Brown and P.M. Allen. 1998. Identification of epitopes of fibronectin attachment protein. FAP-A) of *Mycobacterium avium* which stimulate strong T-cell responses in mice. Infect. Immun. 66(3): 1261-1264.

Hopewell, P.C.. 1992. Impact of human immunodeficiency virus infection on the epidemiology, clinical features, management, and control of tuberculosis. Clin Infect Dis 15(3): 540-547.

Hussain, S., B.S. Zwilling and W.P. Lafuse. 1999. *Mycobacterium avium* infection of mouse macrophages inhibits IFN-gamma Janus kinase-STAT signaling and gene induction by down-regulation of the IFN-gamma receptor. J. Immunol. 163(4): 2041-2048.

Janeway, C.A., Jr. and R. Medzhitov. 2002. Innate immune recognition. Annu Rev Immunol 20: 197-216.

Jouanguy, E., F. Altare, S. Lamhamedi, P. Revy, J.F. Emile, M. Newport, M. Levin, S. Blanche, E. Seboun, A. Fischer and J.L. Casanova. 1996. Interferon-gamma-receptor deficiency in an infant with fatal bacille Calmette-Guerin infection. N Engl J Med 335(26): 1956-1961.

Jouanguy, E., S. Lamhamedi-Cherradi, F. Altare, M.C. Fondaneche, D. Tuerlinckx, S. Blanche, J.F. Emile, J.L. Gaillard, R. Schreiber, M. Levin, A. Fischer, C. Hivroz and J.L. Casanova. 1997. Partial interferon-gamma receptor 1 deficiency in a child with tuberculoid bacillus Calmette-Guerin infection and a sibling with clinical tuberculosis. J. Clin. Invest. 100(11): 2658-2664.

Kamijo, R., J. Le, D. Shapiro, E.A. Havell, S. Huang, M. Aguet, M. Bosland and J. Vilcek. 1993. Mice that lack the interferon-gamma receptor have profoundly altered responses to infection with Bacillus Calmette-Guerin and subsequent challenge with lipopolysaccharide. J. Exp. Med. 178(4): 1435-1440.

Keane, J., S. Gershon, R.P. Wise, E. Mirabile-Levens, J. Kasznica, W.D. Schwieterman, J.N. Siegel and M.M. Braun. 2001. Tuberculosis associated with infliximab, a tumor necrosis factor alpha-neutralizing agent. N Engl J Med 345(15): 1098-1104.

Klein, M.R., S.M. Smith, A.S. Hammond, G.S. Ogg, A.S. King, J. Vekemans, A. Jaye, P.T. Lukey and K.P. McAdam. 2001. HLA-B*35-restricted CD8 T cell epitopes in the antigen 85 complex of *Mycobacterium tuberculosis*. J. Infect. Dis. 183(6): 928-934.

Kumararatne, D.S., A.S. Pithie, P. Drysdale, J.S. Gaston, R. Kiessling, P.B. Iles, C.J. Ellis, J. Innes and R. Wise. 1990. Specific lysis of mycobacterial antigen-bearing macrophages by class II MHC-restricted polyclonal T cell lines in healthy donors or patients with tuberculosis. Clin. Exp. Immunol. 80(3): 314-323.

Kusner, D.J. and J. Adams. 2000. ATP-induced killing of virulent *Mycobacterium tuberculosis* within human macrophages requires phospholipase D. J. Immunol. 164(1): 379-388.

Ladel, C.H., S. Daugelat and S.H.E. Kaufmann. 1995. Immune response to *Mycobacterium bovis* bacille Calmette Guerin infection in major histocompatability complex I- and II-deficient knock-out mice: contribution of CD4+ and CD8+ T cells to acquired resistance. Eur. J. Immunol. 25: 377-384.

Ladel, C.H., G. Szalay, D. Riedel and S.H. Kaufmann. 1997. Interleukin-12 secretion by *Mycobacterium tuberculosis*-infected macrophages. Infect. Immun. 65(5): 1936-1938.

Lalvani, A., R. Brookes, R.J. Wilkinson, A.S. Malin, A.A. Pathan, P. Andersen, H. Dockrell, G. Pasvol and A.V. Hill. 1998. Human cytolytic and interferon gamma-secreting CD8+ T lymphocytes specific for *Mycobacterium tuberculosis*. Proc. Natl. Acad. Sci. USA. 95(1): 270-275.

Lammas, D.A., C. Stober, C.J. Harvey, N. Kendrick, S. Panchalingam and D.S. Kumararatne. 1997. ATP-induced killing of mycobacteria by human macrophages is mediated by purinergic P2Z(P2X7) receptors. Immunity 7(3): 433-444.

Lang, F., M.A. Peyrat, P. Constant, F. Davodeau, J. David-Ameline, Y. Poquet, H. Vie, J.J. Fournie and M. Bonneville. 1995. Early activation of human V gamma 9V delta 2 T cell broad cytotoxicity and TNF production by nonpeptidic mycobacterial ligands. J. Immunol. 154(11): 5986-5994.

Larkin, R., C.D. Benjamin, Y.M. Hsu, Q. Li, L. Zukowski and R.F. Silver. 2002. CD40 ligand. CD154) does not contribute to lymphocyte-mediated inhibition of virulent *Mycobacterium tuberculosis* within human monocytes. Infect. Immun. 70(8): 4716-4720.

Lefford, M.J.. 1975. Transfer by adoptive immunity to tuberculosis in mice. Infect. Immun. 11: 1174-1181.

Lewinsohn, D.M., M.R. Alderson, A.L. Briden, S.R. Riddell, S.G. Reed and K.H. Grabstein. 1998. Characterization of human CD8+ T cells reactive with *Mycobacterium tuberculosis*-infected antigen-presenting cells. J. Exp. Med. 187(10): 1633-1640.

Lewinsohn, D.M., A.L. Briden, S.G. Reed, K.H. Grabstein and M.R. Alderson. 2000. *Mycobacterium tuberculosis*-reactive CD8+ T lymphocytes: the relative contribution of classical versus nonclassical HLA restriction. J. Immunol. 165(2): 925-930.

Li, B., H. Bassiri, M.D. Rossman, P. Kramer, A.F. Eyuboglu, M. Torres, E. Sada, T. Imir and S.R. Carding. 1998. Involvement of the Fas/Fas ligand pathway in activation-induced cell death of mycobacteria-reactive human gamma delta T cells: a mechanism for the loss of gamma delta T cells in patients with pulmonary tuberculosis. J. Immunol. 161(3): 1558-1567.

Li, B., M.D. Rossman, T. Imir, A.F. Oner-Eyuboglu, C.W. Lee, R. Biancaniello and S.R. Carding. 1996. Disease-specific changes in gammadelta T cell repertoire and function in patients with pulmonary tuberculosis. J. Immunol. 157(9): 4222-4229.

Lien, E., T.J. Sellati, A. Yoshimura, T.H. Flo, G. Rawadi, R.W. Finberg, J.D. Carroll, T. Espevik, R.R. Ingalls, J.D. Radolf and D.T. Golenbock. 1999. Toll-like receptor 2 functions as a pattern recognition receptor for diverse bacterial products. J. Biol. Chem. 274(47): 33419-33425.

MacMicking, J.D., R.J. North, R. LaCourse, J.S. Mudgett, S.K. Shah and C.F. Nathan. 1997. Identification of nitric oxide synthase as a protective locus against tuberculosis. Proc. Natl. Acad. Sci. USA. 94(10): 5243-5248.

Mahairas, G.G., P.J. Sabo, M.J. Hickey, D.C. Singh and C.K. Stover. 1996. Molecular analysis of genetic differences between *Mycobacterium bovis* BCG and virulent M. bovis. J. Bacteriol. 178(5): 1274-1282.

Means, T.K., S. Wang, E. Lien, A. Yoshimura, D.T. Golenbock and M.J. Fenton. 1999. Human toll-like receptors mediate cellular activation by *Mycobacterium tuberculosis*. J. Immunol. 163(7): 3920-3927.

Medzhitov, R., P. Preston-Hurlburt and C.A. Janeway, Jr. 1997. A human homologue of the Drosophila Toll protein signals activation of adaptive immunity. Nature 388(6640): 394-397.

Mohagheghpour, N., D. Gammon, L.M. Kawamura, A. van Vollenhoven, C.J. Benike and E.G. Engleman. 1998. CTL response to *Mycobacterium tuberculosis*: identification of an immunogenic epitope in the 19-kDa lipoprotein. J. Immunol. 161(5): 2400-2406.

Molloy, A., P. Laochumroonvorapong and G. Kaplan. 1994. Apoptosis, but not necrosis, of infected monocytes is coupled with killing of intacellular bacillus Calmette-Guerin. J. Exp. Med. 180: 1499-1509.

Morita, C.T., E.M. Beckman, J.F. Bukowski, Y. Tanaka, H. Band, B.R. Bloom, D.E. Golan and M.B. Brenner. 1995. Direct presentation of nonpeptide prenyl pyrophosphate antigens to human gamma delta T cells. Immunity 3(4): 495-507.

Muller, I., S.P. Cobbold, H. Waldmann and S.H. Kaufmann. 1987. Impaired resistance to *Mycobacterium tuberculosis* infection after selective *in vivo* depletion of L3T4+ and Lyt-2+ T cells. Infect. Immun. 55(9): 2037-2041.

Murray, J.F.. 1991. Tuberculosis and human immunodeficiency virus infection during the 1990's. Bull. Int. Union Tuberc. Lung Dis. 66(1): 21-25.

Murray, P.J., L. Wang, C. Onufryk, R.I. Tepper and R.A. Young. 1997. T cell-derived IL-10 antagonizes macrophage function in mycobacterial infection. J. Immunol. 158(1): 315-321.

Mutis, T., Y.E. Cornelisse and T.H. Ottenhoff. 1993. Mycobacteria induce CD4+ T cells that are cytotoxic and display Th1-like cytokine secretion profile: heterogeneity in cytotoxic activity and cytokine secretion levels. Eur. J. Immunol. 23(9): 2189-2195.

Newport, M.J., C.M. Huxley, S. Huston, C.M. Hawrylowicz, B.A. Oostra, R. Williamson and M. Levin. 1996. A mutation in the interferon-gamma-receptor gene and susceptibility to mycobacterial infection. N Engl J Med 335(26): 1941-1949.

Neyrolles, O., K. Gould, M.P. Gares, S. Brett, R. Janssen, P. O'Gaora, J.L. Herrmann, M.C. Prevost, E. Perret, J.E. Thole and D. Young. 2001. Lipoprotein access to MHC class I presentation during infection of murine macrophages with live mycobacteria. J. Immunol. 166(1): 447-457.

North, R.J.. 1998. Mice incapable of making IL-4 or IL-10 display normal resistance to infection with *Mycobacterium tuberculosis*. Clin. Exp. Immunol. 113(1): 55-58.

Noss, E.H., C.V. Harding and W.H. Boom. 2000. *Mycobacterium tuberculosis* inhibits MHC class II antigen processing in murine bone marrow macrophages. Cell. Immunol. 201(1): 63-74.

Noss, E.H., R.K. Pai, T.J. Sellati, J.D. Radolf, J. Belisle, D.T. Golenbock, W.H. Boom and C.V. Harding. 2001. Toll-like receptor 2-dependent inhibition of macrophage class II MHC expression and antigen processing by 19-kDa lipoprotein of *Mycobacterium tuberculosis*. J. Immunol. 167(2): 910-918.

Oddo, M., T. Renno, A. Attinger, T. Bakker, H.R. MacDonald and P.R. Meylan. 1998. Fas ligand-induced apoptosis of infected human macrophages reduces the viability of intracellular *Mycobacterium tuberculosis*. J. Immunol. 160(11): 5448-5454.

Oftung, F., A.S. Mustafa, R. Husson, R.A. Young and T. Godal. 1987. Human T cell clones recognize two abundant *Mycobacterium tuberculosis* protein antigens expressed in Escherichia coli. J. Immunol. 138(3): 927-931.

Orme, I.M. and F.M. Collins. 1983. Protection against *Mycobacterium tuberculosis* infection by adoptive transfer. J. Exp. Med. 158: 74-83.

Ottenhoff, T.H., B.K. Ab, J.D. Van-Embden, J.E. Thole and R. Kiessling. 1988. The recombinant 65-kD heat shock protein of *Mycobacterium bovis* Bacillus Calmette-Guerin/*M. tuberculosis* is a target molecule for CD4+ cytotoxic T lymphocytes that lyse human monocytes. J. Exp. Med. 168(5): 1947-1952.

Pearlman, E., J.W. Kazura, F.E. Hazlett and W.H. Boom. 1993. Modulation of murine cytokine responses to mycobacterial antigens by helminth-induced T helper 2 responses. J. Immunol. 151(9): 4857-4864.

Pechhold, K., D. Wesch, S. Schondelmaier and D. Kabelitz. 1994. Primary activation of V gamma 9-expressing gamma delta T cells by *Mycobacterium tuberculosis*. Requirement for Th1-type CD4 T cell help and inhibition by IL-10. J. Immunol. 152(10): 4984-4992.

Pedrazzini, T., K. Hug and J.A. Louis. 1987. Importance of L3T4+ and Lyt-2+ cells in the immunologic control of infection with *Mycobacterium bovis* strain bacillus Calmette-Guerin in mice. Assessment by elimination of T cell subsets *in vivo*. J. Immunol. 139(6): 2032-2037.

Poquet, Y., P. Constant, F. Halary, M.A. Peyrat, M. Gilleron, F. Davodeau, M. Bonneville and J.J. Fournie. 1996. A novel nucleotide-containing antigen for human blood gamma delta T lymphocytes. Eur. J. Immunol. 26(10): 2344-2349.

Porcelli, S., C.T. Morita and M.B. Brenner. 1992. CD1b restricts the response of human CD4-8- T lymphocytes to a microbial antigen. Nature 360(6404): 593-597.

Ramachandra, L., E. Noss, W.H. Boom and C.V. Harding. 1999. Phagocytic processing of antigens for presentation by class II major histocompatibility complex molecules. Cell. Microbiol. 1(3): 205-214.

Ramachandra, L., E. Noss, W.H. Boom and C.V. Harding. 2001. Processing of *Mycobacterium tuberculosis* antigen 85B involves intraphagosomal formation of peptide-major histocompatibility complex II complexes and is inhibited by live bacilli that decrease phagosome maturation. J. Exp. Med. 194(10): 1421-1432.

Raviglione, M.C., D.E. Snider, Jr. and A. Kochi. 1995. Global epidemiology of tuberculosis. Morbidity and mortality of a worldwide epidemic. JAMA 273(3): 220-226.

Rojas, R.E., K.N. Balaji, A. Subramanian and W.H. Boom. 1999. Regulation of human CD4(+) alphabeta T-cell-receptor-positive (TCR(+)) and gammadelta TCR(+) T-cell responses to *Mycobacterium tuberculosis* by interleukin-10 and transforming growth factor beta. Infect. Immun. 67(12): 6461-72.

Rojas, R.E. and W.H. Boom. 2001. Gamma delta T cells and HIV-1 infection. In: Cellular Aspects of HIV Infection. Wiley-Liss Press, NY, NY. p. 147-205.

Rojas, R.E., M. Torres, J.J. Fournie, C.V. Harding and W.H. Boom. 2002. Phosphoantigen presentation by macrophages to *Mycobacterium tuberculosis*—reactive Vgamma9Vdelta2+ T cells: modulation by chloroquine. Infect. Immun. 70(8): 4019-4027.

Rook, G.A., J. Steele, M. Ainsworth and B.R. Champion. 1986. Activation of macrophages to inhibit proliferation of *Mycobacterium tuberculosis*: comparison of the effects of recombinant gamma-interferon on human monocytes and murine peritoneal macrophages. Immunology 59(3): 333-338.

Rosat, J.P., E.P. Grant, E.M. Beckman, C.C. Dascher, P.A. Sieling, D. Frederique, R.L. Modlin, S.A. Porcelli, S.T. Furlong and M.B. Brenner. 1999. CD1-restricted microbial lipid antigen-specific recognition found in the CD8+ alpha beta T cell pool. J. Immunol. 162(1): 366-371.

Schlesinger, L.S.. 1993. Macrophage phagocytosis of virulent but not attenuated strains of *Mycobacterium tuberculosis* is mediated by mannose receptors in addition to complement receptors. J. Immunol. 150(7): 2920-2930.

Schlesinger, L.S., K.C. Bellinger, N.R. Payne and M.A. Horwitz. 1990. Phagocytosis of *Mycobacterium tuberculosis* is mediated by human monocyte complement receptors and complement component C3. J. Immunol. 144(7): 2771-2780.

Schlesinger, L.S., T.M. Kaufman, S. Iyer, S.R. Hull and L.K. Marchiando. 1996. Differences in mannose receptor-mediated uptake of lipoarabinomannan from virulent and attenuated strains of *Mycobacterium tuberculosis* by human macrophages. J. Immunol. 157(10): 4568-4575.

Schoel, B., H. Gulle and S.H. Kaufmann. 1992. Heterogeneity of the repertoire of T cells of tuberculosis patients and healthy contacts to *Mycobacterium tuberculosis* antigens separated by high-resolution techniques. Infect. Immun. 60(4): 1717-1720.

Serbina, N.V. and J.L. Flynn. 1999. Early emergence of CD8(+) T cells primed for production of type 1 cytokines in the lungs of *Mycobacterium tuberculosis*-infected mice. Infect. Immun. 67(8): 3980-3988.

Serbina, N.V., V. Lazarevic and J.L. Flynn. 2001. CD4(+) T cells are required for the development of cytotoxic CD8(+) T cells during *Mycobacterium tuberculosis* infection. J. Immunol. 167(12): 6991-7000.

Shen, Y., D. Zhou, L. Qiu, X. Lai, M. Simon, L. Shen, Z. Kou, Q. Wang, L. Jiang, J. Estep, R. Hunt, M. Clagett, P.K. Sehgal, Y. Li, X. Zeng, C.T. Morita, M.B. Brenner, N.L. Letvin and Z.W. Chen. 2002. Adaptive immune response of Vgamma2Vdelta2+ T cells during mycobacterial infections. Science 295(5563): 2255-2258.

Sicard, H. and J.J. Fournie. 2000. Metabolic routes as targets for immunological discrimination of host and parasite. Infect. Immun. 68(8): 4375-4377.

Silver, R.F., Q. Li, W.H. Boom and J.J. Ellner. 1998. Lymphocyte-dependent inhibition of growth of virulent *Mycobacterium tuberculosis* H37Rv within human monocytes: requirement for CD4+ T cells in purified protein derivative-positive, but not in purified protein derivative-negative subjects. J. Immunol. 160(5): 2408-2417.

Silver, R.F., R.S. Wallis and J.J. Ellner. 1995. Mapping of T cell epitopes of the 30-kDa alpha antigen of *Mycobacterium bovis* strain bacillus Calmette-Guerin in purified protein derivative (PPD)-positive individuals. J. Immunol. 154(9): 4665-4674.

Skeiky, Y.A., M.J. Lodes, J.A. Guderian, R. Mohamath, T. Bement, M.R. Alderson and S.G. Reed. 1999. Cloning, expression, and immunological evaluation of two putative secreted serine protease antigens of *Mycobacterium tuberculosis*. Infect. Immun. 67(8): 3998-4007.

Smith, S.M., R. Brookes, M.R. Klein, A.S. Malin, P.T. Lukey, A.S. King, G.S. Ogg, A.V. Hill and H.M. Dockrell. 2000. Human CD8+ CTL specific for the mycobacterial major secreted antigen 85A. J. Immunol. 165(12): 7088-7095.

Smith, S.M., A.S. Malin, T. Pauline, Lukey, S.E. Atkinson, J. Content, K. Huygen and H.M. Dockrell. 1999. Characterization of human *Mycobacterium bovis* bacille Calmette-Guerin- reactive CD8+ T cells. Infect. Immun. 67(10): 5223-5230.

Sorensen, A.L., S. Nagai, G. Houen, P. Andersen and A.B. Andersen. 1995. Purification and characterization of a low-molecular-mass T cell antigen secreted by *Mycobacterium tuberculosis*. Infect. Immun. 63: 1710-1717.

Sousa, A.O., R.J. Mazzaccaro, R.G. Russell, F.K. Lee, O.C. Turner, S. Hong, L. Van Kaer and B.R. Bloom. 2000. Relative contributions of distinct MHC class I-dependent cell populations in protection to tuberculosis infection in mice. Proc. Natl. Acad. Sci. USA. 97(8): 4204-4208.

Stenger, S., D.A. Hanson, R. Teitelbaum, P. Dewan, K.R. Niazi, C.J. Froelich, T. Ganz, S. Thoma-Uszynski, A. Melian, C. Bogdan, S.A. Porcelli, B.R. Bloom, A.M. Krensky and R.L. Modlin. 1998. An antimicrobial activity of cytolytic T cells mediated by granulysin. Science 282(5386): 121-125.

Stenger, S., R.J. Mazzaccaro, K. Uyemura, S. Cho, P.F. Barnes, J.P. Rosat, A. Sette, M.B. Brenner, S.A. Porcelli, B.R. Bloom and R.L. Modlin. 1997. Differential effects of cytolytic T cell subsets on intracellular infection. Science 276(5319): 1684-1687.

Sturgill-Koszycki, S., P. Sclesinger, P. Chakraborty, P.L. Haddix, H.L. Collins, A.K. Fok, A.D. Allen, S.L. Gluck, J. Heuser and D.G. Russell. 1994. Lack of acidification in Mycobacterium phagosomes produced by exclusion of the vesicular proton-ATPase. Science 263: 678.

Tan, J.S., D.H. Canaday, W.H. Boom, K.N. Balaji, S.K. Schwander and E.A. Rich. 1997. Human alveolar T lymphocyte responses to *Mycobacterium tuberculosis* antigens: role for CD4+ and CD8+ cytotoxic T cells and relative resistance of alveolar macrophages to lysis. J. Immunol. 159(1): 290-297.

Tanaka, Y., C. Morita, Y. Tanaka, E. Nieves, M.B. Brenner and B.M. Bloom. 1995. Natural and synthetic non-peptide antigens recognized by human gamma delta T cells. Nature 375: 155-158.

Ting, L.M., A.C. Kim, A. Cattamanchi and J.D. Ernst. 1999. *Mycobacterium tuberculosis* inhibits IFN-gamma transcriptional responses without inhibiting activation of STAT1. J. Immunol. 163(7): 3898-3906.

Tsukaguchi, K., K.N. Balaji and W.H. Boom. 1995. CD4+ alpha-beta T cell and gamma delta T cell responses to *Mycobacterium tuberculosis*: similarities and differences in antigen recognition, cytotoxic effector function, and cytokine production. J. Immunol. 154: 1786-1796.

Tsukaguchi, K., B. de Lange and W.H. Boom. 1999. Differential regulation of IFN-gamma, TNF-alpha, and IL-10 production by CD4(+) alphabetaTCR+ T cells and vdelta2(+) gammadelta T cells in response to monocytes infected with *Mycobacterium tuberculosis*-H37Ra. Cell. Immunol. 194(1): 12-20.

VanHeyningen, T.K., H.L. Collins and D.G. Russell. 1997. IL-6 produced by macrophages infected with Mycobacterium species suppresses T cell responses. J. Immunol. 158(1): 330-337.

Vordemeier, H.M., D.P. Harris, E. Roman, R. Lathigra, C. Moreno and J. Ivanyi. 1991. Identification of T cell stimulatory peptides from the 38-kDa protein of *Mycobacterium tuberculosis*. J. Immunol. 147(3): 1023-1029.

Vordermeier, H.M., D.P. Harris, G. Friscia, E. Roman, H.M. Surcel, C. Moreno, G. Pasvol and J. Ivanyi. 1992. T cell repertoire in tuberculosis: selective anergy to an immunodominant epitope of the 38-kDa antigen in patients with active disease. Eur. J. Immunol. 22(10): 2631-2637.

Wiker, H.G. and M. Harboe. 1992. The antigen 85 complex: a major secretion product of *Mycobacterium tuberculosis*. Microbiol Rev 56(4): 648-661.

Wojciechowski, W., J. DeSanctis, E. Skamene and D. Radzioch. 1999. Attenuation of MHC class II expression in macrophages infected with *Mycobacterium bovis* bacillus Calmette-Guerin involves class II transactivator and depends on the Nramp1 gene. J. Immunol. 163(5): 2688-2696.

Worku, S. and D.F. Hoft. 2000. *In vitro* measurement of protective mycobacterial immunity: antigen- specific expansion of T cells capable of inhibiting intracellular growth of bacille Calmette-Guerin. Clin Infect Dis 30 Suppl 3: S257-S261.

Young, D., T. Garbe, R. Lathigra, Z.C. Abou and Y. Zhang. 1991. Characterization of prominent protein antigens from mycobacteria. Bull. Int. Union Tuberc. Lung Dis. 66(1): 47-51.

Young, D.B. and T.R. Garbe. 1991. Lipoprotein antigens of *Mycobacterium tuberculosis*. Res Microbiol 142(1): 55-65.

Zhang, Y., R. Lathigra, T. Garbe, D. Catty and D. Young. 1991. Genetic analysis of superoxide dismutase, the 23 kilodalton antigen of *Mycobacterium tuberculosis*. Mol Microbiol 5(2): 381-391.

From: Tuberculosis: The Microbe Host Interface
Edited by: Larry S. Schlesinger and Lucy E. DesJardin

Chapter 5

New *In Vitro* Models of Mycobacterial Pathogenesis

Frederick D. Quinn,
Luiz E. Bermudez and
Kristin A Birkness

Abstract

Mycobacterium tuberculosis bacilli are inhaled into the lung, eventually reaching the alveoli where the organisms are ingested by alveolar macrophages. If not killed by these macrophages, the bacilli replicate intracellularly and spread to other alveolar and recruited macrophages. Within weeks, a few bacilli can multiply to significant numbers and infect hundreds or thousands of alveolar cells. Subsequent dissemination of these infected macrophages from the alveoli into the lymph and circulatory systems may be critical to the establishment of active and latent disease. This simple view of a complex process describes the diverse environments in which the mycobacteria must compete in order to survive, replicate and ultimately disseminate. This view also illustrates the daunting task facing investigators who are attempting to dissect and analyze each of the steps employed by the pathogen and host. Several model systems are now available to help investigators elucidate various aspects of the *M. tuberculosis* pathogenesis process. Although the animal and tissue culture monolayer models are the most widely used, a third system, complex *in vitro* models are beginning to be utilized. What we describe here are a number of these model systems and how they are or could be used to examine many of the stages of the infectious process.

New *In Vitro* Models of Mycobacterial Pathogenesis

Mycobacterium tuberculosis bacilli are inhaled into the lung, eventually reaching the alveoli where the organisms are ingested by alveolar macrophages. If not killed by the macrophage, the bacilli replicate intracellularly and spread to other alveolar macrophages and to the nonactivated bloodborne macrophages attracted to the site by released bacterial cell debris and chemotactic factors (Schlesinger *et al.*, 1990). Within weeks of a successful aerosol infection by a few *M. tuberculosis* bacilli, one can expect to find a significant increase in the number of microorganisms infecting hundreds or thousands of alveolar cells. Subsequent dissemination of these macrophages from the alveoli into the lymph and circulatory systems may be critical to the establishment of active and latent disease (Poole and Florey, 1970; Mekalanos, 1992). Thus, stages in the disease process might be categorized as: 1) inhalation of the bacilli into the lung alveoli, 2) attachment of the bacilli to alveolar macrophages, 3) internalization and survival of the bacilli in macrophages, 4) dissemination of the bacilli to other organs via infected macrophages, 5) granuloma formation and failure of granuloma to contain infection, 6) multiplication of the bacilli within the upper lungs, and 7) aerosol transmission of the bacilli. This is obviously a simplistic view of a complex interactive process, and there are likely many additional stages in the web of interactions that are as yet unidentified. This simple list, nevertheless, underscores the incredibly diverse environments in which the mycobacteria must compete in order to survive, replicate and ultimately disseminate. Perhaps of equal importance, this list illustrates the daunting task set before investigators trying to dissect and analyze each of the mechanisms employed by both pathogen and host.

Because of our current inability to routinely diagnose early infections, most of what we understand of the human disease process comes from the latest stages: multiplication to high numbers in the lungs with aerosol dissemination. Knowledge of earlier events is based primarily on research using animals and to some extent *in vitro* tissue culture monolayer models. While animal models provide much useful data, the physiology and immune response of most non-primate systems can sometimes make comparisons to human data difficult to interpret (Flynn *et al.,* 1993; Orme, 1993; Rhoades *et al.,* 1995). Human tissue culture cell monolayers, although simpler to manipulate, cannot mimic the wide variety of cell types and environments encountered by the pathogen in the human or animal host. Although these models are widely used and extremely helpful, there are aspects of the infectious process that require human cells and in a more complex array than found in a monolayer.

A potential group of alternative models falls into a third category: the complex *in vitro* models. Such model systems may help to elucidate many aspects of *M. tuberculosis* pathogenesis that are too specific for an animal model or too complex for a tissue culture monolayer. What we describe here are a number of these model systems and how they are being used or could be used to examine many of the described stages of the infectious process.

In Vitro Models

Environmental conditions encountered by pathogens during the course of an infection are responsible for the observed adaptive gene responses and phenotypes. It is well known that events important for a successful infection such as phagocytosis or invasion of non-professional phagocytes can be influenced by a myriad of experimental details (Li *et al.,* 2002). Therefore, it is not surprising that in addition to paying close attention to conditions during the experiment, care also must be taken in choosing cell lines and in preparing the host and pathogen cultures. These concerns are not always obvious in some published reports and can make the resulting data difficult to interpret. The ultimate test of any model system is how closely the conditions encountered by the pathogen mimic those seen in animal and ultimately human disease, thus ensuring the study of the most appropriate responses of the pathogen and host.

Alveolar Bilayer

It is not known precisely how many alveolar macrophages the bacilli might encounter as they enter the alveolar space. It has been estimated that in an average human male there are 1,400 type I pneumocytes, 28,000 type II pneumocytes, and 50-100 alveolar macrophages per alveolus (Crandall and Kim, 1991; Crystal, 1991; Schneeberger, 1991). Thus, the bacilli might well interact initially with epithelial cells lining this space. If the invading bacteria were able to enter these cells, not only would they be protected from the bactericidal activity of the macrophage, they might also be able to pass through these cells into adjacent endothelial cells and thus enter the circulatory system as do other respiratory pathogens such as *Streptococcus pneumoniae* (Cundell and Tuomanen, 1994). *In vitro* studies using cultured pneumocyte monolayers

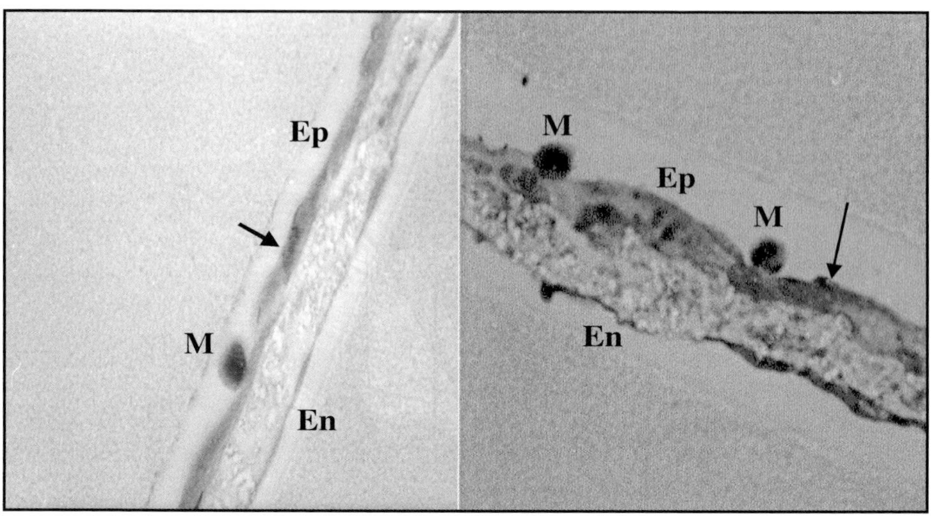

Figure 1. Tissue culture bilayer 4 hours (left) and 24 hours (right) after infection with *M. tuberculosis*. Mononuclear cells (M) have migrated from the lower chamber to the apical surface where they are seen in close association with bacilli (arrows). Microporous membrane separates the apical A549 pneumocyte epithelial layer (Ep) from the basal HULEC (human lung endothelial cell) layer (En). Magnification, x1,000.

have shown that *M. tuberculosis* bacilli are able not only to enter these cells but to multiply intracellularly in far greater numbers than those seen within cultured macrophages, increasing up to 6-fold in 4 days (McDonough and Kress, 1995; Mehta *et al.*, 1996; Li *et al.*, 2002). Similar studies also showed intracellular growth within cultured human lung endothelial cells (Li *et al.*, 2002; Mehta and Quinn, unpublished observations). In these cells, *M. tuberculosis* bacilli replicated intracellularly up to 120-fold by day seven post-infection with large numbers of *M. tuberculosis* bacilli observed within the vacuoles of the cultured cells. Thus, even a few organisms inhaled into the alveolar space could potentially multiply to a much larger number before entering the blood stream.

To examine some fundamental issues about the early stages of infection, an artificial tissue system has been developed that incorporates epithelial and endothelial cell monolayers separated by a microporous membrane in a 6-well tissue culture dish (Birkness *et al.*, 1999; Bermudez *et al.*, 2002). This three-layer system (Figure 1) allows for cell-to-cell communication that influences cell morphology, cell differentiation, cell orientation and cytokine production. It makes possible the examination of microbial attachment, internalization, intracellular multiplication, and intra- or intercellular passage from the epithelial cell surface through the two layers of cells and into the lower chamber. With the addition of peripheral blood lymphocytes and monocytes it is possible to observe the effect of a bacterial infection on the migration of these cells from the lower chamber through the endothelial cells to the epithelial cell surface. One also can observe the effect of these cells on the growth and movement of the invading microbe during the course of infection.

Although primary cell lines can be used, in two recently published studies, human type II pneumocyte epithelial (A549) and human lung endothelial (HULEC) or human umbilical vein/A549 hybrid (Eahy926) cell lines were used (Birkness *et al.*, 1999; Bermudez *et al.*, 2002). In these studies, bacterial attachment, internalization, and passage through the two-cell layers and into the space below were observed. Three *M. tuberculosis* strains and a strain of *M. bovis* BCG all passed at the same low rate through the bilayer even in the presence of a high inoculum and 48 hours of infection (Birkness *et al.*, 1999; Bermudez *et al.*, 2002). Electron micrographs of infected bilayers after 48 hours showed large numbers of bacilli within pneumocytes. The percentage of *M. bovis* BCG bacilli passing into the lower chamber after the first four hours is slightly higher than those of the *M. tuberculosis* strains. However, there is little difference in percentages among the three strains after 48 hours. Unlike the passage of other bacteria such as *Neisseria meningitidis* (Birkness *et al.*, 1995), the observed mycobacterial numbers are all very small percentages of the total innoculum. Pre-culturing bacteria in A549 cells significantly increased the passage rate of *M. tuberculosis* to 15% of the inoculum by 48h (Bermudez *et al.*, 2002).

Peripheral blood mononuclear cells or purified monocytes added to the lower chamber migrated up through the various layers to the epithelial surface in response to the presence of the bacilli. *M. tuberculosis* bacilli also were able to translocate efficiently across the bilayer when inside monocytes, and infected monocytes crossed the barrier with greater efficiency when A549 alveolar cells were infected with *M. tuberculosis* bacilli (Bermudez *et al.*, 2002). Thus, if these observations correlate with early human infection, only a few organisms could multiply within pneumocytes, move to adjacent pneumocytes and macrophages, and eventually gain entry into the circulatory system by passage through the endothelial layer either as free bacteria or within macrophages.

Both studies (Birkness *et al.*, 1999; Bermudez *et al.*, 2002) demonstrated the presence of cytotoxic damage to the host cells or alterations in the structural integrity of the cell layers. Such damage may account for some of the observable tissue damage that is a hallmark of this disease in animals and humans. It also appeared that a majority of the bacteria remained associated with the host cells and did not freely disseminate into the surrounding medium. This observation forms the basis for the monolayer dissemination and cytotoxicity model described in the section entitled *Plaque Dissemination Model* of this chapter.

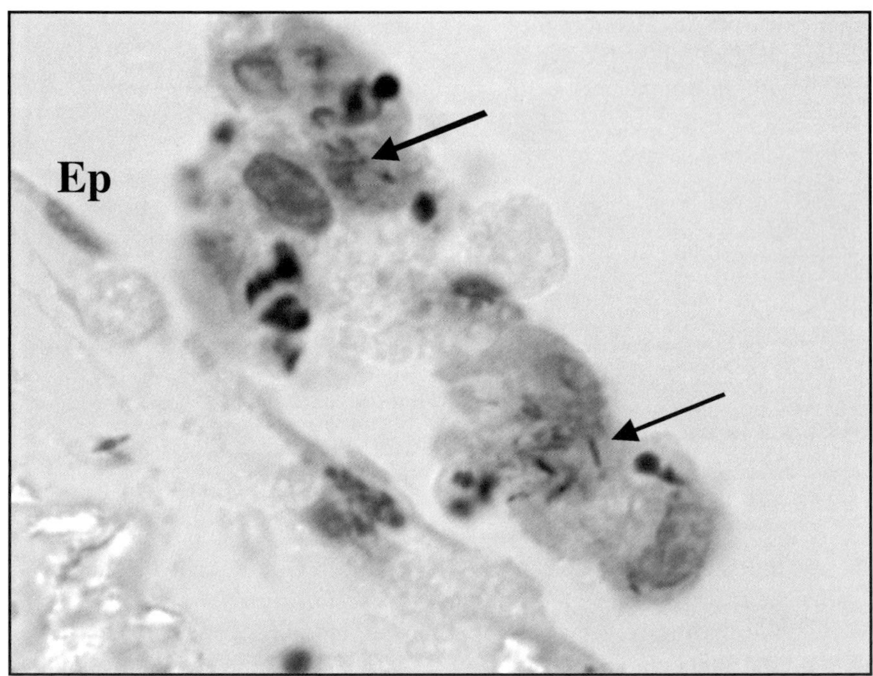

Figure 2. *In vitro* granuloma, composed of human peripheral blood lymphocytes, autologous macrophages and mycobacteria (arrows) attached to apical epithelial cell surface (Ep) of A549/HULEC bilayer. Magnification, x1,000.

The bilayer model offers several directions for further study of mycobacterial pathogenesis. The roles of the various chemokines in leukocyte migration and bacterial passage through the bilayer might be confirmed by blocking the activity of these molecules with the addition of anti-chemokine antibodies. In the presence of anti-MCP-1 antibodies, migration of monocytes across the bilayer was significantly slowed (Bermudez *et al.*, 2002). Another very important component of the host immune response to *M. tuberculosis* infection is the formation of a granuloma, as described in the section entitled *In Vitro Granuloma*. Since one of the places this complex of macrophages, T lymphocytes and fibroblasts is established is within the tissues of the lung (Lurie *et al.*, 1952; McCune and Tompsett, 1956), the best framework for an *in vitro* model of the granuloma may well be the complex of tissues found in the bilayer. There is experimental evidence that granulomas can form on the bilayer surface (Figure 2).

The bilayer is obviously a simple system when compared to human lung alveolus. Substituting primary cells, including both Type I and Type II cells, would add cellular diversity and complexity. It has been suggested that dendritic cells (Henderson *et al.*, 1997) or M cells (Teitelbaum *et al.*, 1999) present in airway epithelium and lung parenchyma may play an important role in the uptake of *M. tuberculosis* bacilli and subsequent priming of T cells to initiate the host immune response. The addition of these cell types to this model might allow us to more clearly identify their role in this response. It would also be of interest to observe what happens as the cell support system is made more complex by the addition of a variety of extracellular matrix components, fibroblast cells and surfactants. In addition, the human pulmonary surfactant protein A (hSP-A), a member of the mammalian collectin family, is thought to play a key defensive role against airborne invading pulmonary pathogens, including *M. tuberculosis*. hSP-A has been shown to promote the uptake and the phagocytosis of pathogenic bacilli through the recognition and the binding of carbohydrate motifs on the invading pathogen surface (Beharka *et al.*, 2002; Sidobre *et al.*, 2002; Weikert *et al.*, 2000). Its addition to the bilayer epithelial surface might also prove interesting. Further refinement of this model will enhance our abilities to elucidate the pathogenic mechanisms of *M. tuberculosis* in the human lung and define the roles of the various alveolar cell types in early infection.

In Vitro Granuloma

In addition to the acute disease, *M. tuberculosis* latently infects perhaps one-third of the world's population (Raviglione *et al.*, 1995). Latent infection begins after inhalation as the bacilli enter the alveoli and are internalized by resident macrophages. The infected phagocytic cells secrete cytokines and chemokines that activate and recruit additional peripheral blood macrophages and T lymphocytes to the site. These cells combine to form a granuloma, the principal defense of the human host against a mycobacterial infection and the likely site of dormant bacilli. This compact, organized collection of activated macrophages, including epithelioid and multinucleated giant cells, surrounded by T lymphocytes, and later by fibroblasts and collagen, serves to isolate the bacteria and may prevent active disease by sequestering the invading organisms. If the granuloma is maintained, these bacteria may remain dormant within the lung for decades.

A number of animal and *in vitro* models have been developed over many decades of research in an attempt to unravel the complex sequence of events involved in granuloma formation and the cytokine cascade that controls the process. The mouse, although the most extensively studied animal model, is much more resistant than humans to *M. tuberculosis* infection and correlates less well in granuloma formation. Granuloma formation is observed in the mouse, but lesions are small, and the cellular organization is somewhat different with a central core of lymphocytes surrounded by epithelioid macrophages (Saunders *et al.*, 1999). Also, the caseating necrosis and tubercule breakdown seen in humans and in primate, rabbit, and guinea pig models is not seen in the mouse. There is an absence of multinucleated giant cells, and the immune response is insufficient to achieve the isolation of granulomas from the surrounding cells (Cardona *et al.*, 2000). *M. tuberculosis* infection in rabbits resembles human disease more closely than any other non-primate animal model and it is the only model in which cavitary disease can be readily produced (McMurray, 1996). The guinea pig is likewise exquisitely susceptible to *M. tuberculosis* infection and, like the rabbit, develops classical granulomas similar to those seen in humans with active tuberculosis; lesions composed of lymphocytes, monocytes, epithelioid cells, and multinucleated giant cells (McMurray, 1996; Orme, 1998; Smith and Wiegeshaus, 1989). However, none of the non-primate animal models exhibit the full complement of the human disease spectrum (Clark-Curtiss, 1998).

While much important information has been gained from these models, there are many limitations. Even in animal models, it is difficult to examine the earliest events in the immune response leading to granuloma formation, so a number of *in vitro* models have been developed using synthetic microparticles (Shikama, 1989) or the eggs of the parasite *Schistosoma mansoni* (Doughty and Phillips, 1982) primarily combined with murine spleen cells. In these models, however, there is no phagocytosis; the mononuclear cells adhere to the particles or the eggs, and there is no evidence of mononuclear cell proliferation or multinucleated giant cell formation. Models using purified protein derivative-coated beads in mice (Chensue *et al.*, 1994; Chensue *et al.*, 1995) do provide some relevant cytokine data but are still limited in that they do not incorporate live organisms and therefore cannot elicit the same immunological response as that seen in human infection. To more closely simulate the process of human granuloma formation during *M. tuberculosis* infection, the subsequent breakdown when host defenses are compromised, and to understand what cytokines play roles in the formation and maintenance of this structure, an *in vitro* granuloma model composed entirely of human cells has been developed (Quinn *et al.*, 2002; Birkness *et al.*, manuscript in preparation).

Human peripheral blood lymphocytes, autologous macrophages and mycobacteria were combined, resulting in the formation of small, rounded aggregate structures within 4 hours (Figure 3A). Microscopic examination of the more mature 9-day aggregates found their composition and architecture to closely resemble the pathology seen in human clinical specimens (Figure 3B). Immunohistochemistry using cell surface markers identified macrophages and T lymphocytes surrounding acid-fast bacilli (Figure 4). To begin to understand how these cells come together and stay together in this structure, the expression patterns of a variety of chemokines and cytokines known to be involved in granuloma formation and maintenance in animals and humans were examined. Assays of granuloma culture supernatants demonstrated cytokine production consistent with findings in studies of patients with clinical tuberculosis and addition of key cytokines to freshly isolated cells enhanced aggregate formation (Birkness *et al.*, manuscript in preparation).

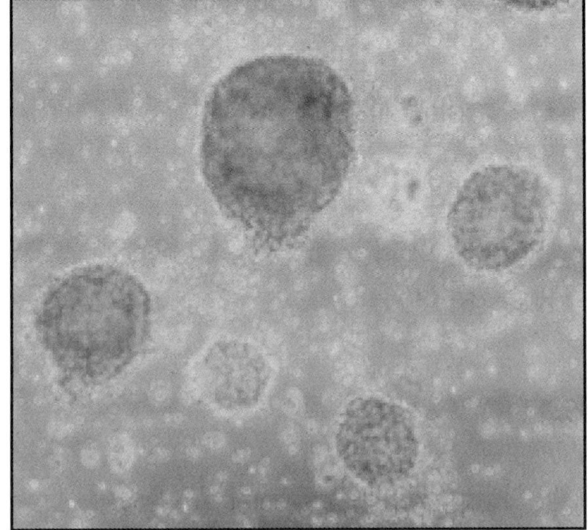

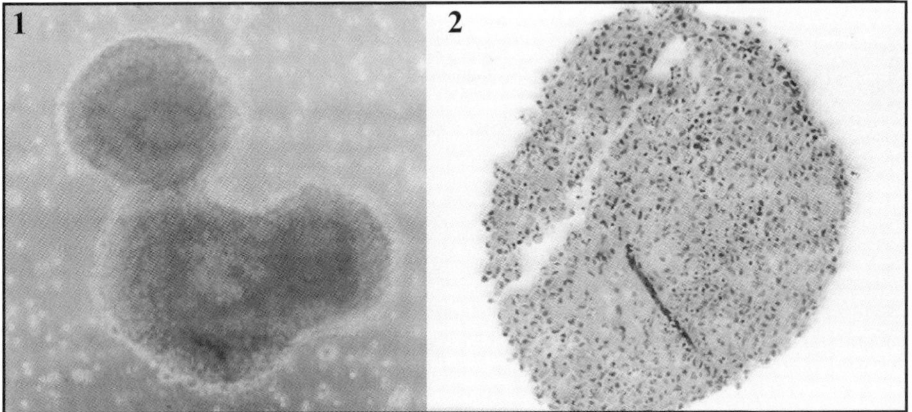

Figure 3. A) *In vitro* granulomas composed of human peripheral blood lymphocytes, autologous macrophages and mycobacteria after 4 hours' incubation. Magnification, x100. B) *In vitro* granulomas composed of human peripheral blood lymphocytes, autologous macrophages and mycobacteria after 9 days' incubation, (1) viable and (2) fixed, sectioned and H&E stained. Magnification, x100 and x200 respectively.

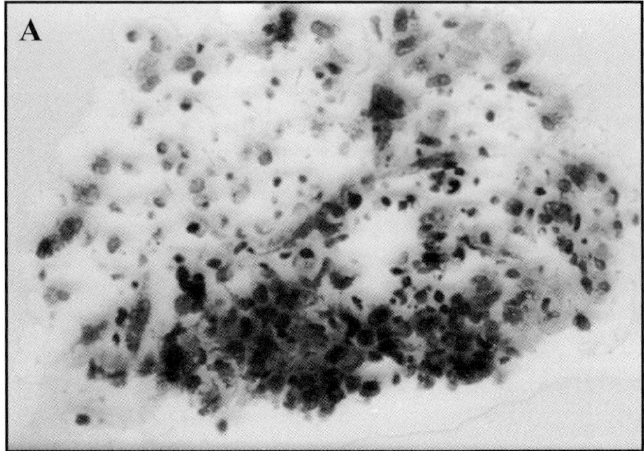

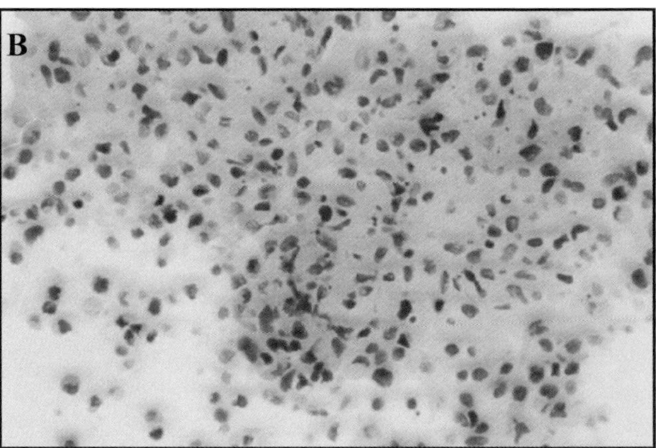

Figure 4. *In vitro* granulomas composed of human peripheral blood lymphocytes, autologous macrophages and mycobacteria after 9 days' incubation, fixed, sectioned and stained for cell surface markers (A) CD68, specific for macrophages, and (B) CD3, specific for T cells. Positive cells stain darkest. Magnification, x400.

The *in vitro* granuloma promises to be a very effective tool for continuing to expand our knowledge of the human immune response to mycobacterial infection. This model may be useful for determining the efficacy of drugs to combat dormant bacilli found in latent infections and, perhaps, for discovering ways of counteracting granuloma collapse and subsequent reactivation tuberculosis during HIV co-infection.

Plaque Dissemination Model

A successful *in vivo* infection begins with successful attachment, phagocytosis and intracellular replication within the macrophage. A number of studies have demonstrated that both *M. tuberculosis* and *M. avium* possess multiple mechanisms for attachment and phagocytosis by macrophages (Bohlson *et al.*, 2001; Castro-Garza *et al.*, 2002; Michell *et al.*, 2003; Middleton *et al.*, 2002; Menozzi *et al.*, 2002; Secott *et al.*, 2001; Schlesinger, 1993). This characteristic may be important to the progress of the infection, since the "secondary" recruited macrophages are likely to be in an activated state following cytokine release by the infected cells and thus possess an enhanced ability to kill the invading microbes. It may be advantageous for the bacteria to have multiple entry points, and thus multiple trafficking pathways through the macrophage, perhaps bypassing some of these bactericidal mechanisms. It also has been observed that once within the phagocytic vesicle of the primary macrophage, the bacteria acquire an "invasive phenotype" that enhances their ability to disseminate and enter secondary macrophages (Bermudez *et al.*, 2002; McDonough *et al.*, 2000). Though the genetic basis for this observation is not known, it again may be advantageous for the mycobacteria to acquire it in order to be more competitive within activated secondary cells.

Recent observations suggest that *M. avium* and likely *M. tuberculosis* bacilli enter adjacent macrophages using different surface receptors, and survive and traffic through a different set of vacuoles compared to the bacteria entering macrophages for the first time after culture on agar plates (Bermudez *et al.*, 1997). Studies have shown that complement and mannose receptors are the main membrane structures used by *M. tuberculosis* and *M. avium* to enter macrophages when the bacteria are grown in culture (Schlessinger *et al.*, 1990; Schlessinger, 1993; Bermudez *et al.*, 1991), while at least the complement receptors are not significantly used by pathogenic mycobacteria when phagocytosed *in vivo* (Bermudez *et al.*, 1999; Hu *et al.*, 2000). When mycobacteria are injected into test animals as opposed to aerosol infection, the intravenous bacteria bypass the mucosal barrier. It is likely that bacteria administered into the blood would enter the "first" monocytes primarily by complement and mannose receptors. Following aerosol administration, the bacteria would interact with the mucosal surface, alveolar macrophages and epithelial cells. Leaving the first cell from this location and subsequent ingestion by the second macrophages would be accomplished by very different mechanisms. Therefore, the ultimate test model for dissemination would include the interaction between mycobacteria and epithelial mucosal cells or phagocytic cells, and should involve entry into alveolar cells with the goal of acquiring an "intracellular phenotype" as the first step towards dissemination.

The role of other alveolar cells in the disease process is less clear, even though pathogenic mycobacteria can enter and replicate in a number of cell types *in vitro*, including alveolar epithelial, endothelial and fibroblast cells

(Mehta *et al.*, 1996). Some *in vivo* evidence for the importance of epithelial cells in disease indicates that the invasion and cytotoxic destruction of human lung epithelial cells by *M. tuberculosis* depends on the presence of surface adhesins, such as heparin binding (Menozzi *et al.*, 1998). An *M. tuberculosis* mutant lacking the heparin-binding hemagglutinin has impaired ability to disseminate in the host when administered by aerosol (Pethe *et al.*, 2001). Once inside alveolar epithelial cells, *M. tuberculosis* bacilli acquire a different phenotype (similar to that observed for macrophages) becoming more invasive and cytotoxic for other epithelial cells and acquiring the ability to invade and translocate across endothelial cells with greater efficiency (Bermudez *et al.*, 2002).

It is known that pneumocytes and macrophages infected with virulent *M. tuberculosis* undergo both necrosis and apoptosis (Dobos *et al.*, 2000; Frantazzi *et al.*, 1999). While apoptosis appears to be the prevalent mechanism of *M. tuberculosis* and *M. avium* killing of pneumocytes and macrophages, it is currently unknown if apoptosis represents a mechanism of host defense, or if it is a phenomenon triggered by the bacterium to facilitate its release from the host cell. *In vitro*, there is a direct correlation between apoptosis of infected cells and the appearance of the invasive phenotype by the intracellular mycobacteria (Bermudez *et al.*, 2002, 1997).

The mechanisms used by *M. tuberculosis* and *M. avium* to invade macrophages and alveolar epithelial cells and spread the infection are only currently beginning to be identified and require investigation. The models used to identify receptors on the host cell surface associated with phagocytosis or internalization certainly have limitations, and novel systems using intracellular bacteria to perform the assays are needed. One potential model to address these issues may be the monolayer dissemination or "plaque" model (Fernandez *et al.*, 1989; Jones and Portnoy, 1994; Byrd *et al.*, 1998; Castro-Garza *et al.*, 2002).

In vitro models have been described for the study of bacterial cell-to-cell spreading through host cell monolayers (Jones and Portnoy, 1994; Sun *et al.*, 1990). Based on a modification of the model presented by Byrd *et al.* (1998) the system described here uses mycobacteria expressing green fluorescent protein (GFP) and the human pneumocyte cell line, A549, overlaid with agarose. We were able to locate individual infected cells and follow the infection process in the monolayer until host cell lysis, bacterial release, and, ultimately, plaque formation were observed.

We developed an *in vitro* tissue-culture model to analyze the process involved in mycobacterial spread through lung epithelial cell monolayers. An A549 cell monolayer was infected with low numbers of viable *M. tuberculosis* bacilli expressing the *gfp* gene. Subsequent addition of a soft agarose overlay prevented the dispersal of the bacilli from the initial points of attachment. By fluorescence microscopy the bacteria were observed to infect and grow within the primary target cells; this was followed by lysis of the infected cells and subsequent infection of adjacent cells. This process repeated itself until an area of clearing (plaque formation) was observed. The addition of amikacin (that kills extracellular bacteria) after initial infection did not prevent intracellular growth; however, subsequent plaque formation was not observed. Plaque formation also was observed after infection with *Mycobacterium bovis* BCG bacilli, but the plaques were smaller than those formed after infection with *M. tuberculosis*. These observations reinforce the possibility that cell-to-cell spreading of *M. tuberculosis* bacilli, particularly early in the course of infection within lung macrophages, pneumocytes, and other cells, may be an important component in the infectious process.

Infection of lung cells by *M. tuberculosis* bacilli and the subsequent dissemination of the bacilli to the bloodstream likely involve bacterial interaction with lung epithelial and endothelial cells in addition to macrophages (Grange, 1998). Development of models to study the interactions between bacteria and host cells, including how the pathogen disseminates from cell to cell, are important components to understanding the virulence process. The model described here permits us to follow the infection of transformed alveolar epithelial cells without the distribution of bacteria and bacterial and host secreted factors throughout the entire monolayer during the course of the experiment as occurs when using standard submerged monolayer systems. In this system, the initial infection and host response remain localized. Also, because the cultures are covered with a semisolid medium, the bacteria can spread by only two routes: extracellularly by lysis and reinfection of closely associated cells, or via intracellular mechanisms as utilized by *Listeria monocytogenes* and *Shigella* sp. (Greiffenberg *et al.,* 1998; Rathman *et al.,* 2000). The precise location of the bacteria in the culture can be determined by detection of GFP expression. In addition, detection of GFP provides enhanced assay sensitivity, allowing detection of a single bacterium or few clustered bacteria in locations throughout the monolayer. This model can also be used to select clones with decreased abilities to attach, internalize or multiply intracellularly as exemplified with *Legionella pneumophila* and *L. monocytogenes* when mutants produced plaques either smaller or larger than those produced by infections with wild-type bacilli (Grange, 1998; Fernandez *et al.,* 1989).

The results of this work suggest that in our *in vitro* system *M. tuberculosis* bacilli predominantly spread following infection of an initial cell by intracellular growth within that cell, lysis of the cell, and subsequent infection of surrounding cells, with repetition of the process ultimately leading to plaque formation. Determining the relevance of these *in vitro* observations to the actual disease process in the lungs is a crucial next step. Inspections of animal alveolar epithelia infected with GFP-producing bacteria in experimental animals within a few days of aerosol infection should permit us to observe and more thoroughly analyze the presence of bacteria within these types of cells.

Biofilm

Highly contagious tuberculosis patients have lung granulomas filled with caseum and viable *M. tuberculosis* bacilli. In late stage tuberculosis, these granulomas rupture releasing their contents into the airways where the bacteria are expelled by the act of coughing. The bacilli travel in small droplets to a healthy host where they are inhaled and deposited on the walls of the terminal bronchi or in the alveolar space, and the infectious cycle begins again. Although animal models and even humans can be used to examine the bacterial responses during late stage disease, a simple *in vitro* model, the biofilm, may mimic and thus permit the study of many of the conditions encountered by the bacteria during this stage of the infection.

Biofilms are dynamic and varied structures composed of cells organized in sessile communities growing as discrete microcolonies within an exopolysaccharide matrix and traversed by a network of open water channels (Lewandowski *et al.*, 1993). Biofilm infections are generally refractory in terms of antibiotic therapy and rapid clearance by the normal immune defense mechanims including antibodies, surfactants and phagocytes (Baltimore and Mitchell, 1980; Jensen *et al.*, 1990). Biofilms persist in the tissue, but generally grow and spread slowly. The body's general strategy in dealing with chronic biofilm infections is to wall off the microcolonies by fibrosis in an attempt to isolate an infection that it cannot resolve — a structural and functional homolog of the granuloma. Some biofilm pathogens are virtually never cleared by antibiotic therapy, as in cystic fibrosis, while other biofilm diseases, such as bacterial endocarditis and perhaps tuberculosis require prolonged courses of treatment (Nickel *et al.*, 1985; Scheld *et al.*, 1978).

When cells in a planktonic (single cell) bacterial population are exposed to an antibacterial agent, all of the bacterial cells are potentially susceptible because they are equally exposed to the drug and existing environment. Clinically, this situation occurs when performing antibiotic susceptibility testing in a clinical laboratory or in the blood of a patient with a bacteremia. In these cases, the antibiotic is expected to kill all of the existing pathogenic organisms.

A biofilm is inherently different from a planktonic population because its component cells grow as microcolonies, each constituting a different microniche of widely varying conditions (Costerton, 1994). An individual bacterium cannot condition its surrounding environment, but the collective community can have a dramatic effect. This phenomenon is referred to as physiological heterogeneity. Thus, the selection of an antibiotic for use in the control of a planktonic infection on the basis of its ability to kill floating single cells in broth medium or on the surface of an agar plate is not sufficient to cover all of the phenotypes of a pathogen during infection.

Direct observations of environmental biofilms using microelectrodes have demonstrated that some loci within a biofilm may be aerobic while others are anaerobic, and some loci are neutral while others have pH values as low as 5.8 (Li and Bishop, 2003; Okabe *et al.,* 1999). Anaerobiosis and acid pH are conditions in which the effectiveness of some antibiotics is reduced (Finelli *et al.,* 2003). When the inherent resistance of sessile bacteria to antibiotics was first noted, this remarkable protection was attributed to a diffusion limitation imposed by the exopolysaccharide matrix that surrounds and encloses the cells within the biofilm (Costerton *et al.,* 1987). Although this fact certainly has some effect on resistance, it is now known that sessile cells in different locations within the biofilm are killed at different times based primarily on the rates of metabolism and phase of growth (Stewart, 1996).

Several groups have published reports on the direct observation of bacteria growing in lung tissue in chronic pulmonary diseases of both humans and animals (Costerton, 2002; Drenkard and Ausubel, 2002). The biofilm mode of growth has been particularly well documented in cystic fibrosis (Lam *et al.,* 1980), Legionnaire's disease (Wright *et al.,* 1989), shipping fever (Morck *et al.,* 1990) and meloidosis (Vorachit *et al.,* 1995) using both light and electron microscopy. Quantitative microbiological experiments on the etiology of these diseases showed that they could be established in experimental animals by the simple installation in the lung of matrix-enclosed microcolonies of the causative organism (Lam *et al.,* 1980). These model biofilm infections, in which the microcolonies closely resemble those seen in the human lung, persist in spite of fully intact host defenses and aggressive antibiotic therapy. As in late stage tuberculosis, the pulmonary surfactants, the mucociliary escalator, and the opsonin-reinforced phagocytosis at the alveolar level are insufficient to clear the chronic biofilm infection once it becomes established. If the chronic pulmonary infection persists for years, as in the case of cystic fibrosis, the patient develops high levels of immune complexes in both the circulation and the lung tissue, and severe tissue damage occurs (Cochrane *et al.,* 1988). In other pulmonary diseases such as meloidosis, the disease can be transformed from its chronic state to an aggressive acute infection with bacteremia, using steroid stress in experimental animals and by environmental stress in humans. In an earlier stage of the disease, the lesions produced by mycobacteria growing in human and animal lungs are characterized as persistent bacterial aggregates surrounded by very large numbers of activated phagocytes. Collapse of the immune system due to conditions such as HIV co-infection or steroid chemotherapy can cause reactivation disease in these latently infected individuals.

The late stage granuloma consists of a mass of replicating bacilli, few viable host cells and a large proportion of necrotic and caseous debris in a polysaccharide matrix (Jacobs *et al.*, 1998) which in some ways mimicks biofilm etiology in the lungs of cystic fibrosis patients (Ernst *et al.*, 1999). A few studies have determined that the environment within the granuloma caseum is slightly acidic, hyperosmolar and, until the moment of the rupture into the airways, anaerobic or microaerophilic, conditions similar to those described for biofilm loci (Dannenberg and Rook, 1994; Poole and Florey, 1970). Hypothetically then the bacteria that escape from the ruptured granuloma may be analogous to the planktonic cells that detach from the biofilm. Comparing bacteria extracted from the planktonic and biofilm culture conditions might give insight into possible biochemical differences that may permit improved survival in aerosol droplets, higher levels of attachment and internalization in alveolar cells, and better survival numbers within the extreme environments such as the granuloma. Quantitative studies on the challenge of animal lungs infected with bacteria have clearly shown that single airborne bacterial cells are rapidly cleared from the lung, but matrix-enclosed biofilm fragments are cleared much more slowly, if at all (Woods *et al.*, 1981). When animals are subjected to crowding, as in shipping fever in cattle (Morck *et al.*, 1990) or to stress, as in meloidosis, bacterial pathogens gain a foothold in the lungs, and the latent disease may take an acute or chronic course. Thus, the combination of a depressed immune response and the presence of the bacterium in an appropriate physiological state may be needed for a successful infection. With current technology, it is possible to monitor the chemical conditions and viability of bacteria within biofilm microniches and thus potentially alter these conditions to improve the outcome for infected patients.

Could the evolutionary response of *M. tuberculosis* to survival within a granuloma be analogous to a mycobacterial response within a biofilm? Biofilms in natural ecosystems are well protected from amoebae and other predators as pathogenic biofilms are resistant to attack by phagocytes and antibodies *in vivo*. When preformed biofilms were placed in the peritoneum of a rabbit immunized against planktonic cells of the same bacterial species, the sessile cells survived and persisted (Ward *et al.*, 1992). Preliminary studies that could be used to confirm this hypothesis might begin by examining further the genetic response of *M. tuberculosis* to pH and anoxia and subsequent internalization by macrophages. Studies by Saviola *et al.* (2003) and Park *et al.* (2003) documented increased expression of a number of genes from *M. tuberculosis* incubated under acidic or microaerophilic conditions, respectively. It also was demonstrated that incubation of virulent strains of *M. tuberculosis* under slightly acidic (pH 6.0) or anoxic conditions increased by several fold the ability of the bacilli to enter macrophages (Li *et al.*, 2002). Interestingly, the bacteria cultured under these conditions enter macrophages using ligands other than the primary complement receptors, CR3 and CR4, or the mannose receptors. Since lipoarabinomannan (LAM) is the adhesin for the mannnose receptor on macrophages, a possible implication is that under acidic conditions the conformation of LAM would be damaged or modified making it incapable of binding to the mannose receptor (Bermudez *et al.*, 1991). Although the precise entry mechanism is currently unknown, it is highly efficient, and upon entry *M. tuberculosis* bacilli do not trigger TNF-α or IL-12 release by the macrophage. If these bacterial responses acquired from various *in vitro* models can be confirmed using a simple biofilm system, we will have demonstrated the utility of a logistically simple model and possibly defined a step in the evolution of this pathogen.

Differentiated Epithelial Organotypic (Raft) Culture

Although most mycobacterial illness and death is caused by tuberculosis, diseases caused by non-tuberculous mycobacteria are having a strong impact on human populations in both developing and industrialized countries (Reichenbach *et al.*, 2001). Many non-tuberculous mycobacterial diseases, such as those caused by *M. avium* complex, are considered opportunistic infections in AIDS patients (Jones *et al.*, 1999), and some research using these species has already been described here. The rates of non-AIDS-associated non-tuberculous mycobacterial infections are also increasing, specifically, disease caused by *M. ulcerans*, *M. marinum*, *M. haemophilum* and other less well characterized yet related species (Fischer *et al.*, 1996; Dobos *et al.*, 1999) has increased in both healthy and immunocompromised patients in the last decade. Of particular concern, Buruli ulcer, caused by *M. ulerans,* is rapidly becoming the third most prevalent mycobacterial disease, with an impact soon to surpass that of leprosy (Dobos *et al.*, 1999).

It is thought that infection with these mycobacteria occurs by inoculation during contact with contaminated water (Dobos *et al.*, 1999). Although this mode of infection has only conclusively been demonstrated for *M. marinum*, all cause dermonecrotic lesions in humans and animals and rarely cause disseminated disease even in immunocompromised individuals. Available research indicates lower optimal growth temperatures (25-33°C) for these species and different target cells are the reasons for limited dissemination. Nevertheless, all species ultimately cause skin ulceration *in vivo*, likely indicating the presence of secreted toxins (Dobos *et al.*, 1999, 2001). Tissue culture monolayers of human adipose cells have proven to be an effective model to differentiate *M. ulcerans* strains based on their cytopathic effects and to assess and compare the effects of potential toxins isolated from *M. ulcerans* cultures. However, to more fully understand the pathogenesis of this disease and how the bacterium is able to cause such severe tissue damage and induce virtually no host cellular immune response, additional *in vivo* and more complex *in vitro* models are needed.

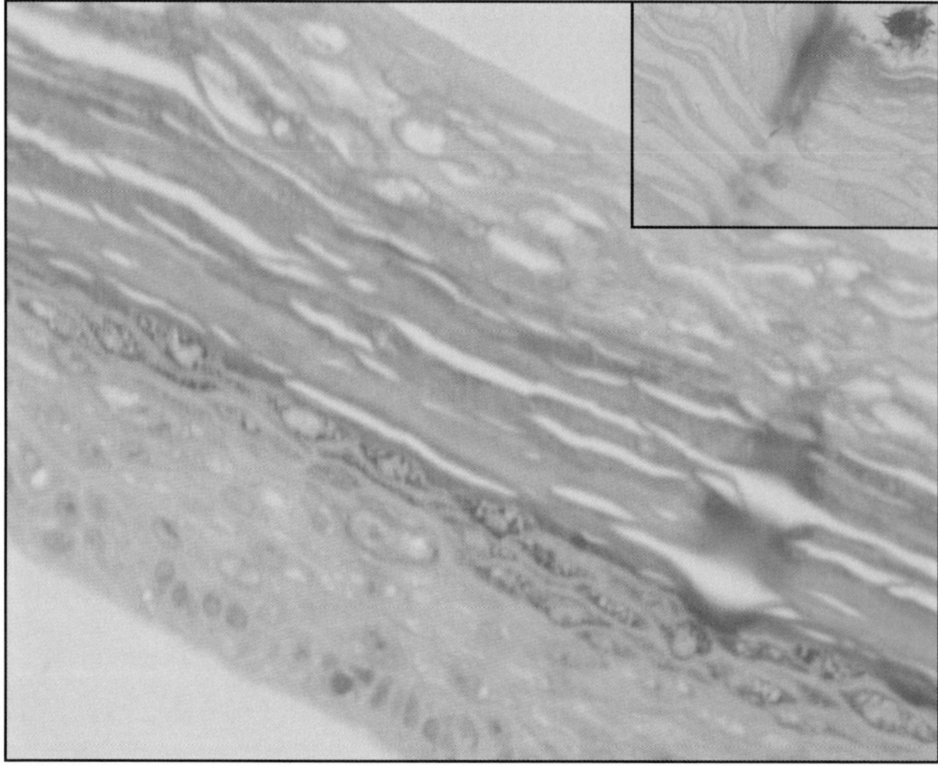

Figure 5. Differentiated layers of epidermal cells in raft skin culture model. Inset: raft 48 hours after surface infection with mycobacteria; several acid fast stained bacilli have migrated through the upper layers. Magnification, x1,000.

Meyers *et al.* (1992) described a differentiated epithelial organotypic (raft) culture that recreates important morphological and physiological skin features. While epithelial cells grown as a monolayer do not differentiate, placing these cells on top of a dermal equivalent and raising them to an air-liquid interface allows for complete differentiation. This epithelial differentiation is accomplished by placing a monolayer of human keratinocytes above a layer of collagen matrix and mouse fibroblasts layered on a rigid but porous support in a liquid growth medium ultimately exposing the developing epidermis to an air-liquid interface. Meyers found that human foreskin keratinocytes showed orderly stratification and differentiation after 9 days with an organized layer of polygonal basal keratinocytes, several layers of spinous cells, several layers of cells containing keratohyalin granules, and flattened, cornified cells with nuclear degeneration.

Preliminary studies using this raft culture model infected with *M. ulcerans* demonstrated extracellular association and penetration through several layers of the skin culture by *M. ulcerans* bacilli as early as 48 hours after addition of the bacteria (Figure 5). This *in vitro* skin raft model can be used to look at a number of factors that play a role in the pathogenesis of Buruli ulcer including determining if a pre-existing lesion is necessary for infection to occur, how inoculum size may influence the course of infection, how the infection progresses through the layers of skin, and at what point ulceration on the skin surface becomes visible. Fibrosis and calcification can be observed, as can the presence of a granulomatous response, at the margin of the ulcer preceding the healing process one sees in animal models. With the incorporation of adipose cells into the model system, it can be determined when the adipose cell necrosis begins, how it progresses, and whether or not bacilli are seen where the necrosis is occurring. If an autologous peripheral blood leukocyte component can be added to the model system, it will be possible to look at the inflammatory response to the infection and to measure cytokine production by RT-PCR or by *in situ* hybrydization. These results may help determine what is preventing host immune response to the mycobacterial infection. The model will be useful to test purified toxin candidates for their ability to induce ulcer development and to determine if antibodies specific to these factors will protect against disease.

Summary

Within the past five years, great strides have been made in the development of model systems and cellular tools for the identification and study of virulence determinants in pathogenic mycobacteria. We are at the threshold of combining knowledge from decades of studies using microbiological, human clinical, animal, and more recently genetic and molecular biological studies towards the goal of understanding the basic mechanisms used by the pathogenic mycobacteria to cause human and animal disease. Adding these novel cellular approaches we hope will contribute to the pace of discovery.

References

Baltimore, R.S., and M. Mitchell. 1980. Immunological investigations of mucoid strains of *Pseudomonas aeruginosa*: Comparison of susceptibility by opsonic antibody in mucoid and nonmucoid strains. J. Infect. Dis. 141: 238-247.

Beharka, A.A., C.D. Gaynor, B.K. Kang, D.R. Voelker, F.X. McCormack, and L.S. Schlesinger. 2002. Pulmonary surfactant protein A up-regulates activity of the mannose receptor, a pattern recognition receptor expressed on human macrophages. J. Immunol. 169: 3565-3573.

Bermudez, L. E., L. S. Young, and H. Enkel. 1991. Interaction of *Mycobacterium avium* complex with human macrophages: roles of membrane receptors and serum proteins. Infect. Immun. 59: 1697-1702.

Bermudez, L. E., and J. Goodman. 1996. *Mycobacterium tuberculosis* invades and replicates within type II alveolar cells. Infect. Immun. 64: 1400-1406.

Bermudez, L. E., A. Parker, and J. R. Goodman. 1997. Growth within macrophages increases the efficiency of *Mycobacterium avium* in invading other macrophages by a complement receptor- independent pathway. Infect. Immun. 65: 1916-1925.

Bermudez, L. E., J. Goodman, and M. Petrofsky. 1999. Role of complement receptors in uptake of *Mycobacterium avium* by macrophages *in vivo*: evidence from studies using CD18-deficient mice. Infect. Immun. 67: 4912-4916.

Bermudez, L. E., F. J. Sangari, P. Kolonoski, M. Petrofsky, and J. Goodman. 2002. The efficiency of the translocation of *Mycobacterium tuberculosis* across a bilayer of epithelial and endothelial cells as a model of the alveolar wall is a consequence of transport within mononuclear phagocytes and invasion of alveolar epithelial cells. Infect. Immun. 70: 140-146.

Birkness, K. A., B. L. Swisher, E. H. White, E. G. Long, E. P. Ewing, Jr., and F. D. Quinn. 1995. A tissue culture bilayer model to study the passage of *Neisseria meningitidis*. Infect. Immun. 63: 402–409.

Birkness, K. A., M. Deslauriers, J. H. Bartlett, E. H. White, C. H. King, and F. D. Quinn. 1999. An *in vitro* tissue culture bilayer model to examine early events in *Mycobacterium tuberculosis* infection. Infect. Immun. 67: 653-658.

Bohlson, S. S., J. A. Strasser, J. J. Bower, and J. S. Schorey. 2001. Role of complement in *Mycobacterium avium* pathogenesis: *in vivo* and *in vitro* analyses of the host response to infection in the absence of complement component C3. Infect. Immun. 69: 77297735.

Byrd, T.F., G.M. Green, S.E. Fowlston, and C.R. Lyons. 1998. Differential growth characteristics and streptomycin susceptibility of virulent and avirulent *Mycobacterium tuberculosis* strains in a novel fibroblast-Mycobacterium microcolony assay. Infect. Immun. 66: 5132-5139.

Cardona, P.J., R. Llatjos, S. Gordillo, J. Diaz, I. Ojanguren, A. Ariza, and V. Ausina. 2000. Evolution of granulomas in lungs of mice infected aerogenically with *Mycobacterium tuberculosis*. Scand. J. Immunol. 52: 156-163.

Castro-Garza, J., C.H. King, W.E. Swords, and F.D. Quinn. 2002. Demonstration of spread by *Mycobacterium tuberculosis* bacilli in A549 epithelial cell monolayers. FEMS Microbiol. Lett. 212: 145-149.

Chensue, S.W., K.S. Warmington, J.H. Ruth, P. Lincoln, and S.L. Kunkel. 1994. Cytokine responses during mycobacterial and schistosomal antigen-induced pulmonary granuloma formation. Production of Th1 and Th2 cytokines and relative contribution of tumor necrosis factor. Amer. J. Pathol. 145: 1105-1113.

Chensue, S.W., K.S. Warmington, J.H. Ruth, P. Lincoln, and S.L. Kunkel. 1995. Cytokine function during mycobacterial and schistosomal antigen-induced pulmonary granuloma formation. Local and regional participation of IFN-gamma, IL-10, and TNF-alpha. J. Immunol. 154: 5969-5976.

Clark-Curtiss, J.E. 1998. Identification of virulence determinants in pathogenic mycobacteria. Curr. Top. Microbiol. Immunol. 225: 57-79.

Cochrane, D.M.G., M.R.W. Brown, H. Anwar, P.H. Weller, K. Lam and J.W. Costerton. 1988. Antibody response to *Pseudomonas aeruginosa* surface protein antigens in a rat model of chronic lung infection. J. Med. Microbiol. 27: 255-261.

Costerton, J.W., D.W. Lambe, Jr., K.J. Mayberry-Carson, and B. Tober-Meyer. 1987. Cell wall alterations in staphylococci growing *in situ* in experimental osteomyelitis. Can. J. Microbiol. 33: 142-150.

Costerton, J.W., Z. Lewandowski, D. DeBeer, D. Caldwell, D. Korber, and G. James. 1994. Biofilms, the customized microniche. J. Bacteriol. 176: 2137-2142.

Costerton, J.W. 2002. Anaerobic biofilm infections in cystic fibrosis. Mol. Cell 10: 699-700.

Crandall, E. D., and K. J. Kim. 1991. Alveolar epithelial barrier properties. In: The Lung: Scientific Foundations. R. J. Crystal and J. B. West, eds. Raven Press, New York, N.Y. p. 273–287.

Crystal, R. J. 1991. Alveolar macrophages. In: The Lung: Scientific Foundations. R. J. Crystal and J. B. West, eds. Raven Press, New York, N.Y. p. 527–538.

Cundell, D. R., and E. I. Tuomanen. 1994. Receptor specificity of adherence of *Streptococcus pneumoniae* to human type-II pneumocytes and vascular endothelial cells *in vitro*. Microb. Pathog. 17: 361–374.

Dannenberg, A. M., Jr., and G. A. W. Rook. 1994. Pathogenesis of pulmonary tuberculosis: an interplay of tissue-damaging and macrophage-activating immune responses—dual mechanisms that control bacillary multiplication. In: Tuberculosis: Pathogenesis, Protection and Control. B. R. Bloom, ed. ASM Press, Washington, D.C. p. 459–483.

Dobos, K.M., F.D. Quinn, D.A.Ashford, C.R. Horsburgh, and C.H. King. 1999. Emergence of a unique group of necrotizing mycobacterial diseases. Emerg. Infect. Dis. 5: 367-378.

Dobos, K.M., E.A. Spotts, F.D. Quinn, and C.H. King. 2000. Necrosis of lung epithelial cells during infection with *Mycobacterium tuberculosis* is preceded by cell permeation. Infect. Immun. 68: 6300-6310.

Dobos, K.M., P.L. Small, M. Deslauriers, F.D. Quinn, and C.H. King. 2001. *Mycobacterium ulcerans* cytotoxicity in an adipose cell model. Infect. Immun. 69: 7182-7186.

Doughty, B.L. and S.M. Phillips. 1982. Delayed hypersensitivity granuloma formation around Schistosoma mansoni eggs *in vitro*. I. Definition of the model. J. Immunol. 128: 30-36.

Drenkard, E., and F.M. Ausubel. 2002. *Pseudomonas* biofilm formation and antibiotic resistance are linked to phenotypic variation. Nature 416: 740-743.

Ernst, R.K., E.C. Yi, L. Guo, K.B. Lim, J.L. Burns, M. Hackett, and S.I. Miller. 1999. Specific lipopolysaccharide found in cystic fibrosis airway *Pseudomonas aeruginosa*. Science 286: 1561-1565.

Fernandez R.C., S.H. Lee, D. Haldane, R. Sumarah, and K.R. Rozee. 1989. Plaque assay for virulent *Legionella pneumophila*. J. Clin. Microbiol. 27: 1961-1964.

Finelli, A., C.V. Gallant, K. Jarvi, and L.L. Burrows. 2003. Use of in-biofilm expression technology to identify genes involved in *Pseudomonas aeruginosa* biofilm development. J. Bacteriol. 85: 2700-2710.

Fischer, L.J., F.D. Quinn, E.H. White, and C.H. King. 1996. Intracellular growth and cytotoxicity of *Mycobacterium haemophilum* in a human epithelial cell line (Hec-1-B). Infect. Immun. 64: 269-276.

Flynn, J.L., J. Chan, K.J. Triebold, D.K. Dalton, T.A. Stewart, and B.R. Bloom. 1993. An essential role for interferon gamma in resistance to *Mycobacterium tuberculosis* infection. J. Exp. Med. 178: 2249-2254.

Frantazzi, C., R. D. Arbeit, C. Carini, M. K. Balcewicz-Sablinka, J. Keane, H. Kornfeld, and H. G. Remold. 1999. Macrophage apoptosis in mycobacterial infections. J. Leuk. Biol. 66: 763-772.

Grange, J.M. 1998. Immunophysiology and immunopathology of tuberculosis. In: Clinical Tuberculosis. Davies, P.D.O., ed. Chapman and Hall, London. p. 129-152.

Greiffenbrg, L., W. Goebel, K.S. Kim, I. Weiglein, A. Bubert, F. Engelbrecht, M. Stins, and M. Kuhn. 1998. Interaction of *Listeria monocytogenes* with human brain microvascular endothelial cells: In1B-dependent invasion, long-term intracellular growth, and spread from macrophages to endothelial cells. Infect. Immun. 66: 5260-5267.

Henderson, R. A., S. C. Watkins, and J. L. Flynn. 1997. Activation of human dendritic cells following infection with *Mycobacterium tuberculosis*. J. Immunol. 159: 635–643.

Hu, C., T. Mayadas-Norton, K. Tanaka, J. Chan, and P. Salgame. 2000. *Mycobacterium tuberculosis* infection in complement receptor 3-deficient mice. J. Immunol.165: 2596-2602.

Jacobs, W., S. Kumar-Singh, J. Bogers, K.Van de Vijver, A. Deelder, and E. Van Marck. 1998. Transforming growth factor-beta, basement membrane components and heparin sulphate proteoglycans in experimental hepatic *Schistosoma mansoni*. Cell Tissue Res. 292: 101-106.

Jensen, E.T., A. Kharazmi, K., K. Lam, and J.W. Costerton. 1990. Human polymorphonuclear leukocyte response to *Pseudomonas aeruginosa* biofilms. Infect. Immun. 58: 2383-2385.

Jones, S. and D.A. Portnoy. 1994. Small plaque mutants. Meth. Enzymol. 236: 526-531.

Jones, J.L., D.L. Hanson, M.S. Dworkin, D.L. Alderton, P.L. Fleming, J.E. Kaplan, and J. Ward. 1999. Surveillance for AIDS-defining opportunistic illnesses, 1992-1997. MMWR CDC Surveill. Summ. 48: 1-22.

Lam, J., R. Chan, K. Lam, and J.W. Costerton. 1980. Production of mucoid microcolonies by *Pseudomonas aeruginosa* within infected lungs in cystic fibrosis. Infect. Immun. 28: 546-556.

Lewandowski, Z., S.A. Altobelli, and E. Fukushima. 1993. NMR and microelectrode studies of hydrodynamics and kinetics in biofilms. Biotechnol. Prog. 9: 40-45.

Li, J., and P.L. Bishop. 2003. Monitoring the influence of toxic compounds on microbial denitrifying biofilm processes. Water Sci. Technol. 47: 211-216.

Li, Y.J., M. Petrofsky, and L.E. Bermudez. 2002. *Mycobacterium tuberculosis* uptake by recipient host macrophages is influenced by environmental conditions in the granuloma of the infectious individual and is associated with impaired production of interleukin-12 and tumor necrosis factor alpha. Infect. Immun. 70: 6223-6230.

Lurie, M. B., S. Abramson, and A. G. Heppleston. 1952. On the response of genetically resistant and susceptible rabbits to the quantitative inhalation of human type tubercle bacilli and the nature of resistance to tuberculosis. J. Exp. Med. 95: 119–137.

McCune, R. M., and R. Tompsett. 1956. Fate of *Mycobacterium tuberculosis* in mouse tissues as determined by the microbial enumeration technique. J. Exp. Med. 104: 737-762.

McDonough, K. A., and Y. Kress. 1995. Cytotoxicity for lung epithelial cells is a virulence-associated phenotype of *Mycobacterium tuberculosis*. Infect. Immun. 63: 4802-4811.

McDonough, K. A., M. A. Florczyk, and Y. Kress. 2000. Intracellular passage within macrophages affects the trafficking of virulent tubercle bacilli upon reinfection of other macrophages in a serum-dependent manner. Tubercle Lung. Dis. 80: 259-271.

McMurray, D.N. 1996. Pathogenesis of experimental tuberculosis in animal models. Curr. Top. Microbiol. Immunol. 215: 157-179.

Mehta, P. K., C. H. King, E. H. White, J. J. Murtagh, Jr., and F. D. Quinn. 1996. Comparison of *in vitro* models for the study of *Mycobacterium tuberculosis* invasion and intracellular replication. Infect. Immun. 64: 2673–2679.

Mekalanos, J. J. 1992. Environmental signals controlling expression of virulence determinants in bacteria. J. Bacteriol. 174: 1-7.

Menozzi, F. D., R. Bischoff, E. Fort, M. J. Brennan, and C. Locht. 1998. Molecular characterization of the mycobacterial heparin-binding hemagglutinin, a mycobacterial adhesin. Proc. Natl. Acad. Sci. 95: 12625-12630.

Meyers C, Frattini MG, Hudson JB, Laimins LA. 1992. Biosynthesis of human papillomavirus from a continuous cell line upon epithelial differentiation. Science 257: 971-973.

Michell, S.L., A.O. Whelan, P.R. Wheeler, M. Panico, R.L. Easton, A.T. Etienne, S.M. Haslam, A. Dell, H.R. Morris, A.J. Reason, J.L. Herrmann, D.B. Young, and R.G. Hewinson. 2003. The MPB83 antigen from *Mycobacterium bovis* contains O-linked mannose and (1, 3)-mannobiose moieties. J. Biol. Chem. 278: 16423-16432.

Middleton, A.M., M.V. Chadwick, A.G. Nicholson, A. Dewar, R.K. Groger, E.J. Brown, T.L. Ratliff, and R. Wilson. 2002. Interaction of *Mycobacterium tuberculosis* with human respiratory mucosa. Tuberculosis 82: 69-78.

Menozzi, F.D., K. Pethe, P. Bifani, F. Soncin, M.J. Brennan, and C. Locht. 2002. Enhanced bacterial virulence through exploitation of host glycosaminoglycans. Mol. Microbiol. 43: 1379-1386.

Morck, D.W., J.W. Costerton, D.O. Bolingbroke, H. Ceri, N.D. Boyd, and M.E. Olson. 1990. A guinea pig model of bovine pneumonic *Pasteurellosis*. Can. J. Vet. Res. 54: 139-145.

Nickel, J.C., I. Ruseska, J.B. Wright, and J.W. Costerton. 1985. Tobramycin resistance of *Pseudomonas aeruginosa* cells growing as a biofilm on urinary catheter material. Antimicrob. Agents Chemother. 27: 619-624.

Okabe, S., T. Itoh, H. Satoh, and Y. Watanabe. 1999. Analyses of spatial distributions of sulfate-reducing bacteria and their activity in aerobic wastewater biofilms. Appl. Environ. Microbiol. 65: 5107-5116.

Orme, I. M. 1993. Immunity to mycobacteria. Curr. Opin. Immunol. 5: 497-502.

Orme, I. M. 1998. The immunopathogenesis of tuberculosis: a new working hypothesis. Trends Microbiol. 6: 94-97.

Park, H.D., K.M. Guinn, M.I. Harrell, R. Liao, M.I. Voskuil, M. Tompa, G.K. Schoolnik, and D.R. Sherman. 2003. Rv3133c/dosR is a transcription factor that mediates the hypoxic response of *Mycobacterium tuberculosis*. Mol. Microbiol. 48: 833-843.

Pethe, K., S. Alonso, F. Biet, G. Delogu, M. J. Brennan, C. Locht, and F. D. Menozzi. 2001. The heparin-binding haemagglutinin of *M. tuberculosis* is required for extrapulmonary dissemination. Nature 412: 190-194.

Poole, J. C., and H. W. Florey. 1970. Chronic Inflammation and Tuberculosis. In *General Pathology*. H. W. Florey, ed. W.B. Saunders Co, Philadelphia.

Quinn, F.D., K.A. Birkness, and P.J. King. 2002. Alpha-crystallin as a potential marker of *Mycobacterium tuberculosis* latency. ASM News 68: 612-617.

Rathman, M., N. Jouirhi, A. Allaoui, P. Sansonetti, C. Parsot, and G. Nhieu Van Tran. 2000. The development of a FACS-based strategy for the isolation of *Shigella flexneri* mutants that are deficient in intracellular spread. Mol. Microbiol. 35: 974-990.

Raviglione, M.C., D.E. Snider, Jr., and A. Kochi. 1995. Global epidemiology of tuberculosis. Morbidity and mortality of a worldwide epidemic. JAMA 273: 220-226.

Reichenbach, J., S. Rosenzweig, R. Doffinger, S. Dupuis, S.M. Holland, and J.L. Casanova. 2001. Mycobacterial diseases in primary immunodeficiencies. Curr. Opin. Allergy Clin. Immunol. 1: 503-511.

Rhoades, E.R., A.M. Cooper, and I.M. Orme. 1995. Chemokine response in mice infected with *Mycobacterium tuberculosis*. Infect. Immun. 63: 3871-3877.

Saunders, B.M., A.A. Frank, and I.M. Orme. 1999. Granuloma formation is required to contain bacillus growth and delay mortality in mice chronically infected with *Mycobacterium tuberculosis*. Immunol. 98: 324-328.

Saviola B., S.C. Woolwine, and W.R. Bishai. 2003. Isolation of acid-inducible genes of *Mycobacterium tuberculosis* with the use of recombinase-based *in vivo* expression technology. Infect. Immun. 71: 1379-1388.

Scheld, W.M., J.A. Vaone, and M.A. Sande. 1978. Bacterial adherence in the pathogenesis of endocarditis. Interaction of bacterial dextran, platelets and fibrin. J. Clin. Invest. 61: 1394-1404.

Schlesinger, L. S., C. G. Bellinger-Kawahara, N. R. Payne, and M. A. Horwitz. 1990. Phagocytosis of *Mycobacterium tuberculosis* is mediated by human monocyte complement receptors and complement component C3. J. Immunol. 144: 2771-2780.

Schlesinger, L. S. 1993. Macrophage phagocytosis of virulent but not attenuated strains of *Mycobacterium tuberculosis* is mediated by mannose receptors in addition to complement receptors. J. Immunol. 150: 2920-2930.

Schneeberger, E. E. 1991. Alveolar type I cells. The Lung: Scientific Foundations. In: R. J. Crystal and J. B. West, eds. Raven Press, New York, N.Y. p. 229–234.

Secott, T.E., T.L. Lin, and C.C. Wu. 2001. Fibronectin attachment protein homologue mediates fibronectin binding by *Mycobacterium avium* subsp. paratuberculosis. Infect. Immun. 69: 2075-2082.

Shikama, Y. 1989. Granuloma formation by artificial microparticles *in vitro*. Macrophages and monokines play a critical role in granuloma formation. Am. J. Pathol. 34: 1189-1199.

Sidobre, S., G. Puzo, and M. Riviere. 2002. Lipid-restricted recognition of mycobacterial lipoglycans by human pulmonary surfactant protein A: a surface-plasmon-resonance study. Biochem. J. 365: 89-97.

Smith, D.W. and E.H. Wiegeshaus. 1989. What animal models can teach us about the pathogenesis of tuberculosis in humans. Rev. Infect. Dis. 11 Suppl. 2: S385-S393.

Stewart, P.S. 1996. Theoretical aspects of antibiotic diffusion into microbial biofilms. Antimicrob. Agents Chemother. 40: 2517-2522.

Sun, A.N., A. Camilli, and D.A. Portnoy. 1990. Isolation of *Listeria monocytogenes* small-plaque mutants defective for intracellular growth and cell to cell spread. Infect. Immun. 58: 3770-3778.

Teitelbaum, R., W. Schubert, L. Gunther, Y. Kress, F. Macaluso, J. W. Pollard, D. N. McMurray, and B. R. Bloom. 1999. The M cell as a portal of entry to the lung for the bacterial pathogen *Mycobacterium tuberculosis*. Immunity 10: 641-650.

Vorachit, M., K. Lam, P. Jayanetra, and J.W. Costerton. 1985. Electron microscopy study of the mode of growth of *Pseudomonas pseudomallei in vitro* and *in vivo*. J. Tropical Med. Hyg. 98: 379-391.

Ward, K.H., M.E. Olson, K. Lam, and J.W. Costerton. 1992. Mechanism of persistent infection associated with peritoneal implants. J. Med. Microbiol. 36: 406-413.

Weikert, L.F., J.P. Lopez, R. Abdolrasulnia, Z.C. Chroneos, and V.L. Shepherd. 2000. Surfactant protein A enhances mycobacterial killing by rat macrophages through a nitric oxide-dependent pathway. Am. J. Physiol. Lung Cell Mol. Physiol. 279: L216-223.

Woods, D.E., D.C. Straus, W.G. Johanson, and J.A. Bass. 1981. Role of fibronectin in the prevention of the adherence of *Pseudomonas aeruginosa* to buccal cells. J. Infect. Dis. 143: 784-790.

Wright, J.B., M.A. Athar, T.M. van Olm, J.S. Wootliff, and J.W. Costerton. 1989. Atypical legionellosis: Isolation of *Legionella pneumophila* serogroup 1 from a patient with aspiration pneumonia. J. Hosp. Infect. 13: 187-190.

From: Tuberculosis: The Microbe Host Interface
Edited by: Larry S. Schlesinger and Lucy E. DesJardin

Chapter 6

Animal Models in the Analysis of Pathogenesis

Andrea M. Cooper

Abstract

The pathogenesis of tuberculosis is a complex process that depends upon components of both the vertebrate host and the mycobacteria. Animal models are required to address the complexity of the interactions between host and pathogen. In their own right they can be used to address hypotheses regarding the role of individual components in pathogenesis but the real strength of the models lies in the use of comparative pathology. Comparing and contrasting the responses of several animal models of tuberculosis can result in the development of novel hypotheses regarding pathogenesis of this disease. This development may lead to a greater understanding of the pathogenic process and thus a greater ability to create effective immune mediated interventions such as vaccines that not only induce anti-bacterial effector functions but also help to limit the pathologic consequences of disease.

Animal Models of Tuberculosis, Are They Useful?

Many complex biological problems have been addressed by the thoughtful and deliberate application of animal models. Investigation of the pathogenesis of tuberculosis and mycobacterial disease is especially amenable to the use of appropriate animal models. The elegance, and ultimately the value, of these models resides in their plasticity to the fundamental tenets of scientific methodology. Few investigators would suggest the results obtained from animal models are directly applicable to the human condition; indeed, that is not their purpose. Instead, the utility of the animal model is to provide an analysis platform for the complex process occurring during the interaction of a vertebrate host and persistent, inflammatory bacilli. Once characterized, this interaction can then provide the basis for extrapolation and further analysis of human data. Despite the wide variety of animal models used in mycobacterial research, a few basic commonalties exist. First, there is little evidence suggesting a vertebrate host can successfully completely clear *Mycobacterium tuberculosis* following infection. Second, most hosts generate some degree of granulomatous cellular response to this persistent pathogen. Beyond these similarities lie many differences in the types and consequences of these cellular responses. Real effort must therefore be devoted to uniform application of techniques and protocols and this, combined with careful appreciation of the differences between models, will allow more rapid progress in the study of the pathogenesis of tuberculosis.

If one considers pathogenesis to be the development of a moribund condition in the host, then it becomes apparent that pathogenesis is a function of not only the bacteria but also the host response. Immunopathogenesis, then, is the component of the disease dependent upon the immune response of the host; that being said, tuberculosis and anthrax were the diseases which resulted in the crystallization of the germ theory of disease and the development of Koch's postulates (Koch, 1884). Tuberculosis, therefore, is a condition that does not occur without the presence of virulent bacteria but for which there is a substantial role for the immunopathogenic response in development of clinical disease. This combination of bacterial and host components in disease progression makes the use of animal models imperative in dissecting the components crucial to pathogenesis of this disease. The various animal models used in the past exhibit varying levels and types of immune response which has been used to suggest that some animals are resistant to disease while others are susceptible. In fact, all the current species of animals used as models of tuberculosis, when infected with a low dose of bacteria, allow similar patterns of bacterial growth, with the most dramatic differences being noted in the cellular and pathological responses (Dannenberg *et al.*, 2001).

The choice of animal model depends almost entirely upon the question to be addressed. The crucial questions are a) what host activities limit bacterial growth, b) which host or bacterial component contributes to pathogenic consequences in tissue, c) how do immunological interventions (i.e. vaccination) alter development of disease. Other factors that contribute to the decision include cost, space, reproducibility and availability of reagents. These latter are instrumental in the pervasive use of the mouse as a model for many infectious diseases. Other animal models useful in TB research include the rabbit, the guinea pig and the macaque monkey. These more costly and less tractable models are used to study diverse aspects of the human disease such as development of progressive pulmonary complications and the efficacy of promising vaccines.

Assessment of Pathogenesis

There are several basic methods of addressing the development of disease within animal models of tuberculosis. In any model of disease however, the measurement of one parameter is insufficient to determine the role of any particular mechanism in protection. The determination of bacterial growth is perhaps the most direct assessment of whether the host can control the bacteria however this technique does not indicate *how* the host is controlling bacterial growth. In addition to bacterial growth, assessment of cellular influx to the site of infection and *ex vivo* cellular responses are important indicators of the level of activation of host defense mechanisms. One aspect of pathogenesis is the induction of a morbid state in the host, which can be assessed by following the animal's weight and demeanor. Indeed upon primary observations of distress such as drop in body weight, inactivity and poor grooming, animals should be humanely euthanized, as rapid deterioration will ensue from this point. Careful analysis of cellular responses, serum cytokine levels and bacterial numbers are required prior to serious signs of illness for correct interpretation of these morbidity studies. Simple "time to death" studies in the absence of more detailed analysis give no indication of the cause of morbidity and mortality.

Bacterial growth in the absence of strong cellular responses can cause significant tissue damage and rapid death and thus determining bacterial growth is one important indicator of disease. Determining bacterial growth throughout the animal is also important, as the expression of protective responses can be different between the organs. An example of this is when bacterial numbers in the lung are similar between two populations but bacterial growth in the spleen and liver are exacerbated in the experimental group (as in (Cooper *et al.*, 2000)).

Determination of bacterial number in the absence of histological examination of the affected tissues can also be misleading as most animal models are capable of controlling bacterial growth over a prolonged period but it is the pathologic sequelae of the protective host immune response that will lead to death. Indeed, bacterial numbers *per se* are not sufficiently detrimental until they reach very high levels as indicated by the enormous numbers of *M. avium* bacteria which can be detected in apparently healthy mice (Florido *et al.*, 1997).

As pathogenesis is the product of interplay between bacterial growth and development of host responses it is essential that the whole animal be considered when assessing pathogenesis. Indeed, to fail to address many components of the host response is to waste the most valuable aspect of the animal model, that being the complexity of the response.

Relative Merits of the Different Animal Species as Models of Tuberculosis

Each investigator can provide a strong rationale for their choice of animal species and most are also aware of the limitations of their own model. For the generalist however, discerning the merits of each model can be confusing and for this reason a brief consideration of the relative merits of each animal species is presented here. The principal merit to any model is its relevance to the human condition. With this in mind, it is generally accepted that while mice are useful in certain studies of the pathogenesis of tuberculosis, more can be learned of the destructive necrosis and caseation evident in human disease by examining the responses in guinea pigs and monkeys. As would be expected, integration of the information gleaned from several models can be more useful than focusing on just one model. Indeed, comparison of the cellular responses between mice and guinea pigs suggests that while murine lymphocytes are located throughout the granuloma, the lymphocyte population in the guinea pig is more peripherally located. It has been hypothesized that the location of murine lymphocytes results in reduced tendency to necrosis in the murine model (Orme, 1998). Additionally, in a recent report, comparative pathology of different mycobacterial diseases within the mouse model has resulted in the postulation that the mediators of the protective response are also instrumental in down-regulating the immune response and thus pathologic consequences of infection (Cooper *et al.*, 2002a).

What then are the key features of the different models? Studies using animal models of tuberculosis began in earnest with Koch at the end of the 19th century (Koch, 1882). Using newly developed methods to culture the bacteria from tubercles taken from a variety of sources, Koch confirmed that the tubercle bacillus was the causative agent of an artificial form of tuberculosis in guinea pigs (Koch, 1881, 1882). Of particular interest to the current discussion was the fact that Koch was able to distinguish between the disease resulting from inoculation and that resulting from spontaneous infection due to co-housing of infected and uninfected guinea pigs (Koch, 1882). He noted that in the artificial disease, progression was rapid and while initially localized to the site of inoculation it disseminated and caused death rapidly. Conversely the disease resulting from spontaneous infection, was localized to the lung and resulted in the development of a "large cheesy mass with extensive decomposition in the center, so that it occasionally resembles the similar processes in the human lung" (from the translation of (Koch, 1882)). In further analysis of the response in the guinea pig, Koch described the rapid inflammatory response resulting from a second infection with the tubercle bacillus (Koch, 1890).

The Guinea Pig

From its early role in the study of the cellular response to tuberculosis described above, the guinea pig has been the corner stone of much of the research into this disease (reviewed in (McMurray, 1994)). It is an exquisitely sensitive host and will succumb to very low doses of bacteria delivered via an aerosol cloud (Ratcliffe *et al.*, 1953; Smith *et al.*, 1966; Wiegeshaus *et al.*, 1970). In a series of reports, Smith and colleagues developed a guinea pig model of aerogenic infection that could be used to address several aspects of human disease. In this model each animal has roughly 3-4 viable bacilli deposited within the lung and dissemination of bacteria to the spleen, lymph nodes and thence to other non-infected lobes of the lung occurs by day 10-14 post infection (Wiegeshaus *et al.*, 1970). Following BCG vaccination of guinea pigs, bacterial growth and the dissemination of bacteria is greatly inhibited; as in humans however, BCG vaccination fails to provide full protection against infection and dissemination (Fok *et al.*, 1976; Harding *et al.*, 1977). The protection that is provided by BCG is, however, dramatic in terms of both survival (Smith *et al.*, 1970) and reduced pathologic sequelae (Smith *et al.*, 1977) in the lung.

A recent body of work from McMurray and colleagues supports a role for malnutrition in increasing susceptibility to TB in the guinea pig model and this has been associated with poor T cell and cytokine responses (Dai *et al.*, 1998; McMurray *et al.*, 1992, 1990). As might be expected from the fact that T cells are compromised, the protective effect of BCG, which depends upon memory T cells, is also compromised by malnutrition (McMurray *et al.*, 1992, 1985, 1986). The protective effect of vaccination in the guinea pig is the most significant evidence that protection in this model is mediated by acquired cellular immunity. The absence of tools with which to examine the cellular immune response in this animal model means that most of discussion of the cellular response in this model involves extrapolation from other animal models. Using the limited antibodies to the CD4 and CD8 T cell markers McMurray has enumerated lymphocytes in the infected guinea pig (Mainali *et al.*, 1992). More recently, the use of immunohistochemistry to locate these lymphocytes in the lesion has indicated that both CD4 and CD8 T cells are recruited to and are distributed throughout the lesion early in granuloma development (Turner,

2002). Of particular interest is the observation that histiocytes dominate the early inflammatory response in this model (Turner, 2002). The reduced necrosis and lesion size observed in the vaccinated guinea pigs may simply be a result of reduced bacterial growth, however, a detailed examination of the granuloma in both naïve and vaccinated animals will be informative in identifying the mechanisms involved in the development of a destructive lesion.

One area where the guinea pig is clearly superior to the murine model for the analysis of cellular immune responses to TB is in the examination of the role of non-traditional T-cell responses. Recently, a family of non-polymorphic genes encoding the potential antigen-presenting molecules CD1 (Porcelli *et al.*, 1998) have been implicated in the response to mycobacterial antigens. Unlike mice, which fail to express the CD1 genes thought to be active in this process, guinea pigs possess a homologue of each human CD1 gene (Dascher *et al.*, 1999) and can be induced to develop a specific response to CD1 restricted mycobacterial lipid antigens (Hiromatsu *et al.*, 2002). The potential for these cells to complement or even orchestrate the response to TB has not yet been addressed but it is in the study of these kinds of populations that keys to the development of novel vaccines may be found.

The Rabbit

In contrast to the guinea pig, which exhibits little genetic variability in its susceptibility to TB (Cohen *et al.*, 1987), the rabbit can exhibit a range of susceptibilities to disease; some develop cavitary disease of the lung while others develop disseminated disease (Lurie, 1941; Lurie *et al.*, 1948, 1950; Wells *et al.*, 1941) (and reviewed in (Dannenberg, 1994)). In some of the early work on rabbits it was shown that a single experimental exposure to droplet nuclei of mycobacteria resulted in a disease profile which was very similar to that demonstrated in rabbits that were naturally infected by co-housing with tuberculous rabbits (Lurie, 1941; Wells *et al.*, 1941). The development of cavitary disease, which is an important aspect of the most dangerous stage of human TB, is only observed in this rabbit model. For this reason the New Zealand White rabbit has been studied recently in an attempt to identify correlates associated with cavitation; correlates identified so far are a strong skin test response and the expression of the proteinase, cathepsin D (Converse *et al.*, 1996, 1998). Further analysis of the causes of cavitation in this model may provide tools for the eventual reduction of this complication of disease in humans.

The rabbit model has also provided a tool with which to examine the devastating tuberculous meningitis seen in children. Using a model of intracerebral inoculation of mycobacteria to produce an acute central nervous system infection, it was shown that although antibiotics reduced bacterial numbers this treatment alone did not reduce the damaging inflammatory response (Tsenova *et al.*, 1998). Reduction in inflammation and increased survival only occurred if an inhibitor of one of the primary inflammatory cytokines, Tumor necrosis factor-α (TNF-α), was included with treatment (Tsenova *et al.*, 1999, 1998).

The role of TNF-α in generating a cavitary lesion in the lungs of the infected rabbit has not as yet been addressed. The role of TNF-α in tuberculous meningitis does however suggest that in the rabbit this potent inflammatory cytokine can behave as a "double-edged" sword. Many of the tools required to study cavitation in the lungs of rabbits are now available (Shigenaga *et al.*, 2001), and it is hoped that progress in this model will be forthcoming.

The Non-Human Primate

The non-human primate model requires a large commitment of space, time and money in order to ensure animals are housed safely and in non-stressful conditions. The macaque has been the non-human primate of choice for most TB work (McMurray, 2000) and is currently being used to test promising vaccines by several groups. Early work resulted in the characterization of the development of disease in non-human primates and demonstrated that macaques could be both infected and that vaccination was protective (Good, 1968). Measures of protection used at the time were X-ray, weight, gross and fine pathology and mortality. A more recent study utilizing a range of doses delivered by intratracheal instillation, demonstrated that the disease which developed was similar to that seen in humans (Walsh *et al.*, 1996). Further to the previous report, and by making use of modern immunological tools, it was shown that while primary disease in two closely related species of macaque was similar, there were clear differences in the disease following vaccination (Langermans *et al.*, 2001). In particular, the cynomolgus monkey (*Macaca fascicularis*) was able to limit bacterial growth in the lung following vaccination whereas the rhesus monkey (*Macaca mulatta*) exhibited no sign of control (Langermans *et al.*, 2001). This difference in ability to control disease occurred despite the induction of an increased proliferative and cytokine response in both types of monkey following vaccination and present prior to challenge. With regard to the discussion here it is important to note that the immunopathogenesis of disease was somewhat different between the two species both at the primary and secondary challenge. The protected cynomolgus monkey exhibited a more controlled lymphocytic granuloma even in the absence of vaccination whereas the rhesus monkey exhibited a more destructive and necrotic lung disease. Even though the rhesus monkey had clearly produced the type of response that is correlated with protection, it was not protected by this response.

In a separate study in which the challenge *M. tuberculosis* dose was approximately 10 fold lower, BCG vaccination was seen to be effective in the rhesus monkey (Shen *et al.*, 2002). This protection was associated with the rapid recall expansion of the dominant γδT cell (Vγ2Vδ2) sub-population (Shen *et al.*, 2002). This γδ T cell population is also the dominant population in humans and can respond in an antigen-specific manner to non-protein antigens derived from a variety of sources including mycobacteria (Bukowski *et al.*, 1999; Morita *et al.*, 1995). While there is no evidence that these cells are protective, there is evidence that γδ T cells modulate the granulomatous response to tuberculosis in the murine model and that this correlates with increased survival (D'Souza *et al.*, 1997).

Further analysis of the development of the response to infection in macaques prior to and following vaccination should allow the relative roles of both γδ T cells and αβ T cells in controlling both bacterial growth and pathogenesis to be addressed. As is seen from the difference in results described above, it is essential that the dose of bacteria and other parameters of infection be carefully considered before comparing the results of any trial.

The Mouse

There are several small animal models of tuberculosis that are economical in terms of space and cost. Infection studies with amphibian models have provided a useful model with which to follow gene expression of mycobacteria within long term granulomas (Ramakrishnan *et al.*, 2000). In addition, studies with *Mycobacterium marinum* in the goldfish have demonstrated that this host develops a granulomatous disease that could be used for the study of mycobacterial pathogenesis (Talaat *et al.*, 1998). On the whole however the mouse is the animal of choice for large-scale preliminary testing of hypotheses regarding the role of specific host or bacterial components in pathogenesis of tuberculosis. Indeed, the mouse has provided a great deal of useful and relevant information regarding the mechanisms of protection in tuberculosis. In the following paragraphs, the benefits of several routes of infection are described along with the caveats that should be considered when interpreting data from each murine model.

High Dose, Intravenous Infection

This model has been a mainstay of tuberculosis research for many years and has provided much useful, relevant information. It is particularly reproducible and provides an immediate challenge to the host and therefore results in large, easily measurable responses. Using this model, the absolute requirement for T cells (Mackaness, 1968; North, 1973), interferon-γ (IFN-γ) (Cooper *et al.*, 1995b; Flynn *et al.*, 1993b), nitric oxide (MacMicking *et al.*, 1997), TNF-α

(Flynn *et al.*, 1995a), and interleukin (IL)-12 (Cooper *et al.*, 1997b; Flynn *et al.*, 1995b) for survival during disseminated disease has been very clearly shown. Essentially, however, this model approximates profound disseminated disease rather than the more chronic natural disease. Care must be taken in interpreting data from this model as mechanisms that are deemed essential to survival when an animal is infected with a large bolus of bacteria, may not be required to survive a low dose. Two examples of this kind of difference were seen when mice lacking either the β2-microglobulin (β2-m) or CD8 gene were infected intravenously or via the aerosol route. Using a high dose intravenous route both of these mice were susceptible to disease and died (Flynn *et al.*, 1993a; Sousa *et al.*, 2000) whereas low dose aerosol infection resulted in moderately increased bacterial growth and altered granuloma formation but not in increased short-term mortality (D' Souza *et al.*, 2000; Mogues *et al.*, 2001; Turner *et al.*, 2001a). Comparison of the data from both the intravenous and aerosol models suggests that the absence of the β2-m molecule may affect cellular recruitment as poor granuloma formation was seen in the lungs if both models (D' Souza *et al.*, 2000; Flynn *et al.*, 1993a).

One important consideration regarding the intravenous model is that bacteria are deposited directly in the spleen, thus allowing a very rapid induction of the acquired response. This rapid induction means that the role of innate immunity and the induction of acquired immunity in the lung are not addressed. Another often-overlooked feature of this model is the location of the bacteria within the lung tissue. Following intravenous infection, bacteria are deposited throughout the interstitium of the lung and inflammatory responses occur throughout the tissue. This is unlike the deposition occurring in natural infection when tuberculosis lesions arise at discrete foci. A corollary to this issue is that the immune lymphocytes and monocytes being recruited to the inflammatory site are trafficking to areas that may not be targeted in natural infection; thus questions about the role of chemokines and cell migration may be compromised. That being said one of the most profound susceptibilities to TB is seen in the intravenously infected mouse lacking the chemokine receptor 2 (CCR2-/-). These mice exhibit defective recruitment of macrophages, dendritic cells and T cells to the lungs following infection and this correlates with a 100 fold increase in bacterial load in this organ (Peters *et al.*, 2001) suggesting an absolute requirement for this receptor in effective control of TB.

As an aside, a major disadvantage of the high dose intravenous model is that the infection technique requires the injection of fairly large doses of bacteria and the potential for an inadvertent generation of an aerosol is ever present, even with the most skilled technician.

Despite the previous limitations, the primary advantage of the intravenous model is that a very precise infectious dose may be administered; thus ensuring reproducible results are observed. One variation of the high dose intravenous model which has been particularly useful is the low dose intravenous challenge used to initiate a chronic infection (Mohan *et al.*, 2001; Scanga *et al.*, 2000; Flynn *et al.*, 1998; Scanga *et al.*, 1999). This model shares many of the advantages and disadvantages of the high dose model but is particular useful due to its high reproducibility. Using this model, recent studies have demonstrated that although TNF-α is crucial for containing bacterial growth, it is also crucial to the maintenance of the protective granulomatous structure (Mohan *et al.*, 2001). In addition, the chronic infection model has been used to confirm that CD4 T cells are required to maintain the long-term control of bacterial growth (Scanga *et al.*, 2000). This role has recently been confirmed in the low dose aerosol model (Saunders *et al.*, 2002).

In summary, the intravenous model can provide useful, reproducible data regarding the requirements for control of bacterial growth and long-term control of chronic infection. It is limited by the fact that bacteria are deposited throughout the body of the host and this limits study of the role of cell trafficking in induction of the protective response.

Low Dose Aerosol Infection

There are several methods available to introduce bacteria into the lungs. As with all models, however, there are advantages and disadvantages to each method. The three commonest methods are intratracheal inoculation, intranasal instillation and exposure of the animal to an aerosol cloud within a confined space. In general, the intratracheal and intranasal infection routes are used when the aerosol route, which requires special equipment and facilities, is unavailable. The intratracheal and intranasal routes require anesthetic and fairly high infectious doses; the intratracheal route is also complicated by the need for surgery. The high doses required to deliver bacteria to the lung using the intra-tracheal or intra-nasal routes may alter several important aspects of pathogenesis. In particular, the deposition of bacteria in the nares and tracheobronchial tissue is a complication not seen with aerosol delivery. With a high dose of bacteria present in the upper respiratory tract, the induction and development of acquired specific responses will be skewed when compared to cases when a small number of bacteria are deposited directly in the alveolar space. Using this model however, an elegant characterization of T cell recruitment was described in the lung, with the production of the protective cytokine IFN-γ being associated with the CD4 T cells expressing the activation phenotype CD44 high/CD45RB low (Lyadova *et al.*, 2000).

The deposition of bacteria in the lower respiratory tract is achieved by the generation of small bacteria-containing aerosolized droplets to which animals are exposed (Middlebrook, 1952; Ratcliffe *et al.*, 1953; Wiegeshaus *et al.*, 1970). The important aspects of the aerosol-generating machine are the generation of the droplets of the correct size and the movement of the droplet cloud throughout the exposure chamber. This aerosol model of infection appears, intuitively, to be more appropriate for the study of tuberculosis as it most closely mimics natural infection. Indeed, as is thought to happen naturally, bacteria are deposited directly in the alveolar space and are not distributed throughout the lung but establish foci of infection at specific sites (Middlebrook, 1952; Ratcliffe *et al.*, 1953).

The development of disease and also the generation of the acquired responses is also different in the aerosol model compared to other means of bacterial delivery. In particular, in this model an acquired cellular response is not immediately induced upon exposure to the pathogen, but takes some time to become established (Cardona *et al.*, 1999; Cooper *et al.*, 1997a; North, 1995). This delay is one of the contributing factors to the increased susceptibility of mice when exposed to via the aerosol route (Cardona *et al.*, 1999; North, 1995). Indeed, if the dose given by the aerosol route is sufficiently high, bacterial growth cannot be controlled by the emerging protective response and even immunocompetent mice will die. The delay in induction of the acquired response also allows the role of innate responses in controlling infection to be addressed. In addition, in this model the induction of the acquired response is achieved via a unique pathway not mimicked by other routes. This induction requires that bacteria and/or bacterial antigens travel from the alveolar space, possibly via dendritic cells (Gonzalez-Juarrero *et al.*, 2001), and are deposited in the draining lymph node. Bacteria may also erode into the blood and spread to spleen. At both locations responses can be detected following aerosol infection by day 15-20 (Cardona *et al.*, 1999; Cooper *et al.*, 1997a).

Granuloma Development in the Murine Model of Pulmonary Tuberculosis

Once the acquired response has been induced, protective cells must be recruited from the blood into the interstitium and then into the granuloma. This recruitment is more easily elucidated in the low dose aerosol model, as there are discreet foci of infection due to the initial deposition of bacilli in the alveolar space. Several reports have supported a crucial role for TNF-α in the induction of the early chemokine response required for recruitment and in the generation of a tightly compact mononuclear granuloma (Rhoades *et al.*, 1995; Roach *et al.*, 2002). In particular, in the absence of TNF-α there is a complete breakdown in the structural integrity of the granuloma which contributes to dramatically increased mortality (Bean *et al.*, 1999) (Mohan *et al.*, 2001)(also true for other mycobacterial infections (Benini *et al.*, 1999; Ehlers *et al.*, 2000; Smith *et al.*, 1997)). In an elegant study making use of bone marrow chimera technology it has also been shown that Lymphotoxin-α (LT-α) is also important in co-ordination of the protective granulomatous response to TB independently of the activity of TNF-α (Roach *et al.*, 2001). An additional and novel experiment

Upper Panel

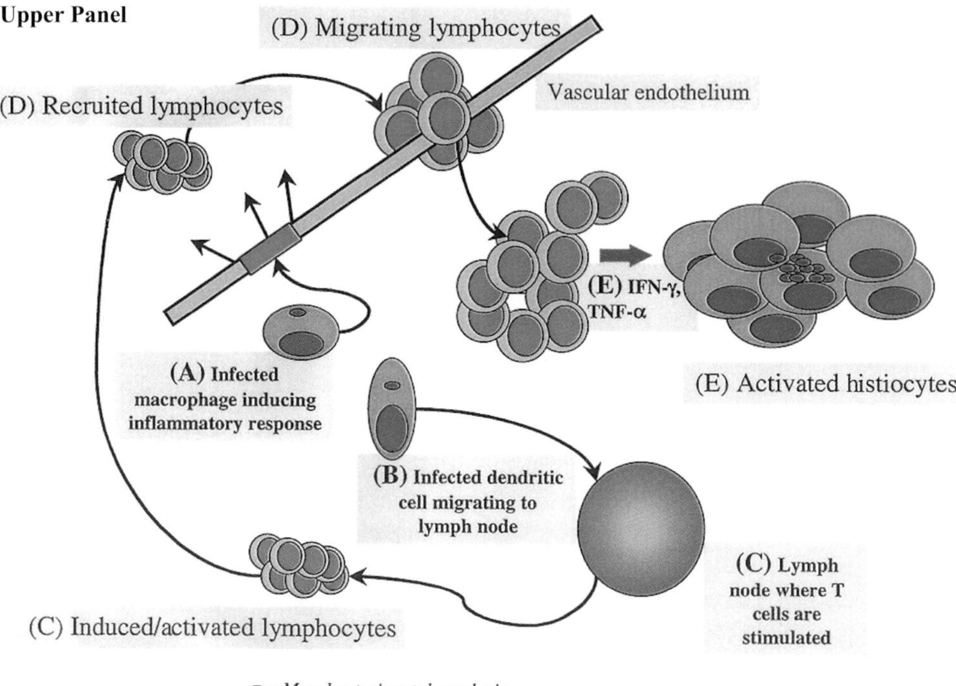

Lower Panel

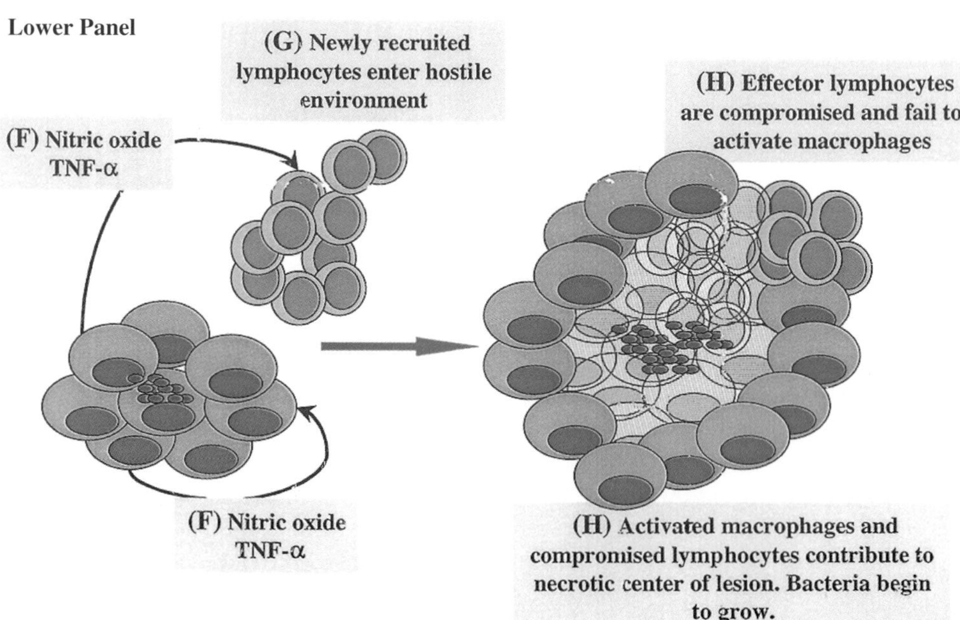

Panel. Development of the pathogenic lesion in TB infected animals.

Upper Panel: In the development of the protective response to TB the vertebrate host must first recognize that it has inhaled a pathogen. Once activated the innate response helps to initiate and orchestrate an inflammatory response (A). At the same time bacteria and bacterial antigens are ingested by immature dendritic cells that then mature and migrate to the lymph node (B). Within the lymph node naïve T cells are exposed to antigen and appropriate stimulatory signals and become activated effector cells (C). Activated lymphocytes then migrate to the site of inflammation and are recruited along a chemokine gradient orchestrated by TNF-α (D). Once within the granuloma the activated lymphocytes activate the infected macrophages via IFN-γ, TNF-α and cessation of bacterial growth ensues (E). **Lower Panel:** As anti-mycobacterial effector molecules such as nitric oxide and TNF-α dominate the lesional site (F) both macrophages and lymphocytes (G) experience a hostile environment (see ref (Cooper *et al.*, 2002a)). As cells begin to die (as a result of apoptosis or necrosis?) then necrotic areas develop in the lesion (H). Cell mediated effector functions are ineffective within the necrotic center and bacteria begin to grow (H). In the mouse the necrotic areas are small and bounded by fibrotic tissue (Orme, 1998) but in the guinea pig, rabbit and macaque, a necrotic center occurs that can consolidate and even progress to cavitation in the rabbit (Converse *et al.*, 1996, 1998).

using mice with a local pulmonary blockade of TNF-α, also demonstrated that this cytokine is required to avoid the development of a neutrophil dominated inflammatory response characterized by high levels of pro-inflammatory cytokines (Smith *et al.*, 2002). These data taken together with the observations made by Flynn in the chronic model (Mohan *et al.*, 2001) suggest that TNF-α is crucial to the development and maintenance of a compact, protective mononuclear granuloma in response to mycobacterial infection.

The relative importance of the granuloma in controlling mycobacterial infection has been discussed throughout the history of tuberculosis research. Studies with the aerosol model of tuberculosis and gene-deleted mice have, however, begun to highlight the role of the host components in this response to infection. In particular, in the absence of the adhesion molecule ICAM-1, mononuclear granulomata fail to coalesce in the lung despite the expression of protective immunity in this organ (Johnson *et al.*, 1998). That the granuloma is crucial to survival is however demonstrated by the increased bacterial growth and generation of destructive neutrophillic lesions following a prolonged period of infection in mice lacking ICAM-1 (Saunders *et al.*, 2000, 1999). Other studies using mice lacking either the CD4 or CD8 co-receptor have also shown that the granuloma breaks down in the absence of an optimal acquired specific T cell response (Saunders *et al.*, 2002; Turner *et al.*, 2001a). Indeed, efficient recruitment of T cells to the lung, while dependent upon TNF-α and chemokine expression is also dependent upon the induction of protective T cells which express the homing molecules needed for them to migrate to sites of inflammation (Feng *et al.*, 1999, 2000). For example, it has been shown that in the absence of T cell activation, such as in IL-12 p40 deficient mice, both

DTH and granuloma formation are compromised (Cooper *et al.*, 2002b, 1997b). Indeed, the addition of IL-12 to mice with defective expression of adhesion markers on T cells resulted in up-regulation of these markers and the increased recruitment of cells to lesions (Cooper *et al.*, 1995a). In other studies looking at the mechanisms behind increased recrudescence in certain strains of mice it was shown that recrudescence was associated with poor expression of adhesion molecules on T cells in the lung (Turner *et al.*, 2001b).

As more murine strains are infected via the aerosol route and the histological consequences examined, it is becoming apparent that we still have a lot to learn about the factors controlling the granulomatous lesion in this model. The knowledge gleaned from this extremely tractable model will hopefully provide the hypotheses to test in the other less simple models.

Current Knowledge Regarding Granuloma Development in Animal Models

The current data from animal models confirms that there is little clearance of *M. tuberculosis* from the lungs of infected animals and that as infection progresses so do the pathologic consequences. It appears that in all models, once the granulomas reach a particular stage (depending upon the animal), bacterial growth is resurgent and the host succumbs to disease (see panel, page 12). This suggests that there is an element of the pathologic response that compromises the expression of immunity. Comparison of the pathologic development in the above animal models may illuminate the elements that limit the expression of the protective response.

In the mouse the lymphocytic nature of the lesion correlates with long-term survival and bacterial growth occurs only when the lymphocytes disappear from the lesions (Rhoades *et al.*, 1997). In contrast, the guinea pig lesion is predominantly histiocytic in character and although we know that CD4 and CD8 T cell are recruited to the lesion, they fail to stop the development of necrosis (Orme, 1998; Turner, 2002). The rabbit also recruits lymphocytes to the lesion site but generates a cavitary lesion which contributes to spread of disease (Converse *et al.*, 1996, 1998). The macaque develops a mononuclear granuloma but if the dose of infection is high then this lesion can be destructive and although effector mechanisms are present their expression may be compromised (Langermans *et al.*, 2001; Shen *et al.*, 2002).

To address the hypothesis that the environment within the granuloma is capable of limiting the efficacy of the protective response several questions must be posed. Firstly, are the known protective lymphocyte responses effectively induced in each animal? The fact that all animal models discussed above can be protected by vaccination suggests that indeed each animal can generate an acquired specific protective response. Secondly, are protective cells effectively recruited to the site of infection? This has not been shown for all the models discussed here but the tools are now available to address this question. Finally, are the recruited lymphocytes able to mediate protection within the granuloma? This is the most difficult issue to address and analysis of this question would benefit from the use of modern tools capable of monitoring both bacterial physiology and macrophage function within the lesion.

Future Directions

There is currently a great deal of excitement in the TB research field. The sequencing of the bacterial genome (Cole *et al.*, 1998) and the development of bacterial mutants with altered virulence (McKinney *et al.*, 2000) have expanded our knowledge of the bacterial role in pathogenesis. The application of newer technologies to the analysis of the host response to infection has also resulted in improved knowledge of the role of host components in pathogenesis. Careful interpretation of the current data base is required for the generation of new hypotheses which can be tested using the presently developing techniques such as laser-capture microscopy (Sun *et al.*, 2001), differential determination of RNA expression by micro-array and dual-photon microscopy (Bousso *et al.*, 2002; Miller *et al.*, 2002). These techniques could be used to compare the responses of vaccinated and unvaccinated animals to determine the cell types, the cytokines/chemokines or the effector molecules most closely related to the development of a self-limiting, non-necrotizing mononuclear granuloma.

In addition to further analyzing the known protective mechanisms, there is hope for the identification of novel mechanisms which, while not activated during natural infection, may provide the anti-mycobacterial activity required for clearance of this persistent bacteria from the vertebrate lung. This goal, while still hypothetical may provide the necessary information for the development of a vaccine capable of not only limiting the spread of TB but also reducing its severity in the host. The ultimate goal is the induction of a rapid response in the lung that kills the bacteria while maintaining the integrity of the host tissue.

References

Bean, A.G.D., Roach, D.R., Briscoe, H., France, M.P., Korner, H., Sedgewick, J.D., and Britton, W.J. 1999. Structural deficiencies in granuloma formation in TNF gene-targeted mice underlie the heightened susceptibility to aerosol *Mycobacterium tuberculosis* infection, which is not compensated for by lymphotoxin. J. Immunol. 162: 3504-3511.

Benini, J., Ehlers, E.M., and Ehlers, S. 1999. Different types of pulmonary granuloma necrosis in immunocompetent vs TNFRp55-gene-deficient mice aerogenically infected with virulent *Mycobacterium avium*. J. Pathol. 189: 127-137.

Bousso, P., Bhakta, N., Lewis, R., and Robey, E. 2002. Dynamics of thymocyte-stromal interactions visualized by two-photon microscopy. Science 296: 1876-1880.

Bukowski, J., Morita, C., and Brenner, M. 1999. Human gamma delta T cells recognize alkylamines derived from microbes, edible plants, and tea: implications for innate immunity. Immunity 11: 57-65.

Cardona, P.-J., Cooper, A.M., Luquin, M., Ariza, A., Filipo, F., Orme, I.M., and Ausina, V. 1999. The intravenous model of murine tuberculosis is less pathogenic than the aerosol model owing to a more rapid induction of systemic immunity. Scand. J. Immunol. 49: 362-366.

Cohen, M., Bartow, R., Mintzer, C., and McMurray, D. 1987. Effects of diet and genetics on *Mycobacterium bovis* BCG vaccine efficacy in in-bred guinea pigs. Infect. Immun. 55: 314-319.

Cole, S.T., Brosch, R., Parkhill, J., Garnier, T., Churcher, C., Harris, D., Gordon, S.V., Eiglmeier, K., Gas, S., Barry, C.E., Tekaia, F., Badcock, K., Basham, D., Brown, D., Chillingworth, T., Connor, R., Davies, R., Devlin, K., Feltwell, T., Gentles, S., Hamlin, N., Holroyd, S., Hornsby, T., Jagels, K., Krogh, A., McLean, J., Moule, S., Murphy, L., Oliver, K., Osbourne, J., Quail, M.A., Rajandream, M.-A., Rogers, J., Rutter, S., Seeger, K., Skelton, J., Squares, R., Squares, S., Sulston, J.E., Taylor, K., Whitehead, S., and Barrell, B.G. 1998. Deciphering the biology of *Mycobacterium tuberculosis* from the complete genome sequence. Nature 393: 537-544.

Converse, P., Dannenberg, A.J., Estep, J., Sugisaki, K., Abe, Y., Schofield, B., and Pitt, M. 1996. Cavitary tuberculosis produced in rabbits by aerosolized virulent tubercle bacilli. Infect. Immun. 64: 4776-4787.

Converse, P., Dannenberg, A.J., Shigenaga, T., McMurray, D., Phalen, S., Stanford, J., Rook, G., Koru-Sengul, T., Abbey, H., Estep, J., and Pitt, M. 1998. Pulmonary bovine-type tuberculosis in rabbits: bacillary virulence, inhaled dose effects, tuberculin sensitivity, and *Mycobacterium vaccae* immunotherapy. Clin. Diag. Lab. Immunol. 5: 871-881.

Cooper, A.M., Adams, L.B., Dalton, D.K., Appelberg, R., and Ehlers, S. 2002a. IFN-γ and NO in mycobacterial disease: new jobs for old hands. Trends Microbiol. 10: 221-226

Cooper, A.M., Callahan, J.E., Griffin, J.P., Roberts, A.D., and Orme, I.M. 1995a. Old mice are able to control low-dose aerogenic infections with *Mycobacterium tuberculosis*. Infect. Immun. 63: 3259-3265.

Cooper, A.M., Callahan, J.E., Keen, M., Belisle, J.T., and Orme, I.M. 1997a. Expression of memory immunity in the lung following re-exposure to *Mycobacterium tuberculosis*. Tuberc. Lung Dis. 78: 67-73.

Cooper, A.M., and Flynn, J.L. 1995b. The protective immune response to *Mycobacterium tuberculosis*. Curr. Op. Immunol. 7: 512-516.

Cooper, A.M., Kipnis, A., Turner, J., Magram, J., Ferrante, J., and Orme, I.M. 2002b. Mice lacking bioactive IL-12 can generate protective, antigen-specific cellular responses to mycobacterial infection only if the IL-12 p40 subunit is present. J. Immunol. 168: 1322-1327.

Cooper, A.M., Magram, J., Ferrante, J., and Orme, I.M. 1997b. IL-12 is crucial to the development of protective immunity in mice intravenously infected with *Mycobacterium tuberculosis*. J. Exp. Med. 186: 39-46.

Cooper, A.M., Pearl, J.E., Brooks, J.V., Ehlers, S., and Orme, I.M. 2000. Expression of nitric oxide synthase 2 gene is not essential for early control of *Mycobacterium tuberculosis* in the murine lung. Infect. Immun. 68: 6879-6882.

D' Souza, C.D., Cooper, A.M., Frank, A.A., Ehlers, S., Turner, J., Bendelac, A., and Orme, I.M. 2000. A novel non-classical β2-microglobulin-restricted mechanism influencing early lymphocyte accumulation and subsequent resistance to tuberculosis in the lung. Am. J. Respir. Cell. Mol. Biol. 23: 188-193.

D'Souza, C.D., Cooper, A.M., Frank, A.A., Mazzaccaro, R.J., Bloom, B.R., and Orme, I.M. 1997. An anti-inflammatory role for γδ T lymphocytes in acquired immunity to *Mycobacterium tuberculosis*. J. Immunol. 158: 1217-1221.

Dai, G., and McMurray, D. 1998. Altered cytokine production and impaired anti-mycobacterial immunity in protein-malnourished guinea pigs. Infect. Immun. 66: 3562-3568.

Dannenberg, A.J. 1994. Rabbit model of tuberculosis. In: Tuberculosis: Pathogenesis, Protection and Control. Bloom, B., ed. American Society for Microbiology, Washington, DC. p. 149-156.

Dannenberg, A.J., and Collins, F. 2001. Progressive pulmonary tuberculosis is not due to increasing numbers of viable bacilli in rabbits, mice and guinea pigs, but is due to a continuous host response to mycobacterial products. Tuberculosis 81: 229-242.

Dascher, C.C., Hiromatsu, K., Naylor, J., Brauer, P., Brown, K., Storey, J., Behar, S., Kawasaki, E., Porcelli, S., Brenner, M., and LeClair, K. 1999. Conservation of a CD1 multigene family in the guinea pig. J. Immunol. 163: 5478-5488.

Ehlers, S., Kutsch, S., Ehlers, E., Benini, J., and Pfeffer, K. 2000. Lethal granuloma disintegration in mycobacteria-infected TNFRp55-/- mice is dependent on T cells and IL-12. J. Immunol. 165: 483-492.

Feng, C.G., Bean, A.G., Hooi, H., Briscoe, H., and Britton, W.J. 1999. Increase in gamma interferon-secreting CD8, as well as CD4, T cells in lungs following aerosol infection with *Mycobacterium tuberculosis*. Infect. Immun. 67: 3242-3246.

Feng, C.G., Britton, W.J., Palendira, U., Groat, N.L., Briscoe, H., and Bean, A.G.D. 2000. Up-regulation of VCAM-1 and differential expression of β integrin expressing lymphocytes is associated with immunity to pulmonary *Mycobacterium tuberculosis*. J. Immunol. 164: 4853-4860.

Florido, M., Appelberg, R., Orme, I.M., and Cooper, A.M. 1997. Evidence for a reduced chemokine response in the lungs of beige mice infected with *Mycobacterium avium*. Immunology 90: 600-606.

Flynn, J., Goldstein, M., Triebold, K., and Bloom, B. 1993a. Major histocompatibility complex class I-restricted T cells are necessary for protection against *Mycobacterium tuberculosis* in mice. Inf. Agents Dis. 2: 259-262.

Flynn, J., Scanga, C., Tanaka, K., and Chan, J. 1998. Effects of aminoguanidine on latent murine tuberculosis. J. Immunol. 160: 1796-1803.

Flynn, J.L., Chan, J., Triebold, K.J., Dalton, D.K., Stewart, T.A., and Bloom, B.R. 1993b. An essential role for interferon gamma in resistance to *Mycobacterium tuberculosis* infection. J. Exp. Med. 178: 2249-2254.

Flynn, J.L., Goldstein, M.M., Chan, J., Triebold, K.J., Pfeffer, K., Lowenstein, C.J., Schreiber, R., Mak, T.W., and Bloom, B.R. 1995a. Tumor necrosis factor-alpha is required in the protective immune response against *Mycobacterium tuberculosis* in mice. Immunity 2: 561-572.

Flynn, J.L., Goldstein, M.M., Triebold, K.J., Sypek, J., Wolf, S., and Bloom, B.R. 1995b. IL-12 increases resistance of BALB/c mice to *Mycobacterium tuberculosis* infection. J. Immunol. 155: 2515-2524.

Fok, J., Ho, R., Arora, P., Harding, G., and Smith, D. 1976. Host-parasite relationships in experimental airborne tuberculosis. V: Lack of hematogenous dissemination of *Mycobacterium tuberculosis* to the lungs in animals vaccinated with Bacille Calmette-Guerin. J. Inf. Dis. 133: 137-144.

Gonzalez-Juarrero, M., and Orme, I. 2001. Characterization of murine lung dendritic cells infected with *Mycobacterium tuberculosis*. Infect. Immun. 69: 1127-1133.

Good, R. 1968. Simian tuberculosis: Immunological aspects. Ann. N.Y. Acad. Sci. 154: 200-213.

Harding, G., and Smith, D. 1977. Host-parasite relationships in experimental airborne tuberculosis. V: Influence of vaccination with Bacille Calmette-Guerin on the onset and/or extent of hematogenous dissemination of virulent *Mycobacterium tuberculosis* to the lungs. J. Inf. Dis. 1977: 439-433.

Hiromatsu, K., Dascher, C., LeClair, K., Sugita, M., Furlong, S., Brenner, M., and Porcelli, S. 2002. Induction of CD1-restricted immune responses in guinea pigs by immunization with mycobacterial lipid antigens. J. Immunol. 169: 330-339.

Johnson, C.M., Cooper, A.M., Frank, A.A., and Orme, I.M. 1998. Adequate expression of protective immunity in the absence of granuloma formation in *Mycobacterium tuberculosis*-infected mice with a disruption in the intracellular adhesion molecule 1 gene. Infect. Immun. 66: 1666-1670.

Koch, R. 1881. Methods for the study of pathogenic organisms. Mitt Kaiser Gesundheit 1: 1-48.

Koch, R. 1882. Die aetiologie der tuberkulose. Berlin Klin Wochenschr 15: 221-230.

Koch, R. 1884. Die aetiologie der tuberkulose. Mitt Kaiser Gesundheit 2: 1-88.

Koch, R. 1890. Weitere mittheilung uber ein heilmittel gegen tuberkulose. Deutsch Medical Wochenschrift 16: 1029-1032.

Langermans, J., Andersen, P., van Soolingen, D., Vervenne, R., Frost, P., van der Laan, T., van Pinxteren, L., van den Hombergh, J., Kroon, S., Peekel, I., Florquin, S., and Thomas, A. 2001. Divergent effect of bacillus Calmette-Guerin (BCG) vaccination on *Mycobacterium tuberculosis* infection in highly related macaque species: implications for primate models in tuberculosis vaccine research. Proc. Natl. Acad. Sci. USA. 98: 11497-11502.

Lurie, M. 1941. Heredity, constitution and tuberculosis: An experimental study. Am. Rev. Tuberc. 44 (suppl.): S1-125.

Lurie, M., and Abramson, S. 1948. Reproduction of human ulcerative pulmonary tuberculosis in rabbits by quantitative natural airborne contagion. Proc. Soc. Exper. Biol. Med. 69: 531-.

Lurie, M., Heppleston, A., Abramson, S., and Swartz, I. 1950. An evaluation of the method of quantitative airborne infection and its use in the study of the pathogenesis of tuberculosis. Am. Rev. Tuberc. 61: 765-797.

Lyadova, I., Eruslanov, E., Khaidukov, S., Yeremeev, V., Majorov, K., Pichugin, A., Nikonenko, B., Kondratieva, T., and Apt, A. 2000. Comparative analysis of T lymphocytes recovered from the lungs of mice genetically susceptible, resistant, and hyperresistant to *Mycobacterium tuberculosis*-triggered disease. J. Immunol. 165: 5921-5931.

Mackaness, G. 1968. The immunology of antituberculous immunity. Am. Rev. Respir. Dis. 97: 337-344.

MacMicking, J.D., North, R.J., LaCourse, R., Mudgett, J., Shah, S.K., and Nathan C.F. 1997. Indentification of NOS2 as a protective locus against tuberculosis. Proc. Natl. Acad. Sci. USA 94: 5243-5248.

Mainali, E., and McMurray, D. 1992. Influence of infection route in the generation of immune lymphocytes by guinea pigs infected with virulent *Mycobacterium tuberculosis*. FASEB J. 6: A1335.

McKinney, J.D., Honer zu Bentrup, K., Munoz-Elias, E., Miczak, A., Chen, B., Chan, W., Swenson, D., Sacchettini, J., Jacobs, W.J., and russell, D.G. 2000. Persistence of *Mycobacterium tuberculosis* in macrophages and mice requires the glyoxylate shunt enzyme isocitrate lyase. Nature 406: 735-738.

McMurray, D. 1994. Guinea Pig models of tuberculosis. In: Tuberculosis: Pathogenesis, Protection and Control. Bloom, B., ed. American Society for Microbiology, Washington, DC. p. 135-147.

McMurray, D. 2000. A nonhuman primate model for preclinical testing of new tuberculosis vaccines. Clin. Infect. Dis. 30: S210-S212.

McMurray, D., and Bartow, R. 1992. Immunosuppression and alteration of resistance to pulmonary tuberculosis in guinea pigs by protein malnourishment. J. Nutr. 122: 738-743.

McMurray, D., Bartow, R., and Mintzer, C. 1990. Protein malnutrition alters the distribution of FcγR+ (Tγ) and FCμR+ (Tμ) T lymphocytes in experimental pulmonary tuberculosis. Infect. Immun. 58: 563-565.

McMurray, D., Carlomango, M., Mintzer, C., and Tetzlaff, C. 1985. *Mycobacterium bovis* BCG vaccine fails to protect protein-deficient guinea pigs against respiratory challenge with virulent *Mycobacterium tuberculosis*. Infect. Immun. 50: 555-559.

McMurray, D., Mintzer, C., Tetzlaff, C., and Carlomango, M. 1986. Influence of dietary protein on the protective effect of BCG in guinea pigs. Tubercle 67: 31-39.

Middlebrook, G. 1952. An apparatus for airborne infection of mice. Proc. Soc. Exper. Biol. Med. 80: 105-110.

Miller, M., Wei, S., Parker, I., and Cahalan, M. 2002. Two-photon imaging of lymphocyte motility and antigen response in intact lymph node. Science 296: 1869-1873.

Mogues, T., Goodrich, M.E., Ryan, L., LaCourse, R., and North, R.J. 2001. The relative importance of T cell subsets in immunity and immunopathology of airborne *Mycobacterium tuberculosis* infection in mice. J. Exp. Med. 193: 271-280.

Mohan, V., Scanga, C., Yu, K., Scott, H., Tanaka, K., Tsang, E., Tsai, M., Flynn, J., and Chan, J. 2001. Effects of tumor necrosis factor alpha on host immune response in chronic persistent tuberculosis: possible role for limiting pathology. Infect. Immun. 69: 1847-1855.

Morita, C., Beckman, E., Bukowski, J., Tanaka, Y., Band, H., Bloom, B., Golan, D., and Brenner, M. 1995. Direct presentation of nonpeptide prenyl pyrophosphate antigens to human gamma delta T cells. Immunity 3: 495-507.

North, J.N. 1995. *Mycobacterium tuberculosis* is strikingly more virulent for mice when given via the respiratory than the intravenous route. J. Inf. Dis. 172: 1550-1553.

North, R.J. 1973. Importance of thymocyte-derived lymphocytes in cell-mediated immunity to infection. Cell. Immunol. 7: 166-176.

Orme, I.M. 1998. The immunopathogenesis of tuberculosis: a new working hypothesis. Trends Microbiol. 6: 94-97.

Peters, W., Scott, H., Chambers, H., Flynn, J., Charo, I., and Ernst, J. 2001. Chemokine receptor 2 serves an early and essential role in resistance to *Mycobacterium tuberculosis*. Proc. Natl. Acad. Sci. USA 98: 7958-7963.

Porcelli, S.A., Segelke, B.W., Sugita, M., Wilson, I.A., and Brenner, M.B. 1998. The CD1 family of lipid antigen-presenting molecules. Immunol. Today 19:362-368.

Ramakrishnan, L., Federspiel, N., and Falkow, S. 2000. Granuloma-specific expression of Mycobacterium virulence proteins from the glycine-rich PE-PGRS family. Science 288: 1436-1439.

Ratcliffe, H., and Palladino, V. 1953. Tuberculosis induced by droplet nuclei infection. J. Exp. Med. 97: 61-68.

Rhoades, E.R., Cooper, A.M., and Orme, I.M. 1995. Chemokine response in mice infected with *Mycobacterium tuberculosis*. Infect. Immun. 63: 3871-3877.

Rhoades, E.R., Frank, A.A., and Orme, I.M. 1997. Progression of chronic pulmonary tuberculosis in mice aerogenically infected with virulent *Mycobacterium tuberculosis*. Tuberc. Lung Dis. 78: 57-66.

Roach, D., Bean, A., Demangel, C., France, M., Briscoe, H., and Britton, W. 2002. TNF regulates chemokine induction essential for cell recruitment, granuloma formation, and clearance of mycobacterial infection. J. Immunol. 168: 4620-4627.

Roach, D., Briscoe, H., Saunders, B., France, M., Riminton, S., and Britton, W. 2001. Secreted Lymphotoxin-α is essential for the control of an intracellular bacterial infection. J. Exp. Med. 193: 239-246.

Saunders, B., Frank, A., Orme, I., and Cooper, A. 2002. CD4 is required for the development of a protective granulomatous response to pulmonary tuberculosis. Cell. Immunol. 216: 65-72.

Saunders, B.M., and Cooper, A.M. 2000. Restraining mycobacteria: role of granulomas in mycobacterial infections. Immunol Cell Biol 78: 334-341.

Saunders, B.M., Frank, A.A., and Orme, I.M. 1999. Granuloma formation is required to contain bacilli growth and delay mortality in mice chronically infected with *Mycobacterium tuberculosis*. Immunology 98: 324-328.

Scanga, C., Mohan, V., Joseph, H., Yu, K., Chan, J., and Flynn, J. 1999. Reactivation of latent tuberculosis: variations on the Cornell murine model. Infect. Immun. 67: 4531-4538.

Scanga, C., Mohan, V., Yu, K., Joseph, H., Tanaka, K., Chan, J., and Flynn, J. 2000. Depletion of CD4(+) T cells causes reactivation of murine persistent tuberculosis despite continued expression of interferon gamma and nitric oxide synthase 2. J. Exp. Med. 192: 347-358.

Shen, Y., Zhou, D., Qiu, L., Lai, X., Simon, M., Shen, L., Kou, Z., Wang, Q., Jiang, L., Estep, J., Hunt, R., Clagett, M., Sehgal, P., Li, Y., Zeng, X., Morita, C., Brenner, M., Letvin, N., and Chen, Z. 2002. Adaptive immune response of Vgamma2Vdelta2+ T cells during mycobacterial infections. Science 295: 2255-2258.

Shigenaga, T., Dannenberg, A., Lowrie, D., Said, W., Urist, M., Abbey, H., Schofield, B., Mounts, P., and Sugisaki, K. 2001. Immune responses in tuberculosis: antibodies and CD4-CD8 lymphocytes with vascular adhesion molecules and cytokines (chemokines) cause a rapid antigen-specific cell infiltration at sites of bacillus Calmette-Guerin reinfection. Immunology 102: 466-479.

Smith, D., Hansch, H., Bancroft, G., and Ehlers, S. 1997. T-cell-independent granuloma formation in response to *Mycobacterium avium:* role of tumour necrosis factor-α and interferon-γ. Immunology 92: 413-421.

Smith, D., and Harding, G. 1977. Animal Model:experimental airborne tuberculosis in the guinea pig. Am. J. Path. 89: 273-276.

Smith, D., McMurray, D., Wiegeshaus, E., Grover, A., and Harding, H. 1970. Host-parasite relationships in experimental airborne tuberculosis. IV: Early events in the the course of infection in vaccinated and nonvaccinated guinea pigs. Am. Rev. Respir. Dis. 102: 937-949.

Smith, D., Wiegeshaus, E., Navalkar, R., and Grover, A. 1966. Host-parasite relationships in experimental airborne tuberculosis. I. Preliminary studies in BCG-vaccinated and nonvaccinated animals. J. Bacteriol. 91: 718-724.

Smith, S., Liggitt, D., Jeromsky, E., Tan, X., Skerrett, S., and Wilson, C. 2002. Local role for tumor necrosis factor alpha in the pulmonary inflammatory response to *Mycobacterium tuberculosis* infection. Infect. Immun. 70: 2082-2089.

Sousa, A.O., Mazzaccaro, R.J., Russell, R.G., Lee, F.K., Turner, O.C., Hong, S., Van Kaer, L., and Bloom, B.R. 2000. Relative contributions of distinct MHC class I-dependent cell populations in protection to tuberculosis infection in mice. Proc. Natl. Acad. Sci. USA 97: 4204-4208.

Sun, B., Xie, W., Yi, G., Chen, D., Zhou, Y., and Cheng, J. 2001. Microminiaturized immunoassays using quantum dots as fluorescent label by laser confocal scanning fluorescence detection. J. Immunol. Meth. 249: 85-89.

Talaat, A., Reimschuessel, R., Wasserman, S., and Trucksis, M. 1998. Goldfish, *Carassius auratus*, a novel animal model for the study of *Mycobacterium marinum* pathogenesis. Infect. Immun. 66: 2938-2942.

Tsenova, L., Bergtold, A., Freedman, V., Young, R., and G, K. 1999. Tumor necrosis factor alpha is a determinant of pathogenesis and disease progression in mycobacterial infection in the central nervous system. Proc. Natl. Acad. Sci. USA 96: 5657-5662.

Tsenova, L., Sokol, K., Freedman, V., and Kaplan, G. 1998. A combination of thalidomide plus antibiotics protects rabbits from mycobacterial meningitis-associated death. J. Inf. Dis. 177: 1563-1572.

Turner, J., D'Souza, C.D., Pearl, J.E., Marietta, P., Noel, M., Frank, A.A., R. , Appelberg, R., Orme, I.M., and Cooper, A.M. 2001a. CD8- and CD95/95 L-dependent mechanisms of resistance in mice with chronic pulmonary tuberculosis. Am. J. Respir. Cell. Mol. Biol. 24: 203-209.

Turner, J., Gonzalez-Juarrero, M., Saunders, B.M., Brooks, J.V., Marietta, P., Ellis, D.L., Frank, A.A., Cooper, A.M., and Orme, I.M. 2001b. Immunological basis for reactivation of tuberculosis in mice. Infect. Immun. 69: 3264-3270.

Turner, O. 2002. In :Microbiology, Immunology and Pathology. PhD Thesis, Dept Microbiology, Immunology and Pathology, Colorado State University, Fort Collins, CO.

Walsh, G., Tan, E., dela Cruz, E., Abalos, R., Villahermosa, L., Young, L., Cellona, R., Nazareno, J., and Horwitz, M. 1996. The Philippine cynomolgus monkey (*Macaca fasicularis*) provides a new nonhuman primate model of tuberculosis that resembles human disease. Nat. Med. 2: 430-436.

Wells, W., and Lurie, M. 1941. Experimental airborne disease:quantitative natural respiratory contagion of tuberculosis. Am. J. Hyg. 34: 21.

Wiegeshaus, E., McMurray, D., Grover, A., Harding, G., and Smith, D. 1970. Host-parasite relationships in experimental airborne tuberculosis III. Relevance of microbial enumerationto acquired resistance in guinea pigs. Am. Rev. Respir. Dis. 102: 422-429.

From: Tuberculosis: The Microbe Host Interface
Edited by: Larry S. Schlesinger and Lucy E. DesJardin

Chapter 7

Analysis of *Mycobacterium tuberculosis* Gene Expression in the Human Host

Josephine E. Clark-Curtiss
and Lucy E. DesJardin

Abstract

Survival and growth within the human host are essential elements of *Mycobacterium tuberculosis* pathogenesis. Thus, understanding the physiology of the tubercle bacilli within the infected host will enable investigators to define the mechanisms of pathogenesis of this organism. In this chapter, we summarize the approaches that have been and are being used to analyze mycobacterial gene expression during its sojourn in the human host, as well as the knowledge that has been gleaned in our understanding of this very successful pathogen.

Introduction

M. tuberculosis has evolved into an extremely successful pathogen by virtue of its ability to survive and grow within its obligate human host, even when confronted with a variety of extreme environments. *M. tuberculosis* can grow as an intracellular parasite in non-activated macrophages, can persist for decades in a quiescent state in the granuloma (reviewed in Kaufmann, 1993; Wayne, 1994), and can survive in an aerosol as it is passed from one host to another host via expulsion from the lung by coughing, singing, or talking (Bates, 1980). The recent completion of the genomic DNA sequences for *M. tuberculosis* strains H37Rv and CDC 1551 demonstrate an unexpectedly large genome size for an obligate parasite, with an estimated 4056 and 3924 genes, respectively (Cole *et al.*, 1998; Camus *et al.*, 2002) (TIGR http://www/tigr.org/tigr-scripts/CMR2). Gene expression in *M. tuberculosis* is predicted to be highly regulated; the genome encodes 13 sigma factors and 100 repressors and activators, indicative of unusually high transcriptional regulation (Cole *et al.*, 1998). Precise control over transcription is predicted to be essential for survival in the various environments encountered by *M. tuberculosis*.

Unlike bacterial pathogens that cause more acute infections of shorter duration and produce virulence determinants such as toxins, capsules or secreted enzymes that interfere with normal host functions, the hallmarks of tuberculosis are the abilities of *M. tuberculosis* to survive and grow within the infected human host. Thus, virulence "determinants" of the tubercle bacilli may be more subtle attributes, such as slow growth, mimicry of host signaling devices, or subterfuge. Understanding the growth and physiology of *M. tuberculosis* within the infected host is essential to defining the mechanisms of pathogenesis of *M. tuberculosis*. The importance of understanding the growth of *M. tuberculosis in vivo* has long been recognized by investigators in the field (reviewed by Segal, 1984; Youmans, 1979; Pierce *et al.*, 1953).

Transmission from the lungs of an infected individual via droplet nuclei to the alveoli of the new host, phagocytosis of the bacilli by alveolar macrophages, and avoidance of normal destructive responses of the macrophages require sophisticated and rapidly acting mechanisms for recognizing the changing environmental conditions and altering gene expression in response. Following successful evasion of the normal phagocytic responses, *M. tuberculosis* must then commence to grow and multiply within the macrophage environment. Does *M. tuberculosis* synthesize all components of its lipids, proteins and nucleic acids *de novo* within the macrophage or is it able to scavenge some precursors from the macrophage? If the latter, how do the bacilli acquire such precursors? What are the sources of carbon, nitrogen, and energy for *M. tuberculosis* within the macrophage? How does the change from active growth to the state of persistence occur *in vivo*?

Understanding physiological processes of *in vivo*-grown *M. tuberculosis* should facilitate development of new antimicrobial agents, especially if they can be designed to disrupt metabolic pathways specific to the tubercle bacilli. Identification of antigens produced by *M. tuberculosis* upon initial interaction within the host and during subsequent growth within an infected individual may aid in development of a more effective vaccine, especially if the modes of presentation of such antigens to the human immune system can be established.

Historically, the occurrence of bacterial physiological processes have been deduced from detection and isolation of specific enzymes for given metabolic pathways or by uptake and conversion of radioactively labeled precursors into specific products, or by isolation of specific components from *in vivo*-grown organisms; an approach also used in the study of *M. tuberculosis*. A caveat of this approach is that these kinds of studies necessitate that the *in vivo*-grown bacteria are grown in animal models rather than the natural host for *M. tuberculosis*, which is humans. Although some studies have been done with *M. tuberculosis* obtained from human patients (e.g., from lung biopsies or surgical resection of lungs), generally the number of bacilli that can be obtained from such specimens is too small to permit isolation of adequate amounts of metabolic enzymes and/or intermediates of metabolic pathways. Early studies by Heplar *et al.* (1954) and Guy *et al.* (1954) comparing metabolic pathways operating in virulent (H37Rv) and avirulent (H37Ra) strains of *M. tuberculosis* grown *in vitro* suggested that the virulence of H37Rv was due to its ability to respire and grow at the lower oxygen tension found in host tissues, compared to the inability of H37Ra to grow in such conditions. Ramakrishnan *et al.* studied utilization of glucose in *in vitro*-grown H37Rv and H37Ra (Ramakrishnan *et al.*, 1962). These investigators demonstrated that the avirulent H37Ra strain utilizes the anaerobic Embden-Meyerhof pathway and the aerobic pentose phosphate pathway equally to catabolize glucose, whereas H37Rv catabolizes glucose predominantly via the Embden-Meyerhof pathway. These studies provided some of the strongest evidence suggesting that *M. tuberculosis* is a facultative aerobe, in contrast to the earlier concept that this organism is a strict aerobe.

Complementing these *in vitro* studies, Segal and Bloch (1956) compared the oxidative responses *in vitro*- and *in vivo* (mouse)-grown H37Rv to a number of substrates and demonstrated significant differences between them. Segal and Bloch recovered *in vivo*-grown *M. tuberculosis* bacilli from the lungs of CF1 (outbred) mice infected 18-23 days previously. These were compared to *M. tuberculosis* grown for 10 days as a submerged culture in Dubos-Middlebrook Tween-albumin broth (Segal and Bloch, 1956). Whereas the *in vitro*-grown bacilli exhibited positive respiratory responses to glycerol, glucose, and a variety of glycolytic intermediates, the *in vivo*-grown bacilli exhibited a much lower rate of metabolic activity, as measured by stimulation of endogenous respiration. Interestingly, there were no differences between the two types of bacilli in respiratory responses to fatty acids such as heptanoic acid, octanoic acid, and oleic acid (Segal and Bloch, 1956).

Several other groups of investigators demonstrated differences between *in vitro-* and *in vivo* (mouse)-grown *M. tuberculosis.* Bekierkunst and Artman detected the presence of an inhibitor of murine lactate dehydrogenase in mice infected with H37Rv, but not in uninfected mice (Bekierkunst and Artman, 1960). These investigators demonstrated that the production of the inhibitor was a consequence of infection with *M. tuberculosis,* although the inhibitor was of host origin. The mechanism of stimulation of production of the inhibitor was not determined in these studies. Other investigators have demonstrated differences in lipid content and in other surface structures of *in vivo-* and *in vitro*-grown *M. tuberculosis* (Kanei, 1967; Segal and Miller, 1965; Pokorny, 1962). Still other investigators detected enzymes of the tricarboxylic acid (TCA) cycle in *in vivo*-grown *M. tuberculosis* (Segal, 1962, Ratledge, 1982), as well as isocitrate lyase, the enzyme that catalyzes the first step of the glyoxalate bypass. These studies suggest that the two pathways may be operating simultaneously, serving both respiratory and biosynthetic functions.

These early pioneering studies have provided the basis for important concepts about growth and survival of *M. tuberculosis* in infected individuals. The concept that *M. tuberculosis* is a facultative aerobe rather than a strict aerobe is consistent with the ability of this organism to persist in foci of the lungs or other parts of the body where oxygen levels are low (Sever and Youmans, 1957).

After a hiatus of 25-30 years, interest in the *in vivo* metabolic capabilities of *M. tuberculosis* has undergone a resurgence with consequent realization of the great gaps in our knowledge regarding the tubercle bacilli. This interest is, in part, fueled by the increase in the incidence of tuberculosis in developing countries. The HIV/AIDS epidemic and associated high proportion of *M. tuberculosis* co-infection, along with an increase in the prevalence of drug resistant strains, often a result of poor compliance associated with long-term, multi-drug therapy, have presented additional challenges in dealing with this disease. A better understanding of the pathogenesis of infection is essential to aid the process of drug discovery and vaccine development. Moreover, recent developments in the areas of genetics and molecular biology have provided additional ways in which to study aspects of *M. tuberculosis* intracellular survival and growth that were not possible in the 1950s, 1960s and 1970s.

Two major philosophical approaches have been employed in these more recent studies: a) analysis of specific genes and gene products, chosen because of hypothetical or real involvement with intracellular survival and/or growth and b) global analyses of gene expression.

The analysis of *M. tuberculosis* specific gene expression and protein synthesis in the host requires basic components. First, it is necessary to define the model system to perform the study. Model systems include tissue culture, either of primary or immortalized cells and typically either of human or murine origin, animal models, typically mice or guinea pigs, and studies of human lung biopsy tissues or sputa from human pulmonary tuberculosis patients. *In vitro* culture models, such as culture under reduced oxygen tension, heat shock, cold shock, oxidative stress, nutrient deprivation and reduced pH are also used as surrogates for the microenvironments that are believed to be encountered by the tubercle bacilli. The relative strengths and complications of these models are discussed in other chapters of this book.

Next, it is necessary to decide if the study is to be qualitative or quantitative, and whether the study is to compare *M. tuberculosis* cultured under two different conditions or to compare expression patterns within a single model, with variations in factors such as time or temperature. It is necessary to select both internal and external controls for *M. tuberculosis* gene expression; a challenging task.

Growth parameters of the mycobacterium to be utilized in these gene expression studies should be carefully examined as the method of culture, stage of growth, strain of *M. tuberculosis*, and choice of medium and additives can have profound effects on *M. tuberculosis* gene expression.

Analysis of Specific Genes and Gene Products

Introduction

Genomic comparisons and sequence homology searches are powerful tools, which have been successfully used for identification of open reading frames (ORFs) that encode potential virulence factors. The rational targeting of genes for further study has been greatly aided by sequencing of the genomes of several species of *Mycobacterium* as well as many other pathogenic bacteria. The genomes of six species of *Mycobacterium* have been or are currently being sequenced, thus enabling comparative genomic studies between human and animal pathogens, as well as between obligate and opportunistic pathogens. Comparative genomics has also allowed presumptive identification of genes encoding enzymes required in the synthesis of both protein and non-protein components of *M. tuberculosis* (Cole *et al.*, 1998; Kato-Maeda *et al.*, 2001). The surface of *M. tuberculosis* contains abundant complex acylated and non-acylated carbohydrates and components recognized as major factors in pathogenicity and host response (Brennan and Nikaido, 1995; Strohmeier and Fenton, 1999; Lee *et al.*, 1996; Aspinall *et al.*, 1995; Ortalo-Magne *et al.*, 1996; Lemassu and Daffé,1994; Vercellone *et al.*, 1998). These complex outermost components of the bacterial cell envelope are the first to contact host cellular constituents and are known to play roles in phagocytosis as well as intracellular events (Brennan and Nikaido, 1995; Chan, 1991).

A number of investigators have conducted elegant studies on specific genes or classes of genes of *M. tuberculosis* whose products are similar to those of proteins that have been shown to be involved in the pathogenesis of other intracellular pathogens. Other investigators have chosen to analyze *M. tuberculosis* genes whose products have been shown to play some role in intracellular growth, or that are proteins involved in the acquisition of essential growth factors such as iron, or are proteins that are involved in essential biosynthetic pathways.

Targeted gene study may involve the analysis of protein products, or specific gene expression by quantification of mRNAs or by the use of reporter constructs. Recent advances in genetic manipulation of *M. tuberculosis* (in particular, advances in the methodology for allelic-exchange mutagenesis) have provided the tools for identification of virulence factors via fulfillment of Koch's molecular postulates (Falkow, 1988). Below, we highlight methods that utilize mRNA, rather than protein, to assess gene expression. The analysis of specific *M. tuberculosis* proteins in the host would utilize similar strategies; however, the relative lack of sensitivity and reagents such as specific antibodies adds an increased level of difficulty to this approach.

Approaches Using Quantitative Measures of Gene Transcription

The *M. tuberculosis* genome encodes 13 sigma factors and 100 repressors and activators (Cole *et al.*, 1998), suggesting unusually high transcriptional regulation of gene expression, particularly for a bacterium that infects only one major host and has no known free-living stage.

Bacteria typically have transcription coupled with translation, with transcriptional regulation the predominant mechanism of obtaining the profile of proteins present within a bacterium at any particular time. Variations in gene expression may also correlate with virulence. *M. bovis* BCG is described as both a deletion and a regulatory mutant (Monahan *et al.*, 2001). Gene expression of six homologous ORFs was decreased in the attenuated *M. tuberculosis* strain H37Ra as compared to its virulent counterpart, H37Rv (Rindi *et al.*, 2001) and alterations in gene expression have been observed that correspond to the drug resistance profile of *M. tuberculosis* strains (Baulard *et al.*, 2000). Bacteria also have relatively short half-lives for mRNAs, estimated at ~3 min, on average. Similarly short half-lives have been observed for some mycobacterial mRNAs, with half-lives of approximately 5 min measured for *fbpB*, and *aroB* mRNAs in log phase cultures of *M. tuberculosis* (DesJardin, unpublished observations). The slow growth rate of the tubercle bacilli leads to a much longer half-life for 16S rRNA and half-lives of many or all RNAs in stationary phase cultures are suspected to be increased (DesJardin, unpublished). Thus, the possibility of post-transcriptional regulation in the form of mRNA stability playing a larger regulatory role for the slow-growing Mycobacterium species cannot be ruled out. In fact, recent studies by Unniraman *et al.* (2002) indicate that stability is important in the regulation of mRNA for DNA gyrase in *M. smegmatis.*

In addition, the proteins specified by many of the H37Rv genome ORFs have not been detected, either due to technical limitations in the methods for detecting proteins, or because of the possibility that not every ORF actually leads to production of an individual protein, thus, measurement of mRNA provides an initial indication that a protein is made. The measurement of mRNA has served as a useful surrogate to follow gene expression and approximate the relative abundance of proteins. This is a particularly important method when samples are limited, as for bacilli within macrophages, and when no antibody to the protein of interest is available.

Quantitative reverse transcriptase-PCR (RT-PCR) was initially laborious and accurate and reproducible values were difficult to obtain, particularly over a wide range of RNA levels. The introduction of machines and methods for quantitative RT-PCR in "real-time" has been instrumental in the practical application of mRNA quantification of very limited numbers in tubercle bacilli in a wide range of samples. These systems have been shown to be accurate and precise over a large dynamic range, allowing for high throughput with no post-amplification processing required, thus minimizing the chance of false-positive values as a result of contamination (Desjardin *et al.*, 1998, 1999). "Real-time" PCR is possible due to the development of fluorescent probes, real-time instruments, and the software necessary to perform the analyses. Several such platforms currently exist (ABI 7700 and ABI 7000 (Applied BioSystems), iCycler (BioRad), SmartCycler (Cephid), LightCycler (Roche). The probes consist of oligonucleotides that are specific for a particular DNA sequence and have either been singly or dually labeled with a fluorophore(s). Probes typically utilize Fluorescence Resonance Energy Transfer (FRET) and include 5'exonuclease assay (e.g. TaqMan), in addition to hybridization probes such as Molecular Beacons and LightCycler probes.

Biological materials contain complex mixtures of RNA from both the host and other microbial sources. This RNA can present a confounding factor in the analysis of specific *M. tuberculosis* mRNA targets. Particular care must be taken to prevent cross-reaction with other molecules that may not be apparent when using RNA from cultured *M. tuberculosis.*

To quantify a particular *M. tuberculosis* mRNA in RNA isolated from sputum samples, the RNA is initially reverse transcribed to cDNA using a primer specific for the target gene (DesJardin, 2000). The use of a specific RT primer results in greater yield of a particular product. The majority of RNA isolated from a biological specimen is likely to be RNA from host cells and other microbes that may be present in the specimen. Additionally, the majority of total RNA is comprised of ribosomal RNA. If random primers are used for the RT reaction, there is competition between different priming events for the limited amount of reverse transcriptase enzyme present, whereas a specific primer can target the enzyme to the specific message.

Sensitivity and specificity are enhanced by use of conditions that favor formation of cDNA from the target message. In some instances, the use of a short RT primer located 50 to 200 bases 3' to the region amplified during subsequent PCR is necessary (DesJardin, unpublished observations). This shorter primer is constructed to have a lower melting temperature (T_m) that allows for added targeting during the RT reaction. A second set of primers that amplify a region 5' to the RT primer is then used in the subsequent PCR of the cDNA. The PCR primers are designed to have a higher T_m than the shorter RT primer. Therefore, the RT primer will not form PCR products at the annealing temperature used during PCR.

To monitor for DNA contamination of the RNA extract, reactions must also be performed which contain all RT reagents except the RT enzyme. RT reactions and mock RT reactions on samples, *M. tuberculosis* control RNA, and *in vitro* transcript RNAs should all be performed simultaneously.

As with any nucleic acid amplification reaction, precautions must be taken to avoid false positive results in addition to those precautions used for handling RNA. PCR is an extremely sensitive and powerful technique, with the potential to amplify a single molecule of input target, resulting in a positive reaction. For this reason, precautions must be taken to avoid contamination during each phase of specimen collection and processing. Protocols for the proper collection and handling of specimens must be generated and followed with frequent checks for false-positive reactions. The preparation of specimen RNA should be done in a biological safety cabinet that is not used for routine culture work. Contact of the specimen with any potential source of *M. tuberculosis* bacteria, such as BACTEC bottles, LJ slants, and other processed or unprocessed samples must be avoided. The use of sterile disposable plastic tubes, disposable gloves and gowns, and aerosol resistant barrier pipette tips is recommended. PCR reactions should be assembled in a dedicated UV-equipped, dead-space cabinet. If at all possible, separate areas should be maintained for sample preparation, PCR master mix preparation, and template addition to PCR reactions. All surfaces and racks should be decontaminated with 10% bleach. Since every

PCR reaction can generate up to 1 billion copies of the target molecule, the UDG decontamination system is highly recommended to help control amplicon contamination from previous reactions (DesJardin, 2000).

Studies defining the role of an iron-responsive DNA-binding protein of *M. tuberculosis*, IdeR, by Gold *et al.* included quantitative RT-PCR analyses using molecular beacon probes with the ABI 7700 system (Gold *et al.*, 2001). IdeR belongs to a family of proteins (the DtxR family), which have been shown to regulate iron acquisition (Gold *et al.*, 2001). The importance of the ability to obtain iron by bacterial pathogens in the iron-limited environment of mammalian hosts is well documented (Bullen, 1981; Ratledge, 1982; Wheeler and Ratledge, 1994) and pathogenic mycobacteria produce siderophores known as mycobactins to aid in iron acquisition. Gold *et al.* (2001) utilized *in silico* analysis to identify a number of genes which possessed sequences homologous to the IdeR binding sequence (IdeR boxes) from the *M. tuberculosis* H37Rv genome sequence (Cole *et al.*, 1998). After initial *in vitro* gel retardation and DNA footprinting experiments demonstrating that IdeR did indeed bind to the IdeR boxes for four of these genes, Gold *et al.* (2001) employed quantitative RT-PCR technology to establish that three of these genes (*mbtB*, *mbtI*, and *rv3402c*) were induced at both 24 and 72 h after infection of the human monocyte-like cell line, THP-1. The genes *mbtB* and *mbtI* encode enzymes involved in the biosynthesis of mycobactin, whereas *rv3402c* encodes a protein of unknown function, but which exhibits significant amino acid similarity (> 40%) to three types of enzymes: an aminotransferase, a dehydratase, and an enzyme involved in perosamine/O antigen biosynthesis (Gold *et al.*, 2001). The results of Gold *et al.* (2001), together with those of DeVoss *et al.* (2000; see below) provide strong evidence that *M. tuberculosis* actively synthesizes mycobactin and acquires iron during growth in human macrophages.

Studies using the ABI Prism 7700 Sequence Detection System have been used successfully to quantify *M. tuberculosis* mRNA in biological samples (Desjardin *et al.*, 1999; Hellyer *et al.*, 1999b), macrophage lysates (Wilkinson *et al.*, 2001), and a variety of *in vitro* culture systems (Hellyer *et al.*, 1999a; Desjardin *et al.*, 2001). Investigators have used quantitative RT-PCR to assess expression of *fbpB* by *in vivo*-grown *M. tuberculosis* (DesJardin *et al.*, 1999; Hellyer *et al.*, 1999b). The *fbpB* gene encodes a mycolyl transferase which is known as Antigen 85B or the α-antigen and which is involved in cell wall biosynthesis. DesJardin *et al.* (1999) and Hellyer *et al.* (1999b) employed quantitative RT-PCR, using the ABI 7700 System and TaqMan probes. In addition to traditional enumeration of acid-fast bacilli (AFB) and viable bacterial CFU, amounts of *M. tuberculosis* transcripts were also quantified in sputum samples of pulmonary TB patients. Specific levels of *M. tuberculosis* mRNA, rRNA and DNA, representing molecules with three different inherent molecular stabilities, were measured to determine which would most appropriately correspond with standard methods to assess response to treatment. Sputum samples were obtained from patients at the time of diagnosis of tuberculosis and at several times after chemotherapy had been initiated (DesJardin *et al.*, 1999). By comparing the levels of *fbpB* mRNA to levels of *M. tuberculosis* DNA (in the form of IS6110) and to viable cell counts, these investigators demonstrated that quantitative determination of the number of RNA transcripts of a gene involved in an essential biosynthetic pathway was an accurate and more rapid way than currently available methods to assess the effectiveness of chemotherapy in the treatment of active tuberculosis.

The use of nested RT-PCR as a measure of viability of *M. tuberculosis* has been used previously (Jou *et al.*, 1997). However, this is the first application of RT-PCR to sputum to measure the bactericidal effect in subjects undergoing chemotherapy for tuberculosis. Quantification showed that the average number of molecules of Antigen 85B mRNA measured in *M. tuberculosis* cells is ~5 molecules/cell at mid-log phase using *in vitro* cultured cells and ~0.1 molecules/cell *in vivo*. Similarly, 16S rRNA levels averaged 5000 molecules/cell *in vitro* and ~800 molecules/cell *in vivo*. The differences between amounts recovered from *in vitro*-grown *M. tuberculosis* and from *M. tuberculosis* recovered from sputum may be due to incomplete recovery of RNA or loss of RNA between the time of sputum collection and processing. It is also possible that there are reduced numbers of ribosomes and certain messenger RNAs in bacteria that are not growing under ideal culture conditions such as an organism would experience in an immunocompetent host.

Wilkinson *et al.* (2001) used the same quantitative RT-PCR approach to assess expression of *fbpB* and another gene, *hspX* (Acr), by *M. tuberculosis* growing in infected human monocytes in culture. The *hspX* gene encodes a 16 kDa protein which has homology to the ubiquitous α-crystallin family of chaperones and which has been shown to be up-regulated for expression during stationary phase growth of *M. tuberculosis* and under *in vitro* growth conditions of low oxygen (DesJardin *et al.*, 2001). Wilkinson *et al.* showed that the amounts of *fbpB* mRNA increased 14.6-fold in *M. tuberculosis* during the first 24 h after infection of human macrophages derived from peripheral blood monocytes (PBMCs). In contrast, the level of *hspX* mRNA decreased 14.3-fold during the same period. The changes in levels of *fbpB* and *hspX* mRNAs were determined relative to the amounts of 16S rRNA present in *M. tuberculosis* at the time of infection and 24 h later (Wilkinson *et al.*, 2001).

Several other groups have employed RT-PCR to identify specific genes that are expressed by *M. tuberculosis* strain Erdman during growth in macrophages or in macrophage-like cell lines. Jensen-Cain and Quinn studied expression of the extracytoplasmic function (ECF) sigma factor, SigE, in *M. tuberculosis* after infection of human PBMC-derived macrophages, compared to expression in *M. tuberculosis* after infection of the human type II pneumocyte cell line, A549 (Jensen-Cain and Quinn, 2001). Although these investigators were able to demonstrate expression of *sigE* by *M. tuberculosis* as early as 6 h after infection of macrophages, they were unable to detect expression of this gene at any time when *M. tuberculosis* was grown in A549 cells. Expression of *sigE* by *M. tuberculosis* was detected throughout the 5-day infection experiments, suggesting that this ECF sigma factor is needed for expression of *M. tuberculosis* genes involved in intracellular growth, although these investigators did not identify the genes regulated by SigE (Jensen-Cain and Quinn, 2001). Manganelli *et al.* (2001) used DNA microarray analysis to define the genes regulated by SigE, by comparing wild-type *M. tuberculosis* to a *sigE* mutant (see Section III.C.5).

Av-Gay *et al.* (1999) studied expression of a gene encoding a serine/threonine protein kinase of *M. tuberculosis (pknB)* in both murine macrophages and in alveolar macrophages from a tuberculosis patient. Determination of the complete nucleotide sequence of the *M. tuberculosis* H37Rv genome has revealed the presence of at least 11 genes encoding serine/threonine protein kinases (Cole *et al.*, 1998). Interestingly, these *M. tuberculosis* protein kinases exhibit significant amino acid similarity, as well as functional similarity, to eukaryotic serine/threonine protein kinases (Av-Gay *et al.,* 1999). Thus, the *M. tuberculosis* proteins must be considered as potentially capable of disrupting host signal transduction mechanisms by competing with host enzymes. In *in vitro* studies, Av-Gay *et al.* initially demonstrated that PknB is a functional kinase that is autophosphorylated on serine/threonine residues and that the protein is able to phosphorylate the eukaryotic peptide substrate, myelin basic protein (Av-Gay *et al.*, 1999). Moreover, by RT-PCR, these investigators found that *pknB* was expressed by *M. tuberculosis* at 24 and 72 h after infection of the murine cell line, J774.2, and more importantly, by *M. tuberculosis* grown for some unspecified period of time in alveolar macrophages obtained by bronchoalveolar lavage of a resected lung from a patient with active tuberculosis (Av-Gay *et al.,* 1999).

Mariani *et al.* (2000) employed RT-PCR to study expression of 14 *M. tuberculosis* strain H37Rv genes, growing in laboratory broth medium (Sauton's) and in human macrophages. These investigators demonstrated that five of the 14 genes were expressed by *M. tuberculosis* growing in both environments (MT10Sa gene, 35 kDa protein gene, *ahpC, sigF,* and *katG*), four genes were only detected in *M. tuberculosis* growing in broth (*fbpB, fbpC, rpoV,* and *esx*) and five genes were only expressed when *M. tuberculosis* was growing in macrophages (*fbpA, rpoB, pab, invA, invB*) (Mariani *et al.*, 2000). It is surprising that these investigators did not detect expression of *fbpB* or *esx* (which encodes the antigen ESAT-6) during growth of *M. tuberculosis* in macrophages, since other investigators have demonstrated expression of *fbpB* in macrophage-grown *M. tuberculosis* (Wilkinson *et al.*, 2001) and since strong immune responses are elicited to both FbpB and ESAT-6 by tuberculosis patients, implying that the genes encoding these proteins are expressed by *M. tuberculosis* during infection. Nevertheless, these studies are notable in that the authors employed the RT-PCR technique to assess expression of more than two or three genes by *M. tuberculosis* grown in human macrophages.

Several of these investigators have expanded these results by studying gene expression by two *M. tuberculosis* strains, H37Rv (a virulent laboratory strain) and CMT97 (a recent clinical isolate), during growth in human PBMC-derived macrophages (Cappelli *et al.*, 2001). Not only did the authors compare gene expression between these two strains of *M. tuberculosis*, but they also compared expression during growth in resting macrophages and in macrophages that had been activated by addition of interferon-gamma (IFN-γ) and LPS to the macrophage culture. In these studies, the investigators assessed expression of 11 *M. tuberculosis* genes by RT-PCR, as well as assessing expression of 8 macrophage genes encoding cytokines. Interestingly, H37Rv expressed more genes (9 of the 11) in resting macrophages than did CMT97 (4 of 11), whereas the opposite occurred in the IFN- γ/LPS-activated macrophages; i.e., the number of genes expressed by CMT97 was greater (10 of 11) than the number expressed by H37Rv (2 of 11) (Cappelli *et al.*, 2001). Although H37Rv did not express many of the genes selected for study during growth in activated macrophages, this strain exhibited a greater than 10-fold increase in cell counts during the period of culture in activated macrophages. Another interesting finding was that activated macrophages infected with H37Rv apparently expressed higher levels of 7 of the 8 cytokine mRNAs than did resting macrophages infected with this strain, whereas the opposite results were obtained when comparing cytokine mRNA levels by activated and resting macrophages infected with the clinical isolate, CMT97. The results reported in this study are intriguing in that they are in disagreement with previously published studies on survival of *M. tuberculosis* in activated vs. resting macrophages. Moreover, in contrast to the results reported by Mariani *et al.* (2000), H37Rv did express *fbpB* and *fbpC* in resting macrophages, although the authors did not detect expression of *esx* by H37Rv in either resting or activated macrophages, in agreement with the previous studies of Cappelli *et al.* (2001).

Approaches Using Qualitative Measures of *M. tuberculosis* Gene Expression

The study of qualitative gene expression in the host often follows the same parameters for quantitative studies. This approach is advantageous when the equipment for quantitative evaluation of gene expression is not available and is typically less expensive than analyses utilizing real-time PCR platforms. Qualitative studies, in the form of *ex vivo* and *in situ* hybridization, can have the distinct advantage of localizing mRNA expression to particular bacilli, can allow for co-localization of both mycobacterial and host transcripts, and can place *M. tuberculosis* expression within the host microenvironment. Studies employing electron microscopy have the potential to define subcellular target location. The disadvantages of qualitative studies largely center upon a decrease in sensitivity and occasionally specificity.

Examination of AFB isolated from lung granulomas of pulmonary tuberculosis patients was pioneered in the early 1950s, with analyses of granuloma types and the ability to culture *M. tuberculosis* from these lung resections (reviewed in Wayne and Sohaskey, 2001). Recently, these original observations have been expanded to the evaluation of *M. tuberculosis* gene expression in the granulomas from tuberculosis patients following surgical resection of the lung (Fenhalls *et al.*, 2002).

Fenhalls *et al.* (2002) performed a comparative analysis of 191 necrotic and non-necrotic granulomas from nine culture-positive pulmonary tuberculosis patients. The presence of AFB by Ziehl-Nielson (ZN) staining, RNA polymerase beta subunit *(rpoB)* mRNA, *M. tuberculosis* DNA (targeting the repetitive PGRS gene family), both by *in situ* hybridization, and the presence of CD68 positive cells (indicating macrophage lineage) by immunohistochemistry, were analyzed in paraffin-embedded sections. The pattern of specific gene expression was found to be compartmentalized within the granuloma, suggesting several metabolic states of *M. tuberculosis* in this environment.

The transcript for *rpoB* was associated with bacilli within macrophages and only found in the periphery of necrotic lesions and throughout non-necrotic granulomas. Earlier studies had found *M. tuberculosis* DNA associated with the alveolar and interstitial macrophages in lung tissue from individuals without histological evidence of tuberculous lesions (Hernandez-Pando *et al.*, 2000).

Fenhalls *et al.* further examined 7 pulmonary tuberculosis patients for mRNA for *icl, rv2557, narX, iniB* and *kasA,* and for *M. tuberculosis* DNA (Fenhalls *et al.*, 2002). All patients had evidence for these targets; however, the number of granulomas with evidence for *narX* expression was very low, with 0-2 granulomas positive per patient out of a total of 122 granulomas examined. These investigators detected mRNAs corresponding to each of the genes in non-necrotic granulomas, but only in one of the zones of the necrotic granulomas (the outer lymphocyte cuff). Only transcripts from *rv2557, kasB,* and *iniB* were detected in the transition zone of necrotic granulomas and no evidence of expression of any of these genes was detected in the central regions of the necrotic granulomas (Fenhalls et al., 2002). It will be interesting to understand the limits of detection of this assay system and to hopefully expand these intriguing observations to confocal microscopy for a 3-dimensional picture.

Use of Reporter Constructs to Identify Promoters Induced During *In Vivo* Growth

Gene reporter constructs measure the abundance of an *M. tuberculosis* gene product by fusing the RNA polymerase-binding region of the promoter of a specific gene to the coding sequence of a substitute protein that can be readily detected, such as β-galactosidase (*lacZ*), green fluorescent protein (GFP), or luciferase. This is a powerful tool for mapping the functional domains of promoters and readily ascertaining induction of gene expression. These studies may also be useful as a mechanism of comparative quantification of *M. tuberculosis* gene expression. To provide accurate information using this approach, the precise identification of the *M. tuberculosis* promoter must be known, the promoter must direct transcription of a single gene, and the promoter must be able to function appropriately when removed from its genomic context. An additional caveat is that the half-lives of reporter proteins may be different from those of the *M. tuberculosis* proteins for which they are acting as markers. A weakness in the design of most of the studies using reporter constructs for analyzing *M. tuberculosis* gene expression is that investigators have cloned the promoters of interest into plasmids harboring the promoterless reporter gene. Although the plasmids used for these constructs were present in low numbers (2-5 copies per chromosome equivalent) in mycobacterial strains, these plasmids were introduced into wild-type strains. Thus, expression of the wild-type gene was occurring simultaneously with expression of the reporter gene from the cloned promoter. Promoter excess may very well titrate out regulatory proteins (transcriptional repressors and/or activators), thereby causing false expression or repression of the gene of interest.

The attractiveness of reporter gene technology is that detection of gene expression is easier and more rapid than conducting functional assays on gene products. Tyagi *et al.* (2000) have recently reviewed the types of reporter constructs that have been used in mycobacteria, most of which have been for analysis of gene expression in *in vitro* conditions. The green fluorescent protein (GFP) of the jellyfish, *Aequoria victoria* (Valdivia *et al.*, 1996) has been the reporter amenable to studying gene expression in mycobacteria growing in cultured macrophages. A number of investigators have constructed plasmids with GFP fusions with promoters of specific genes of *M. tuberculosis* and then introduced the plasmids into *M. bovis* BCG for analysis (Kremer *et al.*, 1995; Dhandayuthapani *et al.*, 1995; Via *et al*, 1996). When these GFP fusions were studied for expression in macrophages, these investigators used the murine cell line, J774A.1. Although this model system is convenient to use, it remains to be seen whether or not expression of *M. tuberculosis* genes is regulated in the same way in *M. tuberculosis* and in BCG (see below). More importantly, one should be exceedingly cautious about interpreting *M. tuberculosis* gene expression during infection of J774A.1 cells, which are an immortalized cell line of murine origin, and in which *M. tuberculosis* grows very differently than in human macrophages. *M. tuberculosis* may express different genes after phagocytosis by J774A.1 cells and regulation of mycobacterial gene expression may be very different in human macrophages.

Springer *et al.* (2001) reported differences in expression patterns of GFP fusions to the promoter of the *M. tuberculosis* alkyl hydroperoxide reductase (*ahpC*) gene in *M. tuberculosis, M. bovis,* and *M. bovis* BCG. These differences occurred during growth of the bacterial strains *in vitro* and in J774A.1 cells. In *M. bovis* BCG, *ahpC* is expressed constitutively, both in broth and in J774A.1 cells, whereas expression of this gene was repressed in virulent *M. tuberculosis* and *M. bovis* strains growing in broth and in *M. tuberculosis* growing in J774A.1 cells (Springer *et al.,* 2001).

McKinney *et al.* (2000) constructed a GFP fusion with the gene encoding isocitrate lyase (*icl*) from *M. tuberculosis* CDC 1551. These investigators demonstrated that the levels of Icl-GFP increased initially after infection of both resting and activated murine bone marrow-derived macrophages but remained higher throughout the infection when the recombinant *M. tuberculosis* expressing Icl-GFP were inside activated murine bone marrow-derived macrophages compared to expression when the bacteria were in resting murine macrophages. Although these studies were done in murine macrophages, the investigators compared expression of *icl-gfp* in both wild-type *M. tuberculosis* and in a Δ*icl* mutant of *M. tuberculosis* and thus obtained additional expression data compared to other studies using GFP reporter constructs (McKinney *et al.,* 2000).

Via *et al.* (1996) constructed a fusion between the promoter of the *mtrA* response regulator gene of *M. tuberculosis* and *gfp* and measured GFP levels in *M. bovis* BCG harboring this fusion after infection of J774 cells. These investigators demonstrated that the *mtrA* promoter was induced in BCG after infection of J774 cells (Via *et al.,* 1996). However, Zahrt and Deretic (2000) used this same P$_{mtrA}$-*gfp* plasmid to compare expression of P$_{mtrA}$-*gfp* in H37Rv and in *M. bovis* BCG after infection of J774A.1 cells (Zahrt and Deretic, 2000). Interestingly, these experiments indicated that *mtrA* is expressed constitutively in *M. tuberculosis*, but that its expression is induced in BCG after phagocytosis by J774A.1 cells (Zahrt and Deretic, 2000). Zahrt and Deretic also analyzed expression of P$_{mtrA}$-*gfp* as a function of time after infection of human PBMC-derived macrophages and demonstrated that *mtrA* is also constitutively expressed by H37Rv in human macrophages. As a control, Zahrt and Deretic monitored expression of P$_{hsp60}$-*gfp* in BCG and in H37Rv after infection of J774A.1 cells and in H37Rv after infection of human macrophages; *hsp60* was constitutively expressed by both bacterial strains and in both kinds of cells (Zahrt and Deretic, 2000).

In another study, Zahrt and Deretic (2001) constructed *gfp* fusions with promoters for 10 putative response regulator genes, identified from the sequence of the *M. tuberculosis* H37Rv genome. These fusions, plus the P_{mtrA}-*gfp* fusion they had previously constructed and studied, were analyzed for patterns of expression during *in vitro* growth of *M. bovis* BCG and *M. tuberculosis* H37Rv in Middlebrook 7H9 broth and during growth in J774A.1 cells and in H37Rv, during growth in human monocytes. The *phoP (rv0757)* and *rv0818* fusions were constitutively expressed by both BCG and H37Rv in J774A.1 cells, although both were expressed at considerably higher levels in BCG than in H37Rv (Zahrt and Deretic, 2001). These two fusions, as well as the P_{mtrA}-*gfp* fusion were also constitutively expressed in H37Rv growing in human monocytes. In agreement with their previous results, *mtrA* expression in BCG growing in J774A.1 cells was induced as a function of time. In BCG, the P_{rv0981}-*gfp* and the P_{rv3143}-*gfp* fusions were also induced as a function of time after infection of J774A.1 cells, but expression of these fusions was not detected in H37Rv growing in either J774A.1 cells or in human monocytes. On the other hand, expression of $P_{rv3133c}$-*gfp* was detected in H37Rv grown for 7 days in human monocytes, but expression of this fusion was not detected at any time in BCG or in H37Rv growing in J774A.1 cells (Zahrt and Deretic, 2001).

These results provide some of the strongest data demonstrating the significant differences between gene expression in BCG compared to virulent *M. tuberculosis* and between mycobacterial gene expression in J774A.1 cells compared to human monocytes. The results of Zahrt and Deretic also suggest that relatively few of the putative two component response regulators appear to be expressed in H37Rv during growth in human monocytes, at least at the times when expression was tested in these studies. However, the proteins of two component regulatory systems would be expected to be present in low amounts in bacterial cells grown under non-inducing conditions and expression of these genes would be expected to be tightly regulated. Since the GFP fusions studied by Zahrt and Deretic (2001) were in *M. tuberculosis* strains that possessed wild type genes for each of the response regulators, one must consider that, if expression of some of the response regulators is regulated by binding of a repressor or an activator, one could get erroneous results (either expression or repression) due to titration of the repressor or activator proteins. This is an issue that must be considered when analyzing gene expression using transcriptional reporter constructs.

Use of reporter gene technology for analyzing gene expression in *M. tuberculosis* (and other bacteria) has several attractive features: the assays are sensitive, reasonably quick, and results can be quantified. Importantly, regulation of expression of a gene of interest may be studied whether or not a functional assay for the gene product is available (Tyagi *et al.*, 2000). On the other hand, there are drawbacks, or potential weaknesses, some of which have been noted above and must be considered when studying genes whose expression is tightly regulated. Another point to be considered is that the stability of the transcript or gene product produced from a reporter fusion construct may be different from the natural transcript or product of the gene, which could affect the function of the gene product in *M. tuberculosis* (Tyagi *et al.*, 2000).

Construction of Genetic Mutants to Assess Roles of Specific Genes

Genomic comparisons and sequence homology searches are powerful and useful tools for selection of gene targets for further study. However, homology alone does not necessarily dictate either function or a role in virulence. Another approach that has been used to evaluate the roles of specific *M. tuberculosis* gene products in survival and growth of the bacilli in the human host has been to mutagenize the genes of interest and to study the effects of such mutations on mycobacterial growth *in vivo*. There are two basic approaches for the generation of genetically altered *M. tuberculosis:* transposon mutagenesis or allelic exchange. Transposon mutagenesis allows for the insertion of small segments of DNA that have been randomly integrated throughout the genome. The transposon insertion typically inactivates the gene into which it has integrated; however, it is possible to increase expression of genes at the site of transposition as well. Libraries of these random mutants may then be studied for altered growth characteristics in the model of choice. Transposon mutant libraries have been constructed in three different strains of *M. tuberculosis* (MT103 [Pelicic *et al.*, 1997], Erdman [Bardarov *et al.*, 1997], and H37Rv [McAdam *et al.*, 2002]), using derivatives of the *M. smegmatis* IS1096 transposon. The second method for generating mutants of *M. tuberculosis* is allelic exchange by homologous recombination. There have been several recent reviews on this subject (Ainsa *et al.*, 2001; Glickman and Jacobs, 2001, Clark-Curtiss and Haydel, 2003). Experimental data derived from these studies are influenced by many of the factors mentioned above, such as test model and culture conditions, in addition to inactivation of the gene of interest without effecting additional gene expression through polar mutations that could either enhance or disrupt additional genes beyond the intended target. It is also essential that any phenotype attributed to allelic-exchange gene inactivation be genetically complemented by introduction of a single copy of the wild-type gene into the mutant strain.

There are many examples of the successful application of these methods, frequently demonstrating that inactivation of a gene leads to attenuation of the mycobacterium. Transposon mutagenesis was used by Bange *et al.* (1996) to demonstrate that *Mycobacterium bovis* BCG, in which the *leuD* gene had been insertionally inactivated by the transposon Tn5367, was unable to grow in the human macrophage-like cell line, THP-1.

In studies characterizing the gene for the 16 kDa α-crystallin chaperone, *hspX (rv2031c,* Acr), Yuan *et al.* (1998) constructed an allelic-exchange deletion mutant of *M. tuberculosis* H37Rv, where the *hspX* gene in the chromosome was replaced with a hygromycin-resistance gene (Hyg[R]). Although the *hspX* mutant grew at rates equivalent to wild type H37Rv in Middlebrook 7H9 broth and was phagocytosed by THP-1 cells as efficiently as wild type, the mutant grew very little in the THP-1 cells during the course of a 10 day infection. Employing an *hspX* promoter-luciferase fusion in wild type H37Rv and measuring luciferase expression as a function of time after infection of THP-1 cells, Yuan *et al.* also demonstrated that the *hspX* promoter was induced as early as one hour after addition of *M. tuberculosis* to the THP-1 cells and maximal induction was achieved by four hours (Yuan *et al.*, 1998). These results suggest that expression of *hspX* may be induced by an environmental signal other than low oxygen levels in the macrophage phagosome. Moreover, these results are in contrast to those of Wilkinson *et al.* (2001) described above. Two points should be noted: (1) the intraphagosomal environment of THP-1 cells (an immortalized cell line) may not be equivalent to that of PBMC-derived macrophages in primary culture and (2) regulation of expression of promoter fusions on plasmids may not be equivalent to that of the native promoter in the context of the chromosome (as discussed above).

Armitage *et al.* (2000) disrupted the *fbpA* and *fbpB* genes in *M. tuberculosis* H37Rv. Although the *fbpB* mutant grew at rates similar to those of its wild type parent in broth media as well as in THP-1 cells, the *fbpA* mutant exhibited little or no growth in THP-1 cells (Armitage *et al.,* 2000). Belisle *et al.* (1997) have shown that the FbpA and FbpC proteins are functionally more efficient than FbpB in broth-grown *M. tuberculosis.* Thus, *fbpA* and/or *fbpC* may be the primary sources of mycolyl transferase during growth in macrophages, as these results suggest for *fbpA.*

De Voss *et al.* (2000) replaced the *mbtB* gene of *M. tuberculosis* H37Rv with a hygromycin-resistance (Hyg[R]) gene by homologous recombination to generate a mutant strain that cannot synthesize an enzyme involved in the biosynthesis of the mycobacterial siderophore, mycobactin. This mutant is unable to acquire iron from the extracellular environment, as evidenced by its restricted growth in iron-poor media. In addition, these investigators demonstrated that the *mbtB* mutant was impaired in growth in THP-1 cells, compared to the parental wild type strain, indicating the importance of iron acquisition to intracellular growth (De Voss *et al.,* 2000).

On the basis of experimental data showing that a gene involved in magnesium ion transport in *Salmonella typhimurium (mgtC)* was necessary for growth of *S. typhimurium* in macrophages and the finding that *M. tuberculosis* possesses a gene encoding a protein whose predicted amino acid sequence exhibits 38% identity with the *S. typhimurium* MgtC protein,

Buchmeier *et al.* (2000) constructed an *mgtC* mutant of *M. tuberculosis* strain Erdman. The *M. tuberculosis mgtC* mutant was constructed by replacement of a 466 bp fragment of the *mgtC* gene with a cassette carrying Hyg^R and Km^R genes. *In vitro* growth of this mutant was impaired in medium with low levels of magnesium and a pH of 6.25. In addition, the *mgtC* mutant grew poorly in human PBMC-derived macrophages, suggesting that the ability to acquire magnesium is essential for growth of *M. tuberculosis* both *in vivo* and *in vitro*, as is the case for *S. typhimurium* (Buchmeier *et al.,* 2000).

Based on studies which had shown that the extracytoplasmic sigma factor *sigF* gene was up-regulated for expression by several external stress conditions (e.g., exposure of *M. tuberculosis* cells to antimycobacterial drugs [Michele *et al.*, 1999], entry into stationary phase growth *in vitro* [DeMaio *et al.*, 1997; Michele *et al.*, 1999]) and during growth in human macrophages (Graham and Clark-Curtiss, 1999; see below) and because of the similarity of the SigF protein to stress/sporulation sigma factors of several other bacteria, Chen *et al.* (2000) constructed a *sigF* mutant of *M. tuberculosis* CDC1551 (a recent clinical isolate). The *sigF* mutant was generated by replacement of a 723 bp intragenic fragment of *sigF* with a Hyg^R gene and introduction of the mutated *sigF* gene into the CDC1551 chromosome by homologous recombination between sequences flanking the *sigF* gene (Chen *et al.*, 2000). The growth rate of the *sigF* mutant was identical to that of the wild type parent in Middlebrook 7H9 broth during exponential growth, although the mutant grew to a three-fold higher density in stationary phase, compared to the wild type (Chen *et al.*, 2000). Interestingly, the authors detected no differences between the mutant and wild type in survival and growth in human PBMC-derived monocytes during the course of an 8-day infection. In these experiments, Chen *et al.*, infected monocytes alone and monocytes in combination with non-adherent peripheral blood lymphocytes. However, when the investigators infected BALB/c mice with the *sigF* mutant and wild type CDC1551, the mutant was less virulent, using time-to-death analysis (Chen *et al.*, 2000). The role of SigF in *M. tuberculosis in vivo* growth remains to be elucidated.

Raynaud *et al.* (2002) disrupted the *ompA (rv0899)* gene of *M. tuberculosis* H37Rv by replacing part of the coding sequence with a Km^R cassette, followed by introduction of the disrupted gene into the H37Rv chromosome by homologous recombination. Analysis of the *ompA* mutant revealed that loss of *ompA* greatly impaired the ability of the strain to grow at reduced pH. The mutant strain also exhibited reduced permeability to several small water-soluble substances such as serine, glucose and glycerol (Raynaud *et al.*, 2002). These investigators also demonstrated that the *M. tuberculosis ompA* mutant grew significantly less well than its wild-type parent in both murine bone marrow-derived macrophages and THP-1 cells. The data presented in this paper suggest that OmpA functions both as a pore-forming protein and in enabling *M. tuberculosis* to respond to reduced environmental pH (Raynaud *et al.* 2002).

Global Gene Expression Approaches

Introduction

The second philosophical approach that has been used by investigators in recent years has been to utilize techniques that allow analysis of expression of a large number of genes during growth in a specific environment (global expression analysis techniques). The underlying philosophy of these approaches is to let *M. tuberculosis* show the investigator what gene products are needed for growth and survival in a given environment, rather than having the investigator predict which gene products he/she thinks are important for *M. tuberculosis*. Such approaches are more comprehensive than analyzing specific genes. Because of that, tremendous amounts of data are and will be generated, which no single investigator or research group could realistically thoroughly investigate alone. Thus, global expression analyses potentially provide avenues of research for multiple investigators and should promote cooperative scientific endeavors.

Identification of Proteins Produced by *M. tuberculosis* by 2-Dimensional Gel Electrophoresis.

The most straightforward way to determine the proteins that are important for survival and growth of *M. tuberculosis* in macrophages would be to identify all of the proteins present in the bacilli at different times after infection of macrophages. However, with a potential coding capacity for nearly 4,000 proteins in the *M. tuberculosis* genome, the task of identifying each of them is formidable. Moreover, techniques for detecting tiny amounts of proteins from small numbers of bacteria are not presently available, although there are methods that can provide some information. Two-dimensional (2-D) gel electrophoresis is a method that separates large numbers of proteins by molecular mass in one dimension and by isoelectric point (pI) in the second dimension (O'Farrell, 1975). Advances in image scanning for quantification and analysis, mass spectrometry for protein identification, and peptide database searching, coupled with high resolution 2-D gel electrophoresis have greatly augmented identification of bacterial proteins.

Lee and Horwitz (1995) utilized 2-D gel electrophoresis to analyze proteins of *M. tuberculosis* Erdman grown for approximately 80 h in THP-1 cells. Proteins were labeled during the last 20 h of infection by the addition of [^{35}S]-methionine to the culture medium. Cyclohexamide was also added to the culture medium to preclude protein synthesis by the THP-1 cells. Lee and Horwitz demonstrated that at least 16 *M. tuberculosis* proteins were induced (at least two-fold, compared to *M. tuberculosis* grown in broth), while 28 proteins were repressed (at least a two-fold decrease). Six of the proteins that were induced in THP-1-grown *M. tuberculosis* were absent from 2-D gels prepared from *M. tuberculosis* grown under normal conditions in Middlebrook 7H9 broth or in broth simulating several kinds of stress conditions (low pH, increased temperature, or presence of hydrogen peroxide). The GroE (Rv3417c) and DnaK (Rv0350) proteins and members of the Antigen 85 complex were among the most abundant proteins expressed by *M. tuberculosis* grown in THP-1 cells (Lee & Horwitz, 1995). The six proteins apparently only expressed by *M. tuberculosis* during growth in THP-1 cells were not identified by Lee and Horwitz.

Recently, two groups (Jungblut *et al.*, 1999; Mattow *et al.*, 2001a, 2001b; Rosenkrands *et al.,* 2000) have combined 2-D gel electrophoresis with mass spectrometry to compile sets of data identifying proteins of *M. tuberculosis* that are produced during growth of the organism under different environmental conditions. This is a powerful approach that will provide invaluable data. The data generated thus far have been on *in vitro*-grown *M. tuberculosis*, with some alterations mimicking aspects of *in vivo* growth, but no data have been generated from macrophage-grown *M. tuberculosis*.

Two-dimensional gel electrophoresis necessitates use of fairly large numbers of *M. tuberculosis* in order to have sufficient amounts of each protein to be detected. This, in turn, requires large-scale macrophage cultures, since low multiplicities of infection must be used to simulate natural infection. Thus, 2-D gel analysis of *M. tuberculosis* proteins produced *in vivo* will likely have to be done using a human macrophage-like cell line, as did Lee and Horwitz (1995). It is not known whether *M. tuberculosis* responds physiologically in the same way in immortalized cell lines as it does in primary human macrophages. Another consideration regarding 2-D gel electrophoresis experiments is the use of cyclohexamide in cell culture media to preclude macrophage protein synthesis during labeling of proteins. Burns-Keliher *et al.* demonstrated significant differences in 2-D gel protein profiles of *S. typhimurium* when the bacteria were grown in Int-407 cells in the presence or absence of cyclohexamide (Burns-Keliher *et al.*, 1997).

Analysis of Global Mycobacterial Gene Expression Through Transcript Analysis

Whole genome analysis of *M. tuberculosis* gene expression is a vibrant and growing area, with new methodologies being developed at a tremendous pace. Analysis of gene expression within the host was initially virtually impossible due to the necessity for large amounts of *M. tuberculosis* RNA, coupled with the requirement that this RNA is free of host RNA contamination. Several innovative approaches have since coupled RNA amplification (RT-PCR) with analysis of mRNA expression profiles and several new strategies amenable to high throughput analyses have recently emerged.

With all the advances in discovery of differentially expressed genes, it still requires a tremendous amount of additional work to prove that differential expression of an individual mRNA molecule results in a specific phenotype. This is illustrated in *M. tuberculosis* by the relatively limited number of recognized virulence factors, in comparison to the large number of differentially regulated genes and proteins described. Advancements in *M. tuberculosis* genetics underscore the need for less labor-intensive, and more standardized *in vitro* models to screen candidate virulence factors.

Detection of Gene Expression by cDNA Subtractive Hybridization

The usual way to differentiate between mycobacterial and host gene expression is to selectively lyse macrophages at various times after infection, recovering the mycobacteria and isolating RNA from the bacteria. This approach has been used by several investigators who have employed complementary DNA (cDNA) subtractive hybridization techniques. The objective of subtractive hybridization is to identify genes that are expressed when bacteria are growing in a specific environment but not when they are growing in another. This is accomplished by isolating RNA from bacteria grown in the environment of interest (e.g., within macrophages), converting the RNA to cDNA, and then subtracting out cDNA corresponding to genes that are expressed in both the environment of interest and in another environment (e.g., broth laboratory medium). The common cDNA molecules can be subtracted out by hybridization with cDNA generated from RNA isolated from bacteria grown in the second environment or by hybridization with the RNA directly. The hybrids are then removed from the reaction mixture, leaving behind cDNA molecules representing genes that are up-regulated for expression or that are uniquely expressed by the bacteria when they are growing in the environment of interest.

Successful subtractive hybridization necessitates a step to remove a large proportion of ribosomal RNA (rRNA) from the total RNA preparation. Many investigators have hypothesized that rRNA molecules would self-anneal during the hybridization reaction and thus would not be involved in subsequent analyses of the cDNA molecules. While self-annealing of rRNA certainly does occur during the hybridization reaction, if these hybrids are not removed from the reaction mixture, the proportion of cDNA molecules representing mRNA will remain small and, during subsequent labeling procedures, will take up a correspondingly small proportion of the label. Moreover, the presence of large amounts of rRNA (or cDNA corresponding to rRNA) will interfere with hybridization between labeled mRNA-derived cDNA and plasmid or cosmid DNA on filters, if this approach is being used to analyze the cDNA molecules (see Hou *et al.*, 2002). If the cDNA molecules recovered after hybridization are cloned to generate cDNA libraries, the investigator will have to analyze many individual clones, because a large percentage of the clones will contain cDNA inserts corresponding to rRNA.

Kinger and Tyagi (1993) were the first to use cDNA subtractive hybridization to identify differentially expressed genes of mycobacteria, although the studies were conducted on broth-grown bacteria. Subtractive hybridization between cDNA molecules generated from *M. avium* grown in human PBMC-derived macrophages and cDNA molecules from *M. avium* grown in Middlebrook 7H9 broth led to the identification of a gene designated *mig* by Plum and Clark-Curtiss (1994). Although this paper describes gene expression by an opportunistic pathogenic mycobacterium, this is the only published paper in which cDNA subtractive hybridization studies were done on mycobacteria grown in primary human macrophages (Plum and Clark-Curtiss, 1994). After three rounds of subtractive hybridization, the remaining cDNA molecules from macrophage-grown *M. avium* were labeled and used as a probe in dot blot hybridization experiments with an *M. avium* cosmid genomic library. Cosmids bearing the *mig* gene hybridized very strongly with the subtracted cDNA probe and were further characterized, resulting in the identification of *mig* (Plum and Clark-Curtiss, 1994). This was not the only gene of *M. avium* that was up-regulated for expression in macrophage-grown *M. avium*, but subsequently, a better method for identifying up-regulated genes was devised (see Section III.C.4). Characterization of *mig* revealed that this gene is present only in *M. avium* (Plum and Clark-Curtiss, 1994; Beggs *et al.*, 2000), although since it encodes an acyl-Coenzyme A synthase (Morsczek *et al.*, 2001), *M. tuberculosis* may produce a protein with similar activity from a gene with significantly different nucleotide sequence. Alternatively, the ability to synthesize Mig may allow *M. avium* to utilize fatty acid substrates that *M. tuberculosis* may not be able to use.

Li *et al.* (2001) studied mycobacterial gene expression using a cDNA-RNA subtractive hybridization approach. These investigators prepared cDNA from *M. bovis* BCG 24 h after infection of the THP-1 monocyte-like cell line. After subtraction of cDNA representing commonly expressed genes, the remaining cDNA molecules were hybridized with clones of a *M. tuberculosis* H37Rv genomic plasmid library. Because this was a plasmid library, it was necessary to include more than 18,000 individual clones on the filters used for each hybridization). Li *et al.* characterized 19 of the strongest-hybridizing clones, which represented 15 genes that were apparently up-regulated for expression in BCG growing in THP-1 cells. These investigators also conducted northern blotting and primer extension experiments with 5 of these genes, which confirmed that they were up-regulated (Li *et al.,* 2001). Among the genes identified by these investigators were *mas* (which encodes mycocerosic acid synthase, which is involved in the synthesis of multi-methylated long chain fatty acids), *furB*, *groEL-2*, *rplE*, and *fadD28* (Li *et al.,* 2001).

Detection of Gene Expression by Dfferential Display PCR

A second method that has been used to study gene expression by organisms growing in different environments or by two closely related organisms is differential display PCR (DD-PCR). This approach has been used by two groups to compare gene expression between *M. tuberculosis* strains growing in or on laboratory media (Rivera-Marrero *et al.*, 1998; Rindi *et al.*, 1999). No studies have been published in which this approach was used to study gene expression in *in vivo*-grown mycobacteria.

DD-PCR appears to be a technically challenging technique that has given rise to a large number of apparently differentially expressed genes upon initial analysis. However, many of these turned out to be non-specific PCR products with little homology to the genome of *M. tuberculosis* (Rivera-Marrero *et al.*, 1998; Rindi *et al.*, 1999). Moreover, subsequent analysis by RT-PCR using gene-specific primers has indicated that a high percentage of the identified genes were not differentially expressed (Rivera-Marrero *et al.*, 1998; Rindi *et al.*, 1999).

Detection of Gene Expression by DECAL (Differential Expression Using Customized Amplification Libraries)

Alland *et al.* (1998), described an approach for analyzing gene expression by *M. tuberculosis* that involved hybridization of cDNA molecules to a library of *M. tuberculosis* genomic DNA. The library of genomic DNA did not include the ribosomal RNA genes (hence the name, Customized Amplification Library: CAL) and consisted of fragments within a narrow size range (400-1500 bp). Total RNA was isolated from *M. tuberculosis* H37Rv cultures grown *in vitro* in the presence and absence of isoniazid. During reverse transcription of the RNA, biotinylated primers led to the incorporation of biotin into the cDNA molecules. The mixture of cDNA molecules was hybridized to the CAL, the hybrids were separated from non-reacting sequences by mixing with streptavidin-coated beads, and the recovered hybrids were PCR-amplified. The amplified products were labeled and used as probes in hybridizations with the individual clones of the CAL. Theoretically, the DECAL approach could be used to study gene expression by *M. tuberculosis* growing in macrophages, but the studies published by Alland *et al.* compared gene expression by *M. tuberculosis* grown in broth in the presence and absence of isoniazid (Alland *et al.*, 1998).

Gene Expression Detected by SCOTS (Selective Capture of Transcribed Sequences)

Graham and Clark-Curtiss (1999) developed a method for analyzing gene expression by *M. tuberculosis*, which is called SCOTS. A significant difference between SCOTS and the methods described above is that the mycobacteria were not separated from the infected macrophages prior to isolation of RNA. Because there are few data concerning the half-lives of mRNAs in *M. tuberculosis*, but knowing that half-lives of at least some mRNAs in other bacteria can be quite short (less than 2 minutes), the SCOTS method was designed to minimize the possibility of losing some mRNAs due to normal turnover during the time it takes to lyse macrophages, separate the mycobacteria, and isolate RNA (Graham and Clark-Curtiss, 1999). Thus, Graham and Clark-Curtiss isolated total RNA from human PBMC-derived macrophages infected with *M. tuberculosis* H37Rv and converted the RNA to cDNA, by priming with random nonamers coupled to a specific linker, in the presence of reverse transcriptase. *M. tuberculosis* chromosomal DNA was biotinylated and pre-hybridized with *M. tuberculosis* rRNA genes prior to addition of the cDNA mixture. The cDNA mixture was hybridized with the *M. tuberculosis* chromosomal DNA; only cDNA molecules derived from *M. tuberculosis* should hybridize to (i.e., be captured by) the chromosomal DNA. The hybrids were separated from the macrophage cDNA and non-hybridized cDNA (from rRNA) by reaction with streptavidin-coated beads. The cDNA molecules were released from the hybrids by denaturation and were PCR-amplified using primers specific for the linkers added during preparation of the cDNA. The amplified cDNA molecules were subjected to two more rounds of SCOTS to prepare high quality cDNA mixtures (Graham and Clark-Curtiss, 1999). To enrich for cDNA molecules corresponding to genes that were up-regulated for expression or that were uniquely expressed by *M. tuberculosis* during growth in macrophages, the cDNA mixtures were subjected to three additional rounds of SCOTS with *M. tuberculosis* chromosomal DNA that had been pre-hybridized with both rRNA genes and with cDNA prepared from H37Rv grown in Middlebrook 7H9 broth (Graham and Clark-Curtiss, 1999). After the rounds of enrichment, the cDNA molecules were cloned into the plasmid, pBluescript, to generate cDNA libraries.

Libraries of cDNA molecules were prepared from *M. tuberculosis* grown for 18 h, 48 h, and 110 h in human PBMC-derived macrophages. Individual cDNA clones were chosen at random, the *M. tuberculosis* inserts sequenced, and the genes corresponding to the inserts identified by comparison of the cDNA sequences to the H37Rv genome sequence. In these studies, Graham and Clark-Curtiss identified 11 genes of H37Rv that were up-regulated for expression (*sigA* and *sigH*) or were apparently uniquely expressed during growth in macrophages (*icl [aceA], ponA, pks2, uvrA, ctpV, mceD, rv0903c, rv3070,* and *rv3483c*; Graham & Clark-Curtiss, 1999). Further analyses have identified 29 additional genes that are expressed by *M. tuberculosis* during growth in human macrophages (Graham & Clark-Curtiss, unpublished; Haydel & Clark-Curtiss, unpublished). It should be noted that the genes thus far identified do not represent all of the genes that are expressed by *M. tuberculosis* during growth in human macrophages. Sequence determination and identification of additional genes is currently ongoing in the Clark-Curtiss lab.

Hou *et al.* (2002) applied the SCOTS approach to identify genes of *M. avium* that are expressed by this opportunistic pathogen during growth in human macrophages. A total of 54 cDNA molecules that represented 46 genes that are expressed by *M. avium* during growth in macrophages were identified and included genes that encoded enzymes for several biosynthetic pathways (mycobactin, polyketides, pyrimidines), enzymes involved in intermediary metabolism, energy metabolism (TCA cycle, glyoxalate shunt), and nitrogen metabolism (Hou *et al.*, 2002). Genes encoding several regulatory proteins were identified, as well as genes that encoded proteins of unknown function. Among the last group were genes that code for proteins similar to members of the PPE family of *M. tuberculosis* proteins and proteins similar to the Mce proteins of *M. tuberculosis* (Hou *et al.*, 2002).

Comparisons between genes expressed by *M. tuberculosis* and *M. avium* after infection of human macrophages have revealed several genes and classes of genes that are expressed by both mycobacteria. These included genes from two of the *mce* operons (*mceD* in *M. tuberculosis*; *lprM* in *M. avium*), genes encoding enzymes involved in polyketide synthesis, genes encoding members of the PPE family of proteins, nitrite extrusion protein genes, the gene encoding isocitrate lyase, and genes encoding proteins of a two-component regulatory system (Graham & Clark-Curtiss, 1999; Hou *et al.* 2002, Haydel & Clark-Curtiss, unpublished). The products of some of these genes (e.g., *mceD*) have been implicated by others to be important for mycobacterial survival in macrophages (Arruda *et al.*, 1993). The products of other genes have been hypothesized to play roles in intracellular growth, although the precise roles have yet to be determined (e.g., polyketide synthases, PPE proteins; Cole *et al.*, 1998). As noted above, isocitrate lyase has been shown to be important for long-term survival of *M. tuberculosis* in the infected mouse model (McKinney *et al.*, 2000). Thus, comparisons of gene expression by two mycobacterial pathogens growing within human macrophages are likely to provide insights as to mycobacterial functions that are essential to intracellular survival and growth.

The SCOTS approach has the potential for providing more global information about gene expression than many of the approaches discussed above. A drawback of the SCOTS approach is the somewhat labor-intensive necessity of analyzing individual clones from the cDNA libraries constructed from cDNA mixtures prepared by SCOTS. It is difficult to predict how many genes will be identified from any cDNA library prepared from a single time point after infection, because relatively little is known even now about *M. tuberculosis* metabolism during growth in human macrophages. The SCOTS technique may also result in recovery of cDNA molecules corresponding to the more abundant classes of mRNA in the mycobacterial cell at the time of RNA extraction. However, several features of the SCOTS technique, such as the use of random priming to generate cDNA molecules of a limited size range (200-400 bp) and amplification for a limited number of cycles, which has been shown to amplify sequences without significant bias (Ko *et al.*, 1990), serve to minimize the likelihood that only cDNA molecules representing the most abundant mRNAs will be recovered (Graham and Clark-Curtiss, 1999). Further experiments are needed to verify this supposition, such as quantitative RT-PCR to determine levels of expression for a number of the genes identified from cDNAs recovered by SCOTS. Another drawback of the SCOTS technique is that only genes that are expressed can be identified; genes that are repressed will not be detected.

The SCOTS approach has several benefits that are not currently available with other methods for analyzing gene expression. Of significant importance is that large amounts of RNA are not required; in the experiments described by Graham and Clark-Curtiss, and Hou et al., we have recovered cDNA mixtures using SCOTS from as few as 8×10^5 to 1×10^6 mycobacteria in infected macrophages (corresponding to total mycobacterial RNA of 200-250 ng, of which approximately 10-12.5 ng was mRNA). Another benefit is that generation of cDNA libraries from bacterial grown under a variety of environmental conditions, both *in vitro* and *in vivo*, is relatively easy, providing the investigator with a plethora of material to further study.

Detection of Gene Expression in Promoter-probe Libraries

Several groups have constructed libraries of random fragments of *M. tuberculosis* chromosomal DNA cloned upstream of promoterless reporter genes, such as firefly luciferase (*Fflux*), *lacZ, inhA,* or *gfp.*

Marston and Shinnick (1996) identified nine recombinant strains that exhibited at least five-fold more luciferase activity after three days of growth in THP-1 cells, compared to growth in broth, but the genes have not yet been described. Triccas *et al.* (1999) constructed an *M. tuberculosis* promoter-probe library with *gfp* as the reporter gene. These investigators identified 7 *M. tuberculosis* promoters that were up-regulated in recombinant *M. bovis* BCG that were grown for six days in murine bone marrow-derived macrophages. The promoters were from *tkt* (*rv1449c*, which encodes a transketolase of the pentose phosphate pathway), *fadB4* (*rv3141*, which encodes a protein homologous to 3-hydroxyl-coenzyme A dehydrogenase), two genes that code for proteins with similarity to proteins involved in sulfur metabolism (*sseA [rv3283]* and *cysD [rv1285]*), *rv0834c* (which encodes a protein of the PE-PGRS family), and two genes that code for proteins of unknown function (Triccas *et al.*, 1999).

Hobson *et al.* (2002) constructed a promoter-probe library in which fragments of *M. tuberculosis* H37Rv DNA were cloned upstream of the *lacZ* gene. This library was introduced into *M. bovis* BCG for initial screening, which was exposure to acidified sodium nitrite. Recombinant *M. bovis* BCG clones that were induced for expression of β-galactosidase were further evaluated after infection of THP-1 cells. Three of the 13 genes identified in these screens were then evaluated for induction of expression by real-time RT-PCR using RNA from *M. tuberculosis* H37Rv grown for 24 h in THP-1 cells. Two of the three genes, *rv1265* and *rv2711 (ideR)* were induced after infection of THP-1 cells (Hobson *et al.*, 2002).

Dubnau *et al.* (2002) constructed a promoter-probe library using *inhA* as a reporter, to identify *M. tuberculosis* genes induced after infection of THP-1 cells. These investigators identified 43 genes from the library and evaluated 13 of them by real-time RT-PCR. Eight of the 13 genes exhibited two-fold or higher induction in *M. tuberculosis* grown for 24 h in THP-1 cells, compared to bacilli grown in Middlebrook 7H9 broth. These genes included *icl* (encodes isocitrate lyase), three genes thought to be involved in fatty acid metabolism (*echA19, fadA4,* and *pckA*), a gene that codes for a possible potassium channel protein (*rv3237c*), a gene that codes for a possible DNA-binding protein, and two genes that code for proteins of unknown function (Dubnau *et al.*, 2002).

These promoter-probe library approaches afford another means to identify a number of genes that are up-regulated for expression in response to the macrophage environment. The limitations associated with the use of reporter gene constructs discussed in Section II.D apply to the promoter-probe libraries as well.

Gene Expression Detected by DNA Microarray Analysis

The most desirable way to assess gene expression in *M. tuberculosis* (or any organism) growing in a specific environment would be to analyze every possible open reading frame (ORF) in the organism's genome. Microarray analysis allows for the simultaneous detection of RNA transcripts from every ORF in the *M. tuberculosis* genome (deSaizieu *et al.*, 1998; Wilson *et al.*, 1999). Moreover, this kind of analysis enables the investigator to semi-quantitatively compare global gene expression in an organism grown under two different conditions (e.g., in the presence and absence of an antimicrobial agent). A third advantage of the DNA microarray analysis technique is that repression of gene expression, as well as up-regulation of expression, can be determined. The keys to successful application of this technology are robust and quality procedures for *M. tuberculosis* RNA isolation, reverse transcription and labeling of cDNA products, and a means of normalizing expression patterns. These conditions must be optimized to limit both intra- and inter-assay variability. As with any RT-PCR protocol, it is essential that the PCR amplification is uniform throughout the range of cDNAs. This is a new area of investigation and there are many considerations in instrumentation, production of reproducible microarray chips, and availability of appropriate software to consider.

Several investigators have applied comparative microarray analysis to studies of *M. tuberculosis* gene expression in various environments or during different phases of growth. In the first *M. tuberculosis* DNA microarray study, Wilson *et al.* identified genes of *M. tuberculosis* H37Rv that are differentially expressed when the bacteria are growing in the presence and absence of isoniazid (Wilson *et al.*, 1999). These investigators identified genes that encode proteins physiologically relevant to the mode of action of isoniazid as being up-regulated when *M. tuberculosis* was grown in the presence of isoniazid (Wilson *et al.*, 1999).

Talaat *et al.* demonstrated that specific genome-directed primers were superior to random priming for generating *M. tuberculosis* cDNA probes to compare gene expression during logarithmic phase and stationary phase growth in broth medium (Talaat *et al.*, 2000).

Fisher *et al.* (2002) utilized a system of subjecting actively growing *M. tuberculosis* to acidic medium for several different periods of time in order to mimic the initial acidification of the macrophage phagosomal compartment. The authors were able to identify 81 genes that were differentially expressed, with the largest induction observed for those genes that showed homology with non-ribosomal peptide synthatases/polyketide synthases.

Manganelli *et al.* (2001) employed microarray technology to analyze differential gene expression during exponential growth in Middlebrook 7H9 broth by wild type *M. tuberculosis* H37Rv and a mutant in which the *sigE* (*rv1221*) gene had been insertionally inactivated. In a second study utilitilizing DNA microarrays, Manganelli *et al.* (2002) compared global gene expression in wild-type *M. tuberculosis* H37Rv and in a *sigH* mutant following a 60 min exposure to diamide (a thiol-specific oxidizing agent). Forty-eight genes of wild-type *M. tuberculosis* were up-regulated for expression under these oxidative stress conditions. However, in the *sigH* mutant, 39 of these genes were not induced under these same stress conditions, suggesting that SigH regulates expression of these genes, either directly or indirectly (Manganelli *et al.*, 2002).

Another group of investigators has also used DNA microarray analysis to compare gene expression between wild-type *M. tuberculosis* CDC1551 and a *sigH* mutant of this strain (Kaushal *et al.*, 2002). These investigators evaluated gene expression in the two strains at four stages of growth in Middlebrook 7H9 broth and identified 31 genes that were apparently regulated by SigH (Kaushal *et al.*, 2002). Both Manganelli *et al.* (2002), and Kaushal *et al.* (2002) showed that genes encoding proteins involved in thiol metabolism (e.g., thioredoxin and thioredoxin reductase) were regulated by SigH.

Another application of DNA microarray technology was the study comparing *M. tuberculosis* H37Rv gene expression during growth under reduced oxygen levels to that of H37Rv grown at normal oxygen levels (Sherman *et al.*, 2001). These investigators identified more than 100 genes that were differentially expressed in response to hypoxic growth conditions (Sherman *et al.*, 2001). Many of the genes whose expression was induced during hypoxic growth code for proteins of unknown function (Sherman *et al.*, 2001). Interestingly, the *devRdevS* two-component regulatory system genes were among the hypoxia-induced genes identified in this study. These genes had been shown by Kinger and Tyagi (1993) to be differentially expressed in *M. tuberculosis* H37Rv and H37Ra strains. Sherman *et al.* (2001) disrupted the *devRdevS* operon and demonstrated that normal regulation of expression of the *hspX* gene was also altered, suggesting that the *devRdevS* two-component system controls gene expression during exposure to hypoxic conditions.

Stewart *et al.* (2002) used DNA microarray analysis to compare gene expression between wild-type *M. tuberculosis* H37Rv and several mutant strains after exposure to a 30 min heat shock of 45°C. These investigators identified two heat-shock regulons and further demonstrated overlapping regulation of expression of the genes in the two regulons (Stewart *et al.*, 2002).

All of the DNA microarray studies have generated large amounts of data and have amply demonstrated the benefits of applying DNA microarray analysis to studies of gene expression in response to environmental conditions. However, DNA microarray analysis at present is not amenable for studying gene expression of *M. tuberculosis* in human macrophages, because of the requirement for large amounts of total RNA: approximately 1-2 µg of mRNA (which comprises 5% or less of total RNA) is necessary to generate the cDNA probes (Talaat *et al.*, 2000). Investigators using DNA microarray technology have employed RNA isolated from pure cultures of bacteria, although Talaat *et al.* compared hybridization to DNA microarrays of cDNA from a pure culture of *M. tuberculosis* to mixtures of mammalian cDNA and *M. tuberculosis* cDNA simulating a mock infection (Talaat *et al.*, 2000). There have been preliminary reports of investigators using DNA microarray analysis to analyze *M. tuberculosis* gene expression in mouse lungs (Talaat *et al.*, 2001) and in murine bone marrow-derived macrophages (Schnappinger *et al.*, 2001)

Analysis of Gene Expression by Signature-Tagged Mutagenesis

Signature-tagged mutagenesis (STM) is an approach whereby a library of mutants of a given bacterial species is generated by insertion of transposons carrying unique DNA sequence tags that are randomly inserted into the bacterial chromosome (Hensel *et al.*, 1995). Thus, each mutant can be distinguished from the others by hybridization to detect the tags. Pools of tagged mutants are introduced into a host organism and are allowed to multiply. At some point after infection, the mutants are recovered from the host and analyzed by hybridization with the tags back to filters prepared from master plates that include each of the individual mutants. Attenuated mutants are those that are unable to multiply in the infected host (Hensel *et al.*, 1995).

This approach has been used by Camacho *et al.* (1999) to generate a signature-tagged transposon library of mutants of *M. tuberculosis* strain MT103 (a clinical isolate). The STM library was used to infect BALB/c mice by the intravenous route. The animals were euthanized three weeks later, and the pools of *M. tuberculosis* mutants recovered were analyzed by hybridization to filters containing DNA from each of the original mutants. These investigators identified 16 attenuated mutants of *M. tuberculosis*. Interestingly, most of the mutated genes encoded proteins involved in lipid metabolism and in transport processes. Another significant finding was that four of the mutations were located in a cluster of genes in a 50 kb region of the *M. tuberculosis* chromosome (Camacho *et al.*, 1999).

Cox *et al.* (1999) also generated an STM library, but in *M. tuberculosis* strain Erdman. These investigators identified 14 attenuated transposon mutants following infection of C57BL/6 mice. Many of these mutants had transposons inserted into genes of unknown function, but Cox *et al.* identified three mutants with insertions in a region of the *M. tuberculosis* chromosome containing genes encoding enzymes involved in the synthesis of the cell wall-associated lipid, pthiocerol dimycocerosate (PDM). These genes were in the same region of the chromosome as those identified by Camacho *et al.* (1999), reinforcing the importance of lipid transport and biosynthesis in *M. tuberculosis* pathogenicity.

To date, the STM approach has only been used to identify *M. tuberculosis* genes that may be important for survival of the bacilli in the infected mouse model, although theoretically, one should be able to infect human macrophages in culture with subsets of an STM library and identify mutants unable to survive.

Conclusion

This is an exciting time to be studying the pathogenesis of *M. tuberculosis*. With the development of a diversity of experimental approaches, coupled with the availability of the complete nucleotide sequence for *Mycobacterium tuberculosis* and other mycobacterial genomes, there are many opportunities to formulate hypotheses and to test such hypotheses. Knowledge gained from studying gene expression by *M. tuberculosis* during its interactions with macrophages and other cells of the human host should enable us to gain a more thorough understanding of the mechanisms whereby *M. tuberculosis* is able to evade normal macrophage functions and grow within these phagocytic cells. Not only will we learn more about the metabolic activities of *M. tuberculosis* during intracellular growth, but there is also the potential to learn more about macrophage biology. Understanding *M. tuberculosis* physiology during intracellular growth should facilitate rational design of new antimicrobial agents, identification of potential antigens that could be incorporated into vaccines, and identification of genes that could be potential targets for inactivation to generate attenuated mutants of *M. tuberculosis*.

References

Ainsa, J.A., Martin, C., and Gicquel, B. 2001. Molecular approaches to tuberculosis. Mol. Microbiol. 42: 561-570.

Alland, D., Kramnik, I., Weisbrod, T.R., Ohtsubo, L., Cerny, T., Miller, L.P., Jacobs, W.R., Jr., and Bloom, B.R. 1998. Identification of differentially expressed mRNA in prokaryotic organisms by customized amplification libraries (DECAL): the effect of isoniazid on gene expression in *Mycobacterium tuberculosis*. Proc. Natl. Acad. Sci. USA 95: 13227-13232.

Armitage, L.Y., Jagannath, C., Wagner, A.R., and Norris, S.J. 2000. Disruption of the genes encoding Antigen 85A and Antigen 85B of *Mycobacterium tuberculosis* H37Rv: effect on growth in culture and in macrophages. Infect. Immun. 68: 767-778.

Arruda, S., Bonfim, G., Knights, R., Huima-Byron, T., and Riley, L.W. 1993. Cloning of an *M. tuberculosis* DNA fragment associated with entry and survival inside cells. Science 261: 1454-1457.

Aspinall, G.O., Chatterjee, D., and Brennan, P.J. 1995. The variable surface glycolipids of mycobacteria: Structures, synthesis of epitopes, and biological properties. Adv. Carbohydr. Chem. Biochem. 51:169-242.

Av-Gay, Y., Jamil, S., and Drews, S.J.. 1999. Expression and characterization of the *Mycobacterium tuberculosis* serine/threonine protein kinase PknB. Infect. Immun. 67: 5676-5682.

Bange, F-C., Brown, A.M., and Jacobs, W.R., Jr. 1996. Leucine auxotrophy restricts growth of *Mycobacterium bovis* BCG in macrophages. Infect. Immun. 64: 1794-1799.

Bardarov, S., Kriakov, J., Carriere, C., Yu, S., Vaamonde, C., McAdam, R.A., Bloom, B.R., Hatfull, G.F., and Jacobs, W.R., Jr. 1997. Conditionally replicating mycobacteriophages: a system for transposon delivery to *Mycobacterium tuberculosis*. Proc. Natl. Acad. Sci. USA 94: 10961-10966.

Baulard, A.R., Betts, J.C., Engohang-Ndong, J., Quan, S., McAdam, R.A., Brennan, P.J., Locht, C., Besra, G.S. 2000. Activation of the pro-drug ethionamide is regulated in mycobacteria. J. Biol. Chem. 275: 28326-28331.

Bates, J.H. 1980. Transmission and pathogenesis of tuberculosis. Clin. Chest Med. 1:167-174.

Beggs, M.L., Stevanova, R., and Eisenach, K.D. 2000. Species identification of *Mycobacterium avium* complex isolates by a variety of molecular techniques. J. Clin. Microbiol. 38: 508-512.

Bekierkunst, A. and Artmen, M. 1960. Studies on *Mycobacterium tuberculosis* H37Rv grown *in vivo*: inhibitor of lactic acid dehydrogenase in normal and infected mice. Proc. Soc. Exptl. Biol. Med. 105: 605-609.

Belisle, J.T., Vissa, V.D., Sievert, T., Takayama, K., Brennan, P.J., and Besra, G.S. 1997. Role of the major antigen of *Mycobacterium tuberculosis* in cell wall biogenesis. Science 276: 1420-1422.

Brennan, P.J. and Nikaido,H. 1995. The envelope of mycobacteria. Annu. Rev. Biochem. 64:29-63.

Buchmeier, N., Blanc-Potard, A., Ehrt, S., Piddington, D., Riley, L., and Groisman, E.A. 2000. A parallel intraphagosomal survival strategy shared by *Mycobacterium tuberculosis* and *Salmonella enterica*. Mol. Microbiol. 35: 1375-1382.

Bullen, J.J. 1981. The significance of iron in infection. Rev. Infect. Dis. 3: 1127-1138.

Burns-Keliher, L., Portteus, A., and Curtiss, R. III. 1997. Specific detection of *Salmonella typhimurium* proteins synthesized intracellularly. J. Bacteriol. 179: 3604-3612.

Camacho, L.R., Ensergueix, D., Perez, E., Gicquel, B., and Guilhot, C. 1999. Identification of a virulence gene cluster of *Mycobacterium tuberculosis* by signature-tagged transposon mutagenesis. Mol. Microbiol. 34: 257-267.

Camus, J-C., Pryor, M.J., Medigue, C.M., and Cole, S.T.. 2002. Re-annotation of the genome sequence of *Mycobacterium tuberculosis* H37Rv. Microbiology 148: 2967-2973.

Capelli, G., Volpe, P., Sanduzzi, A., Sacchi, A., Colizzi, V., and Mariani, F. 2001. Human macrophage gamma interferon decreases gene expression but not replication of *Mycobacterium tuberculosis*: analysis of the host-pathogen reciprocal influence on transcription in a comparison of strains H37Rv and CMT97. Infect. Immun. 69: 7262-7270.

Chan, J., Fan, X., Hunter, S.W., Brennan, P.J. and Bloom, B.R. 1991. Lipoarabinomannan, a possible virulence factor involved in persistence of *Mycobacterium tuberculosis* within macrophages. Infect.Immun. 59:1755-1761.

Chen, P., Ruiz, R.E., Li, Q., Silver, R.F., and Bishai, W.R. 2000. Construction and characterization of a *Mycobacterium tuberculosis* mutant lacking the alternate sigma factor gene, *sigF*. Infect. Immun. 68: 5575-5580.

Clark-Curtiss, J.E. and Haydel, S.E. 2003. Molecular genetics of *Mycobacterium tuberculosis* pathogenesis. Annu. Rev. Microbiol. 37: 517-549.

Cole, S.T., Brosch, R., Parkhill, J., Garnier, T., Chrurcher, C., Harris, D., Gordon, S.V., Eiglmeier, K., Gas, S., Barry, C.E. III, Tekaia, F., Badcock, K., Basham, D., Brown, D., Chillingworth, T., Connor, R., Davies, R., Devlin, K., Feltwell, T., Gentles, S., Hamlin, N., Holroyd, S., Hornsby, T., Jagels, K., Krogh, A., McLean, J., Moule, S., Murphy, L., Oliver, K., Osborn, J., Quail, M.A., Rajandream, M.-A., Rogers, J., Rutter, S., Seeger, K., Skelton, J., Squares, R., Squares, S., Sulston, J.E., Taylor, K., Whitehead, S., and Barrell, B.G. 1998. Deciphering the biology of *Mycobacterium tuberculosis* from the complete genome sequence. Nature 293: 537-544.

Cox, J.S., Chen, B., McNeil, M., and Jacobs, W.R., Jr. 1999. Complex lipid determines tissue-specific replication of *Mycobacterium tuberculosis* in mice. Nature 402: 79-83.

Dhandayuthapani, S., Via, L.E., Thomas, C.A., Horwitz, P.M., Deretic, D., and Deretic, V. 1995. Green fluorescent protein as a marker for gene expression and cell biology of mycobacterial interactions with macrophages. Mol. Microbiol. 17: 901-912.

DesJardin, L.E. 2000. Isolation of *M. tuberculosis* RNA from sputum. In: Methods in Molecular Medicine: Antibiotic Resistance Methods. S. Gillespie. ed. Humana Press, Totowa, New Jersey.

Desjardin, L.E., Chen, Y., Perkins, M.D., Teixiera, L., Cave, M.D. and Eisenach, K.D. 1998. Comparative use of the ABI 7700 (TaqMan) and competitive PCR for quantification of *IS6110* DNA in sputum during the treatment of tuberculosis. J. Clin.Microbiol. 36:1964-1968.

DesJardin, L.E., Hayes, L.G., Sohaskey, C.D., Wayne, L.G., and Eisenach, K.D. 2001. Microaerophilic induction of the alpha-crystalline chaperone protein homologue (*hspX*) mRNA of *Mycobacterium tuberculosis*. J. Bacteriol. 183: 5311-5316.

Desjardin, L.E., Perkins, M.D., Wolski, K., Haun, S., Teixiera, L., Johnson, J.L., Ellner ,J.J., Dietze, R., Bates, J., Dave, M.D., and Eisenach, K.D. 1999. Measurement of sputum *Mycobacterium tuberculosis* mRNA as a surrogate for response to chemotherapy. Am. J. Respir. Crit. Care Med. 160: 203-210.

DeMaio, J., Zhang,Y., Ko, C., and Bishai, W.R. 1997. A stationary phase stress-response sigma factor from *Mycobacterium tuberculosis*. Proc. Natl. Acad. Sci. USA 93: 2790-2794.

de Saizieu, A., Certa, U., Warrington, J., Gray, C., Keck, W., and Mous, J. 1998. Bacterial transcript imaging by hybridization of total RNA to oligonucleotide arrays. Nat. Biotechnol. 16: 45-48.

De Voss, J.J., Rutter, K., Schroeder, B.G., Su, H., Zhu, Y., and Barry, C.E. III. 2000. The salicylate-derived mycobactin siderophores of *Mycobacterium tuberculosis* are essential for growth in macrophages. Proc. Natl. Acad. Sci. USA. 97: 1252-1257.

Dubnau, E., Fontan, P., Manganelli, R., Soares-Appel, S., and Smith, I. 2002. *Mycobacterium tuberculosis* genes induced during infection of human macrophages. Infect. Immun. 70: 2787-2795.

Falkow, S. 1988. Molecular Koch's postulates applied to microbial pathogenicity. Rev. Infect. Dis. 10 (2): S274-S276

Fenhalls, G., Stevens, L., Moser, L., Bezuidenhout, J., Betts, J.C., van Helden, P., Lukey, P.T., and Duncan, K. 2002. *In situ* detection of *Mycobacterium tuberculosis* transcripts in human lung granulomas reveals differential gene expression in necrotic lesions. Infect. Immun. 70: 6330-6338.

Fisher, M.A., Plikaytis, B.B., and Shinnick, T.M. 2002. Microarray analysis of the *Mycobacterium tuberculosis* transcriptional response to the acidic conditions found in phagosomes. J. Bacteriol. 184: 4025-4032.

Gold, B., Rodriguez, G.M., Marras, S.A., Pentecost, M., and Smith, I. 2001. The *Mycobacterium tuberculosis* IdeR is a dual function regulator that controls transcription of genes involved in iron acquisition, iron storage, and survival in macrophages. Mol. Microbiol. 42: 851-865.

Glickman, M.S. and Jacobs, W.R., Jr. 2001. Microbial pathogenesis of *Mycobacterium tuberculosis*: dawn of a new discipline. Cell 104: 477-485.

Graham, J.E. and Clark-Curtiss, J.E. 1999. Identification of *Mycobacterium tuberculosis* RNAs synthesized in response to phagocytosis by human macrophages by selective capture of transcribed sequences (SCOTS). Proc. Natl. Acad. Sci. USA 96: 11554-11559.

Guy, L.R., Raffel, S., and Clifton, C.E. 1954. Virulence of the tubercle bacillus. II. Effects of oxygen tension upon growth of virulent and avirulent bacilli. J. Infect. Dis. 94: 99-106.

Hellyer, T., Desjardin, L.E., Hehman, G.L., Cave, M.D. and Eisenach, K.D. 1999a. Quantitative analysis of mRNA as a marker for viability of *Mycobacterium tuberculosis*. J. Clin. Microbiol. 37: 290-295.

Hellyer, T.J., DesJardin, L.E., Teixeira, L., Perkins, M.D., Cave, M.D., and Eisenach, K.D. 1999b. Detection of viable *Mycobacterium tuberculosis* by reverse transcriptase-strand displacement amplification of mRNA. J. Clin. Microbiol. 37: 518-523.

Hensel, M., Shea, J.E., Gleeson, C., Jones, M.D., Dalton, E., and Holden, D.W. 1995. Simultaneous identification of bacterial virulence genes by negative selection. Science 269: 400-403.

Heplar, J.Q., Clifton, C.E., Raffel, S., and Futrelle, C.M. 1954. Virulence of the tubercle bacillus: I. Effect of oxygen tension upon respiration of virulent and avirulent bacilli. J. Infect. Dis. 94: 90-98.

Hernandez-Pando, R., Jeyanathan, M., Mengistu, G., Aguilar, D., Orozco, H., Harboe, M., Rook, G.A., and Bjune, G. 2000. Persistence of DNA from

Mycobacterium tuberculosis in superficially normal lung tissue during latent infection. Lancet 356: 2133-2138.

Hobson, R.J., McBride, A.J.A., Kempsell, K.E., and Dale, J.W. 2002. Use of an arrayed promoter-probe library for identification of macrophage-regulated genes in *Mycobacterium tuberculosis*. Microbiology 148: 1571-1579.

Hou, J.Y., Graham, J.E. and Clark-Curtiss, J.E. 2002. *Mycobacterium avium* genes expressed during growth in human macrophages detected by selective capture of transcribed sequences (SCOTS). Infect. Immun. 70: 3714-3726.

Jensen-Cain, D.M. and Quinn, F.D. 2001. Differential expression of *sigE* by *Mycobacterium tuberculosis* during intracellular growth. Microb. Pathogen. 30: 271-278.

Jou, N.T., Yoshimori, R.B., Mason G.R., Louie, J.S., and Liebling, M.R. 1997. Single-tube, nested, reverse transcriptase PCR for detection of viable *Mycobacterium tuberculosis*. J. Clin. Microbiol. 35:1161-5.

Jungblut, P.R., Schaible, U.E., Mollenkopf, H.J., Zimny-Arndt, U., Raupach, B., Mattow, J., Halada, P., Lamer, S., Hagens, K., and Kaufmann, S.H.E.. 1999. Comparative proteome analysis of *Mycobacterium tuberculosis* and *Mycobacterium bovis* BCG strains: towards functional genomics of microbial pathogens. Mol. Microbiol. 33: 1103-1117.

Kanei, K. 1967. Detection of host-originated acid phosphatase on the surface of *in vivo*-grown tubercle bacilli. Japan J. Med. Sci. Biol. 20: 73-90.

Kato-Maeda, M., Rhee, J.T., Gingeras, T.R., Salamon, H., Drenkow, J, Smittipat, N., and Small, P.M. 2001. Comparing genomes within the species *Mycobacterium tuberculosis*. Genome Res. 11: 547-554.

Kaufmann, S.H.E. 1993. Immunity to intracellular bacteria. Annu. Rev. Immunol. 11:129-163.

Kaushal, D., Schroeder, B.G., Tyagi, S., Yoshimatsu, T., Scott, C., Ko, C., Carpenter, L., Mehrotra, J., Manabe, Y.C., Fleishmann, R.D., and Bishai, W.R. 2002. Reduced immunopathology and mortality despite tissue persistence in a *Mycobacterium tuberculosis* mutant lacking alternative σ factor, SigH. Proc. Natl. Acad. Sci. USA 99: 8330-8335.

Kinger, A.K. and Tyagi, J.S. 1993. Identification and cloning of genes differentially expressed in the virulent strain of *Mycobacterium tuberculosis*. Gene 131: 113-117.

Ko, M.S. 1990. An "equalized cDNA library" by the reassociation of short double-stranded cDNAs. Nucl. Acids Res. 18: 5705-5711.

Kremer, L., Baulard, A., Estquier, J., Poulain-Godefroy, O., and Locht, C. 1995. Green fluorescent protein as a new expression marker in mycobacteria. Mol. Microbiol. 17: 913-922.

Lee, B-Y. and Horwitz, M.A. 1995. Identification of macrophage and stress-induced proteins of *Mycobacterium tuberculosis*. J. Clin. Invest. 96: 245-249.

Lee, R.E., Brennan, P.J., and Besra, G.S. 1996. *Mycobacterium tuberculosis* cell envelope. Curr. Top. Microbiol. Immunol. 215:1-27.

Lemassu, A. and Daffé, M. 1994. Structural features of the exocellular polysaccharides of *Mycobacterium tuberculosis*. Biochem. J. 297:351-357.

Li, M-S., Monahan, I.M., Waddell, S.J., Mangan, J.A., Martin, S.L., Everett, M.J., and Butcher, P.D. 2001. cDNA-RNA subtractive hybridization reveals increased expression of mycocerosic acid synthase in intracellular *Mycobacterium bovis* BCG. Microbiology 147: 2293-2305.

Manganelli, R., Voskuil, M.I., Schoolnik, G.K., and Smith, I. 2001. The *Mycobacterium tuberculosis* ECF sigma factor σ^E: role in global gene expression and survival in macrophages. Mol. Microbiol. 41: 423-437.

Manganelli, R., Voskuil, M.I., Schoolnik, G.K., Dubnau, E., M. Gomes, and I. Smith. 2002. Roles of the extracytoplasmic-function σ factor σ^H in *Mycobacterium tuberculosis* global gene expression. Mol. Microbiol. 45: 365-374.

Mariani, F., Capelli, G., Riccardi, G., and Colizzi, V. 2000. *Mycobacterium tuberculosis* H37Rv comparative gene expression analysis in synthetic medium and human macrophage. Gene 253: 281-291.

Marston, B.J. and Shinnick, T.M. 1996. Differentially expressed genes of *Mycobacterium tuberculosis*. Annals NY Acad. Sci. 797: 32-41.

Mattow, J., Jungblut, P.R., Muller, E.C., and Kaufmann, S.H.E. 2001a. Identification of acidic, low molecular mass proteins of *Mycobacterium tuberculosis* strain H37Rv by matrix-assisted layer desorption/ionization and electrospray ionization mass spectrometry. Proteomics 1: 494-507.

Mattow, J., Jungblut, P.R., Schaible, U.E., Mollenkopf, H.J., Lamer, S., Zimny-Arndt, U., Hagens, K., Muller, E.C., and Kaufmann, S.H.E. 2001b. Identification of proteins of *Mycobacterium tuberculosis* missing in attenuated *Mycobacterium bovis* BCG strains. Electrophoresis 22: 2936-2946.

McAdam, R.A., Quan, S., Smith, D.A., Bardarov, S., Betts, J.C., Cook, F.C., Hooker, E.U., Lewis, A.P., Everett, M.J., Lukey, P.T., Bancroft, G.J., Jacobs, W.R., Jr., and Duncan, K. 2002. Characterization of a *Mycobacterium tuberculosis* H37Rv transposon library reveals insertions in 351 ORFs and mutants with altered virulence. Microbiology 148: 2975-2986.

McKinney, J.D., Honer zu Bentrup, K., Munoz-Elias, E.J., Miczak, A., Chen, B., Chan, W.-T., Swenson, D., Sacchettini, J.C., Jacobs, W.R., Jr., and Russell, D.G.. 2000. Persistence of *Mycobacterium tuberculosis* in macrophages and in mice requires the glyoxalate shunt enzyme isocitrate lyase. Nature 406: 735-738.

Michele, T., Ko, C., and Bishai, W.R.. 1999. Antibiotic exposure induces expression of the *Mycobacterium tuberculosis sigF* gene: implications for chemotherapy against mycobacterial persisters. Antimicrob. Agents Chemother. 43: 218-225.

Monahan, I.M., Betts, J., Bannerjee, D.K., and Butcher, P.D. 2001. Differential expression of mycobacterial proteins following phagocytosis by macrophages. Microbiology 147: 459-471.

Morsczek, C., Berger, S., and Plum, G. 2001. The macrophage-induced gene (*mig*) of *Mycobacterium avium* encodes a medium-chain acyl-coenzyme A synthetase. Biochim. Biophys. Acta 1521: 59-65.

O'Farrell, P.H. 1975. High-resolution two-dimensional gel electrophoresis of proteins. J. Biol. Chem. 250: 4007-4021.

Ortalo-Magné, A., Andersen, A.B., and Daffé, M. 1996. The outermost capsular arabinomannans and other mannoconjugates of virulent and avirulent tubercle bacilli. Microbiology. 142: 927-935.

Pierce, C.H., Dubos, R.J., and Schaefer, W.B. 1953. Multiplication and survival of tubercle bacilli in the organs of mice. J. Exp. Med. 97: 189-206.

Pelicic, V., Jackson, M., Reyrat, J.M., Jacobs, W.R., Jr., Gicquel, B., and Guilhot, C. 1997. Efficient allelic exchange and transposon mutagenesis in *Mycobacterium tuberculosis*. Proc. Natl. Acad. Sci. USA. 94: 10955-10960.

Plum, G. and Clark-Curtiss, J.E. 1994. Induction of *Mycobacterium avium* gene expression following phagocytosis by human macrophages. Infect. Immun. 62: 476-483.

Pokorny, J. 1962. Cultivation of mycobacteria *in vivo*. II. Study of cellular lipids. Rozhl. Tuberk. 22: 511-517.

Ramakrishnan, T., Indira, M., and Muller, R.K. 1962. Evaluation of the route of glucose utilization in virulent and avirulent strains of *Mycobacterium tuberculosis*. Biochim. Biophys. Acta 59: 529-532.

Ratledge, C. 1982. Nutrition, growth, and metabolism. In: The Biology of the Mycobacteria, Vol. 1. C. Ratledge and J. Stanford, eds. Academic Press, New York, NY. p. 186-271.

Raynaud, C., Papavinasasundarum, K.G., Speight, R.A., Springer, B., Sander, P., Böttger, E.C., Colston, M.J., and Draper, P. 2002. The functions of OmpATb, a pore-forming protein of *Mycobacterium tuberculosis*. Mol. Microbiol. 46: 191-201.

Rindi, L., Lari, N., and Garzelli, C. 1999. Search for genes potentially involved in *Mycobacterium tuberculosis* virulence by mRNA differential display. Biochem. Biophys. Res. Commun. 258: 94-101.

Rindi, L., Lari, N., and Garzelli, C. 2001. Genes of *Mycobacterium tuberculosis* H37Rv downregulated in the attenuated strain H37Ra are restricted to *M. tuberculosis* complex species. New Microbiol. 24:289-294.

Rivera-Marrero, C.A., Burroughs, M.A., Masse, R.A., Vannberg, F.O., Leimbach, D.L., Roman, J., and Murtagh, J.J., Jr. 1998. Identification of genes differentially expressed in *Mycobacterium tuberculosis* by differential display PCR. Microb. Pathogen. 25: 307-316.

Rosenkrands, I., Weldingh, K., Jacobsen, S., Hansen, C.V., Florio, W., Gianetri, I., and Andersen, P. 2000b. Mapping and identification of *Mycobacterium tuberculosis* proteins by two-dimensional gel electrophoresis, microsequncing, and immunodetection. Electrophoresis 21: 935-948.

Schnappinger, D., Ehrt, S., Voskuil, M., Mangan, J., Monahan, I., Wilson, M., Butcher, P., Nathan, C., and Schoolnik, G.K. 2001. Microarray expression analysis of the *Mycobacterium tuberculosis* macrophage interaction.

Abstract 153, Keystone Meeting on Molecular and Cellular Aspects of Tuberculosis Research in the Post Genome Era, Taos, NM

Segal, W. 1962. Differential succinic dehydrogenase activity in intact cells of *Mycobacterium tuberculosis* grown *in vitro* and *in vivo*. Bacteriol. Proc. 62: 96.

Segal, W. 1984. Growth dynamics of *in vivo* and *in vitro* grown mycobacterial pathogens. In: The Mycobacteria: A Sourcebook, Part A. G.P. Kubica and L.G. Wayne, eds. Marcel Dekkar, Inc., New York, NY. p. 547-573.

Segal, W. and Bloch, H. 1956. Biochemical differentiation of *Mycobacterium tuberculosis* H37Rv grown *in vivo* and *in vitro*. J. Bacteriol. 72: 132-141.

Segal, W. and Miller, W. 1965. Comparative study of *in vivo* and *in vitro* grown *M. tuberculosis*. III. Lipid composition. Proc. Soc. Exp. Biol. Med. 118: 613-616.

Sever, J.L. and Youmans, G.P. 1957. The relation of oxygen tension to virulence of tubercle bacilli and to acquired resistance to tuberculosis. J. Infect. Dis. 101: 193-202.

Sherman, D.R., Voskuil, M., Schnappinger, D., Liao, R., Harrell, M.I., and Schoolnik, G.K. 2001. Regulation of the *Mycobacterium tuberculosis* hypoxic response gene encoding α-crystallin. Proc. Natl. Acad Sci. USA 98: 7534-7539.

Springer, B., Master, S., Sander, P., Zahrt, T., McFalone, M., Song, J., Papavinasasundarum, K.G., Colston, M.J., Bottger, E., and Deretic, V. 2001. Silencing of oxidative stress response in *Mycobacterium tuberculosis*: expression patterns of *ahpC* in virulent and avirulent strains and effect of *ahpC* inactivation. Infect. Immun. 69: 5967-5973.

Stewart, G.R., Wernisch, L., Stabler, R., Mangan, J.A., Hinds, J., Liang, K.G., Young, D.B., and Butcher, P.D. 2002. Dissection of the heat-shock response of *Mycobacterium tuberculosis* using mutants and microarrays. Microbiology 148: 3129-3138.

Strohmeier, G.R. and Fenton, M.J. 1999. Roles of lipoarabinomannan in the pathogenesis of tuberculosis. Microbes Infect. 1:709-717.

Talaat, A.M., Hunter, P., and Johnston, S.A. 2000. Genome-directed primers for selective labeling of bacterial transcripts for DNA microarray analysis. Nat. Biotechnol. 18: 679-682.

Talaat, A.M., Lyons, R., and Johnston, S.A. 2001. Genome expression profile of *M. tuberculosis* grown in mice lungs. Abstract 147, Keystone Meeting on Molecular and Cellular Aspects of Tuberculosis Research in the Post Genome Era. Taos, NM.

Tyagi, A.K., Das Gupta, S.K., and Jain, S. 2000. Gene expression: reporter technologies. In: Molecular Genetics of Mycobacteria. G.F. Hatfull and W.R. Jacobs, Jr., eds. American Society for Microbiology Press, Washington, DC. p. 131-147.

Unniraman, S., Chatterji, M., and Nagaraja, V. 2002. A hairpin near the 5' end stabilizes the DNA gyrase mRNA in *Mycobacterium smegmatis*. Nucl. Acids Res. 30: 5376-5381.

Valdivia, R.H., Hromokyj, A.E., Manack, D., Ramakrishnan, L., and Falkow, S. 1996. Applications for green fluorescent protein (GFP) in the study of host-pathogen interactions. Gene 173: 47-52.

Vercellone, A., Nigou, J. and Puzo, G. 1998. Relationships between the structure and the roles of lipoarabinomannans and related glycoconjugates in tuberculosis pathogenesis. Frontiers in Bioscience. 3: 149-163.

Via, L.E., Curcic, R., Mudd, M.H., Dhandayuthapani, S., Ulmer, R.J., and Deretic, V. 1996. Elements of signal transduction in *Mycobacterium tuberculosis*: *in vitro* phosphorylation and *in vivo* expression of the response regulator MtrA. J. Bacteriol. 178: 3314-3321.

Wayne, L.G. 1994. Dormancy of *Mycobacterium tuberculosis* and latency of disease. Eur. J. Clin. Microbiol. Infect. Dis. 13: 908-914.

Wayne, L.G. and Sohaskey, C.D. 2001. Nonreplicating persistence of *Mycobacterium tuberculosis*. Annu. Rev. Microbiol. 55: 39-63.

Wheeler, P.R. and Ratledge, C. 1994. Metabolism of *Mycobacterium tuberculosis*. In: Tuberculosis: Pathogenesis, Protection and Control. B.R. Bloom, ed. American Society for Microbiology Press, Washington, DC. p. 353-385.

Wilkinson, R.J., DesJardin, L.E., Islam, N., Gibson, B.M., Kanost, R.A., Wilkinson, K.A., Poelman, D., Eisenach, K.D., and Toosi, Z. 2001. An increase in expression of a *Mycobacterium tuberculosis* mycolyl transferase gene (*fbpB*) occurs early after infection of human monocytes. Mol. Microbiol. 39: 813-821.

Wilson. M.A., DeRisis, J., Kristensen, H.-H., Imboden, P., Rane, S., Brown, P.O., and Schoolnik, G.K. 1999. Exploring drug-induced alterations in gene expression in *Mycobacterium tuberculosis* by microarray hybridization. Proc. Natl. Acad. Sci. USA 96: 12833-12838.

Youmans, G.P. 1979. Tuberculosis. W.B. Saunders, Philadelphia, PA

Yuan, Y., Crane, D.D., Simpson, R.M., Zhu, Y.Q., Hickey, M.J., Sherman, D.R., and Barry, C.E., III. 1998. The 16 kDa α-crystallin (Acr) protein of *Mycobacterium tuberculosis* is required for growth in macrophages. Proc. Natl. Acad. Sci. USA 95: 9578-9583.

Zahrt, T.C. and Deretic, V. 2000. An essential two-component signal transduction system in *Mycobacterium tuberculosis*. J. Bacteriol. 182: 3832-3838.

Zahrt, T.C. and Deretic, V. 2001. *Mycobacterium tuberculosis* signal transduction system required for persistent infections. Proc. Natl. Acad. Sci. USA 98: 12706-12711.

From: Tuberculosis: The Microbe Host Interface
Edited by: Larry S. Schlesinger and Lucy E. DesJardin

Chapter 8

Analysis of Latency

John Chan and JoAnne Flynn

Abstract

Reactivation of latent tuberculous infection plays a significant role in the pathogenesis of tuberculosis. The mechanisms involved in the establishment of latent tuberculosis and subsequent recrudescence are, however, not well understood. Studies designed to characterize these mechanisms have been hampered by the lack of knowledge of the precise immunologic response of the host as well as the physiologic state of the tubercle bacilli during latent and reactivation tuberculosis, thus rendering modeling of these two phases of tuberculous infection difficult. Nonetheless, results obtained through the use of various *in vitro* and *in vivo* models have shed considerable light on the host-bacillus interaction in latent and reactivation tuberculosis.

Introduction

Mycobacterium tuberculosis has the unique ability to persist in an infected host in a dormant state whose nature remains to be characterized (Flynn and Chan, 2001; McKinney *et al.,* 1998; Parrish *et al.,* 1998). A host harboring dormant tubercle bacilli, in general, does not exhibit clinical signs or symptoms, but is at risks for developing reactivation disease at a later time that could be decades after the primary infection. It is generally assumed that individuals with a positive tuberculin reaction without prior BCG vaccination — that is, those who have been exposed to and infected with *M. tuberculosis* — harbor latent bacilli, and have a 10% chance of developing reactivation disease during their lifetime (Styblo, 1980). Given the estimation that one-third of the world's population is infected with the tubercle bacillus (Dye *et al.*, 1999), understanding the mechanisms by which *M. tuberculosis* establishes a latent infection and subsequently reactivates may lead to improved prevention and treatment strategies against tuberculosis. The goal of this monograph is to review the *in vivo* and *in vitro* methods and models that are commonly used for the study of tuberculous latency and reactivation.

In Vitro Models of Tuberculous Latency

A logical approach to gain insight into the mechanisms by which *M. tuberculosis* adapt to the host immune response to maintain persistence is to design *in vitro* systems that mimic the environment the tubercle bacillus is exposed to in infected tissues. Obviously, it is desirable that these *in vitro* systems are amenable to experimental manipulation. Although the precise host environments encountered by the tubercle bacillus from the time of entry into the lungs to the point of active infection or establishment of latency remain to be defined, various physiological states, based on existing knowledge of the physiology of both the host and the bacillus, as well as clinicopathological studies of tuberculosis, have been used as experimental conditions for the establishment of *in vitro* models of *M. tuberculosis* persistence. In this review, we will focus on the modeling of anaerobiosis and nutrient deprivation.

Modeling Anaerobiosis

Although formal studies designed to evaluate the oxygen tension in a tuberculous granuloma have not been carried out, it is generally thought that the granulomatous tissues harboring dormant bacilli are anaerobic. This notion, though not substantiated by formal experimental evidence, is not an unreasonable one considering the following observations. Based on the known metabolism of mononuclear phagocytic cells within which *M. tuberculosis* resides, Loebel *et al.*, (1933) estimated that the center of a granuloma with a radius of 0.075 to 0.35 mm is totally anaerobic. Caseous necrosis, a common occurrence in the granuloma of tuberculosis patients (Haas and des Perez, 1995), is likely to be relatively anaerobic due to the destruction of tissues including the oxygen-supplying vasculature. Viable bacilli had been cultured from surgical lung specimens with no patent bronchial access from a tuberculous patient whose sputum had been negative for *M. tuberculosis* for at least 9 months (Wayne and Salkin, 1956). Thus, much research effort has been directed at understanding the biology of the tubercle bacillus under anaerobic conditions in attempt to characterize the latent state of the tubercle bacillus.

Wayne *et al.* have established *in vitro* models of *M. tuberculosis* latency that generates non-replicating bacilli via adaptive growth in conditions of different levels of anaerobicity (Wayne, 1976, 1977; Wayne and Sramke, 1979, 1994; Wayne and Lin, 1982; Wayne and Hayes, 1996). An earlier model involved incubating mycobacterial cultures in a sealed container without agitation (Wayne, 1976, 1977; Wayne and Sramek, 1979, 1994; Wayne and Lin, 1982). As a result, as the bacilli deplete the culture of oxygen and settle to the bottom, a self-generated anaerobic environment gradually forms. The bacilli in the sediment are in a non-replicating state, but are able to resume normal growth in a synchronized fashion upon replenishment of oxygen (Wayne, 1977). These bacteria have adapted to the microaerophilic conditions encountered in the bottom of the culture, and thus exhibit enhanced tolerance to anaerobicity compared to organisms grown in normal aerated conditions. Antigenic analysis of the microaerophilic bacilli demonstrated differences from those grown in normal culture condition (Wayne and Sramek, 1979; Wayne and Lin, 1982). Sepcifically, there was a marked increase in the expression of a mycobacterial component designated URB (unbound antigens unique to resting bacilli) (Wayne and Sramek, 1979; Wayne and Lin, 1982). During the adaptation to microaerophilic conditions, the bacilli also become resistant to certain antituberculous drugs including isoniazid and rifampin, with concomitant development of susceptibility to metronidazole (Wayne and Sramek, 1994), an antimicrobial effective against anaerobic organisms. In addition, the non-replicating bacilli display a distinctive metabolism. For example, the activity of isocitrate lyase, an enzyme of the glyoxylate cycle, is enhanced in bacilli in the sediment of the microaerophilic cultures (Wayne and Lin, 1982). This microaerophilic non-replicating state is also associated with enhanced expression of glycine dehydrogenase (as assessed by NADH oxidation during reductive amination of glyoxylate), the enzyme presumably responsible for maintaining nicotinamide adenine dinucleotide (NAD) levels in relatively anaerobic conditions (Wayne and Lin, 1982). The gene responsible for this activity in *M. tuberculosis*, however, remains to be determined. Recent evidence suggests that this activity may be attributable to the action of *M. tuberculosis* alanine dehydrogenase (Usha *et al.*, 2002).

One drawback of the above-described model of anaerobic dormancy is the non-agitated nature of the culture system, which results in the generation of an oxygen gradient that decreases toward the bottom of the medium. Because of the existence of such a gradient, the bacteria generated by this culture system are not physiologically homogeneous (Wayne and Hayes, 1996). Cognizant of this drawback, Wayne *et al.* developed a second model of anaerobic dormancy that produces *M. tuberculosis* bacilli with a homogenous physiological state (Wayne and Hayes, 1996). In this refined model of anaerobic dormancy, *M. tuberculosis* is cultured in sealed containers containing a fixed ratio of the volume of media and head-space air called HSR (head space ratio). Homogeneity of culture conditions is assured by constant gentle stirring using a magnetic bar. This produces an environment in which the oxygen levels diminish over time. The HSR, the rigor with which the organisms are mixed, and the size and configuration of the culture container are important experimental parameters that can affect the growth kinetics and the physiology of the cultured bacilli. Using this system, it has been demonstrated that as oxygen is being depleted in the culture medium, the bacteria enter into a non-replicating persistent state, but can resume growth upon replenishment of oxygen. The entry into this non-replicating state occurs first at a microaerophilic environment. This microaerophilic non-replicating persistence state (NRP stage 1) is characterized by cessation of bacterial growth and DNA synthesis associated with increase in culture turbidity. This phenomenon of growth-culture turbidity dissociation is due to an increase in the size of tubercle bacilli when grown in microaerophilic conditions. The microaerophilic non-replicating state is also associated with enhanced expression of glycine dehydrogenase (Wayne and Hayes, 1996).

As the oxygen saturation diminishes further, *M. tuberculosis* enters into a state, designated anaerobic (NRP stage 2), during which the increase in culture turbidity terminates, and the activity of glycine dehydrogenase decreases. Although DNA synthesis is apparently absent during the NRP 1 and NRP 2 phases of non-replicating persistence, pulse-labeling studies indicate that RNA synthesis is maintained (Wayne and Hayes, 1996). Bacilli from both the NRP stages develop resistance to specific antimycobacterial agents including isoniazid. In the late NRP stage 2, the non-replicating *M. tuberculosis* becomes susceptible to metronidazole. This latter susceptibility is, however, not a property of the NRP stage 1 bacilli. The stage 1 and stage 2 non-replicating *M. tuberculosis* also differ in their growth pattern upon replenishment of oxygen. Thus, with re-aeration, bacteria in the NRP stage 2 (anaerobic) replicate in a synchronous fashion, while those in the NRP stage 1 (microaerophilic) do not (Wayne and Hayes, 1996). These data, together with those provided by drug susceptibility testing, as well as the analysis of glycine dehydrogenase expression and the relationship between culture turbidity and time, strongly suggest that the biological properties of the NRP stage 1 and NRP stage 2 bacteria are different.

As interest in mycobacterial latency and reactivation rekindle, results generated by the Wayne system of anaerobic dormancy have become the basis for studying the persistent state induced by anaerobic conditions. Importantly, data generated by the more recent studies designed to characterize the physiology of *M. tuberculosis* in anaerobic conditions support and extend some of the original observations reported by Wayne *et al.* Thus, it has been shown that under relatively anaerobic conditions, the mycobacterial cell wall thickens, with deposition of α-crystallin, a 16-kD chaperon of *M. tuberculosis* (Cunningham and Spreadbury, 1998). Recently, the expression of α-crystallin has been shown to be markedly enhanced under relatively anaerobic conditions (Yuan *et al.,* 1996), and in a modified Wayne model, has been identified by immunoreactivity as the URB antigen (Desjardin *et al.,* 2001), the mycobacterial component shown by Wayne *et al.* a decade earlier to be highly expressed during hypoxia (Wayne and Sramek, 1979; Wayne and Lin, 1982). These results are reminiscent of those obtained using the Wayne model of anaerobic dormancy demonstrating an increase in size of the non-replicating tubercle bacillus in the NRP stage 1 growth phase (Wayne and Hayes, 1996). Importantly, *M. tuberculosis* α-crystallin knockout mutant has been shown to be deficient for growth in macrophages (Yuan *et al.,* 1998), although not in mice (J. Flynn, unpublished). However, the biological function of α-crystallin and its essentiality for the survival of the tubercle bacilius in the presumed anaerobic environment of the human granuloma remains to be determined.

Isocitrate lyase (Icl), a key component of the glyoxylate cycle, has been a focus for investigation with regard to its role in promoting persistence of the tubercle bacillus in the host. The expression of Icl was enhanced in non-replicating *M. tuberculosis* in microaerophilic condition generated by the original Wayne model (Wayne and Sramek, 1994). The glyoxylate cycle, present in prokaryotes, lower eukaryotes, and plants, facilitates the utilization of fatty acids as the sole carbon source in biosynthesis. It has been shown previously that Icl activity is enhanced in aged cultures of *M. tuberculosis* (Murthy *et al.,* 1973). Purified *M. tuberculosis* Icl exhibits isocitrate lyase activity and the production of this enzyme is enhanced under minimal growth condition when supplemented with acetate or palmitate (zu Bentrup *et al.,* 1999). Using genetic approaches, recent studies have demonstrated that *M. tuberculosis icl* is essential for persistence in the mouse (McKinney *et al.,* 2000). These data support the utility of the Wayne *in vitro* anaerobic model of tuberculous latency for the characterization of some *M. tuberculosis* components that are essential for survival within the infected host.

The Wayne Model Revisited: Analysis of *M. tuberculos*is Anaerobiosis by Microarray and Proteomics

Availability of sequence information of the entire *M. tuberculosis* genome has greatly facilitated studies designed to understand the biology of the tubercle bacillus (Cole *et al.,* 1998). In particular, this sequence information has made possible the use of the microarray technology to determine the gene expression profile of *M.* tuberculosis under specific environmental conditions, as well as that of mutants (Fisher *et al.,* 2002; Betts *et al.,* 2002; Kaushal *et al.,* 2002;

Sherman *et al.*, 2001; Manganelli *et al.*, 2001; Wilson *et al.*, 1999). For example, this method has recently been exploited to profile gene expression of *M. tuberculosis* under hypoxic condition (Sherman *et al.*, 2001). Microarray-based comparison of gene expression using RNAs extracted from *M. tuberculosis* cultured in air and those shifted to a hypoxic atmosphere has demonstrated that relative anaerobicity up-regulates the expression of 47 open reading frames (Sherman *et al.*, 2001). Interestingly, among those up-regulated is the gene encoding α-crystallin (*acr*), whose expression has been shown to be enhanced in the original Wayne model (Wayne and Sramek, 1979; Wayne and Lin, 1982), and subsequently reported to be induced by hypoxia (Yuan *et al.*, 1996; Desjardin *et al.*, 2001). More important, this study also identified enhanced expression of a two-component response regulator *Rv3123c*, which has been shown, by genetic analysis, to be essential for the up-regulation of *acr* (Sherman *et al.*, 2001). Thus, *Rv3133c* may be responsible for the regulation of the hypoxia-induced regulon of *M. tuberculosis*, an important piece in the puzzle of how the tubercle bacillus survives in the anaerobic environment of the granuloma.

Sequence information of the complete *M. tuberculosis* genome has also facilitated the application of proteomics to characterizing the protein expression patterns of different strains of mycobacteria, as well as to examine the mechanisms by which the tubercle bacillus response to specific environmental conditions (Rosenkrands *et al.*, 2002; Jungblut *et al.*, 2001; Florczk *et al.*, 2001; Covert *et al.*, 2001; Mattow *et al.*, 2001; Rosendrands *et al.*, 2000; Mollenkopf *et al.*, 1999). Using metabolic labeling, two-dimensional gel electrophoresis (2D-GE) in conjunction with matrix-assisted laser desorption ionization-time of flight (MALDI-TOF), and protein signature peptide liquid chromatography-electronspray ionization mass spectroscopy (LC-MS) analysis of total bacterial lysates, Rosenkrands *et al.* have identified 5 proteins that are more abundant in *M. tuberculosis* grown under hypoxic conditions compared to control bacterial cultured in normal atmospheric oxygen tension (Rosenkrands *et al.*, 2002). Two additional proteins were found in increased amounts in the culture filtrates of bacilli grown under relatively anaerobic environment compared to controls. Among these proteins is α-crystallin, whose expression has been shown to be hypoxia-induced in a number of experimental anaerobic *in vitro* systems (Wayne and Sramek, 1979; Wayne and Lin, 1982; Yuan *et al.*, 1996; Desjardin *et al.*, 2001). Of particular interest is the demonstration that protein signature peptide LC-MS analysis of total mycobacterial lysates was able to confirm the increased abundance of proteins as assessed by 2D-GE and MALDI-TOF. Protein signature peptide LC-MS involves tryptic digestion of total bacterial lysates into peptides, which are separated by LC prior to MS analysis. Accurate assignment depends on the analysis of unique signatures of peptides derived from digestion of the specific proteins. Relative quantification is achieved using a predictably stable protein, in this case Rp1L, as an internal control (Rosenkrands *et al.*, 2002; Ji *et al.*,

2000). Although the specificity and precision of signature peptide LC-MS remains to be determined, these latter results demonstrate the feasibility of this efficient method to directly use total bacterial lysates for the characterization of differential protein expression of the tubercle bacillus.

Modeling Starvation

In the early 1930's, Loebel *et al.* proposed that in order to understand how the tubercle bacillus survives within the host, it is important to determine the metabolic characteristics of *M. tuberculosis* and the physical and chemical properties of the tissue environment where the bacteria reside (Loebel *et al.*, 1933, 1933a). A starvation model was established to study the influence of various nutrients on the metabolism of *M. tuberculosis* (Loebel *et al.*, 1933). Using this model, experiments were carried out to study the level of respiration of the tubercle bacillus upon transfer from a Long's medium culture to saline phosphate, using a modified Barcroft-Warburg monometer (Loebel *et al.*, 1933). In addition, Loebel *et al.* also examined the "persistence of the proliferative capacity" of the starved bacilli when transferred to medium containing adequate nutrients (Loebel *et al.*, 1933, 1933a). The results of these studies demonstrated that the level of respiration of *M. tuberculosis* diminishes significantly after a few days of starvation. This starvation-induced decrease in the level of respiration can be restored through addition of nutrients even after prolonged deprivation. The starved bacilli also maintain their capacity to proliferate upon replenishment of nutrients. Loebel *et al.* surmised that these attributes of *M. tuberculosis* may be "the factors contributing to the hardiness of the tubercle bacillus" (Loebel *et al.*, 1933, 1933a).

Although the degree of nutrient restriction within the tuberculous granuloma remains to be determined, the Loebel's starvation model, designed to examine the metabolism of starved bacilli, whose capability to proliferate upon replenishment of nutrients is maintained, is a reasonable system of *M. tuberculosis* persistence. Betts *et al.* has recently developed an *in vitro* system, based on the Loebel starvation model, to characterize the effects of starvation on *M. tuberculosis* expression of genes and proteins (Betts *et al.*, 2002). In this model, nutrient-deprived condition is achieved by shifting log-phase bacilli grown in nutrient-rich media to phosphate-buffered saline (PBS). The PBS culture is carried in sealed bottles at 37°C. This culture system generates apparently non-replicating bacilli that exhibit a low level of respiration, as assessed by methylene blue decolorization (while cultures in nutrient-rich media decolorize methylene blue completely 9 days after inoculation, PBS cultures of the same age do not). This result is in keeping with that of experiments by Loebel *et al.* (1933), who demonstrated, using a Warburg manometer, that the level of respiration of stringently starved *M. tuberculosis* drops significantly by 4 days after inoculation. The starved bacteria exhibit no loss of viability in the nutrient-deprived condition for up to 6 weeks. Similar to *M. tuberculosis* grown in hypoxic condition in the Wayne model, starved bacilli develop resistance to isoniazid and rifampin. However, unlike the oxygen-depletion, starvation is not associated with the acquirement of susceptibility to metronidazole (Betts *et al.,* 2002). Since metronidazole requires anaerobic conditions for activity, the fact that nutrient-deprived bacilli do not develop sensitivity to this antimcirobial further supports the finding based on mythelene blue decoloarization that the starved *M. tuberculosis* culture is not hypoxic. As a result, this system has the ability to distinguish between the response of *M. tuberculosis* to nutrient depletion and hypoxia.

Using the modified Loebel model of starvation, in conjunction with gene microarray and proteome analysis, the effects of nutrient deprivation on *M. tuberculosis* has been studied in detail (Betts *et al.,* 2002). The results of this study provide evidence in support of the observations that starved *M. tuberculosis* is likely non-replicating while exhibiting a diminished level of respiration. Thus, subsets of genes responsible for energy metabolism, lipid biosynthesis, cell division, and transcription are down-regulated. In addition, starvation induces stress response genes such as the *M. tuberculosis* α-crystallin.

In Vivo Models of Tuberculous Latency

Research into the mechanisms, both host and bacterial, important in latent and reactivation tuberculosis has been hampered by lack of a precise experimental animal model. Tractable animal models that genuinely represent latent tuberculous infection in humans do not exist. As discussed above, human latent tuberculosis remains an enigma, in terms of the status of the tubercle bacillus, the specific host factors that prevent reactivation, and the local environment of the granuloma. The lack of a definition of human latent *M. tuberculosis* infection in biochemical and molecular terms contributes to the difficulty in developing models that mimic this state. Other factors that complicate development and use of animal models of latent tuberculosis are the need for Biosafety Level 3 containment, the slow growth rate of the organism, and the cost of lengthy chronic infection studies.

Murine Models

Rodents are the most commonly used animals for tuberculosis studies. The advantages of using mouse models, such as cost, ease of manipulation (including the feasibility of doing large-scale aerosol infections), availability of reagents, inbred strains, and ability to contain the mice under BSL3 conditions, make studies in this model desirable. Mice infected with a low dose (20-100 bacilli) of virulent *M. tuberculosis*, either via the intravenous or aerosol route, do not initially control the infection, and bacterial numbers increase to ~10^5 to 10^6 in the lungs and spleen. By 4 weeks post-infection, the adaptive immune response gains control over the infection, and the bacterial numbers in the lung and spleen stabilize or decrease slightly. However, even though a robust cell-mediated response develops, the organisms are not eliminated and persist at fairly high levels in the lungs and spleens. This chronic infection can be maintained for many months (~1 year) with no apparent ill effects on the mouse (Flynn *et al.,* 1998). As mice age, resurgence of the persistent infection can result in death due to tuberculosis (Orme, 1988). If the relevant antimycobacterial immune response is compromised prior to infection (for example, in certain gene-knockout mice), establishment of the chronic state does not occur, supporting the importance of specific innate and adaptive immune response in controlling infection (reviewed in Flynn and Chan, 2001; Flylnn and Ernst, 2000). In addition, as detailed below, specific immune deficiencies induced during the chronic phase can lead to reactivation of the infection and death of the mice. This low-dose chronic persistent model mimics human latent tuberculosis in that the immune system is essential for maintaining the infection, and there are no clinical signs of illness in the mice for many months. The best and most useful description of this model is that it represents a steady state, clinically latent infection characterized by a stable equilibrium between the host and the microbe for an extended period of time. Although the level of bacteria maintained in the mouse lungs may not accurately reflect the human latent tuberculous infection, there is evidence that in this phase of steadily maintained bacillary burden, replication of the tubercle bacillus may be minimal (Rees and Hart, 1961).

Another murine model of latent tuberculosis was first established in the 1950s, by McCune, McDermott and colleagues at Cornell University (McCune *et al.,* 1966, 1966a). The "Cornell" model is based on administration of antimycobacterial drugs to infected mice to reduce the bacterial burden. In the original Cornell model, *M. tuberculosis*-infected mice were treated with isoniazid and pyrazinamide for 3 months to induce an apparently "sterile" state. After discontinuation of antibiotics for 3 months, reactivation or exacerbation of the presumably latent infection could occur spontaneously or through administration of immunosuppressive agents (McCune *et al.,* 1966, 1966a). A distinct advantage of the Cornell model, therefore, is its resemblance of human latent tuberculous infection in the sense that the bacterial burden of the infected host is low. An obvious weakness of this model, however, is the interference of the interaction between the host immune response and the bacilli by the anti-tuberculous agents needed to induce the persistent infection, thus differing from the naturally established latent human infection. A recent study of modifications of the Cornell model indicated that this model can be technically difficult to use and unpredictable for studies on the immunologic basis of latency and reactivation (Scanga *et al.,* 1999).

Despite the shortcomings, murine models of latent tuberculosis derived from the low-dose and the Cornell systems have provided information pertinent to human immunological mechanisms involved in the prevention of disease recrudescence. The results of these *in vivo* studies of murine persistent and reactivation tuberculosis are summarized in the next section.

The Utility of the Mouse Model of Latent Tuberculosis

Most studies in animal models of tuberculosis have been focused on the acute or initial stages of infection. Using aerosol or intravenous infection murine models, certain immune components essential for control of acute tuberculosis have been identified, including interferon (IFN)-γ, tumor necrosis factor (TNF)-α, interleukin (IL)-12, CD4 and CD8 T cells, and inducible nitric oxide synthase (NOS2) (reviewed in Flynn and Chan, 2001; Flynn and Ernst, 2000). IFN-γ is a key factor in control of tuberculosis, and mice deficient in this cytokine are the most susceptible mouse strain reported to date (Flynn *et al.,* 1993; cooper *et al.,* 1993). The requirement for IL-12 is linked to the importance of this cytokine in inducing IFN-γ production by T cells (Cooper *et al.,* 1997). Macrophage activation is impaired in the absence of IFN-γ, and bacterial replication proceeds apparently unimpeded (Flynn *et al.,* 1993). The importance of IFN-γ (Jouanguy *et al.,* 1996; Newport *et al.,* 1996), and IL-12 (Altare *et al.,* 1998; de Jong *et al.,* 1998) in host defense against mycbacoterial infection has recently been demonstrated in humans. CD4 T cells are clearly important in control of tuberculosis, as suggested by the enhanced susceptibility of AIDS patients to tuberculosis. This was substantiated in murine models with deficiencies in CD4 T cells (Caruso *et al.,* 1999; Tascon *et al.,* 1998). IFN-γ production and macrophage activation were delayed in CD4 T cell deficient mice infected with *M. tuberculosis,* resulting in increased bacterial growth and death of the mice. The requirement for CD8 T cells in control of acute infection has been demonstrated in a number of murine models (Orme and Collins, 1984; Flynn *et al.,* 1992; Behar *et al.,* 1999; Feng *et al.,* 1999), and these cells both produce cytokines and function as cytotoxic cells during tuberculous infection (Tascon *et al.,* 1998; Feng *et al.,* 1999; Serbina *et al.,* 1999; Serbina *et al.,* 2000; Lewinsohn *et al.,* 1998). TNF-α appears to play multiple roles, including macrophage activation and granuloma formation (Flynn *et al.,* 1995; Bean *et al.,* 1999; Adams *et al.,* 1995). Macrophage activation is pivotal to control of *M. tuberculosis* infection; in mouse models, production of reactive nitrogen intermediates (RNI) by activated macrophages via NOS2 is a major antimycobacterial mechanism (reviewed in Chan and Flynn, 1999; Nathan and Shiloh, 2000). Mice deficient in the ability to produce RNI are quite susceptible to acute infection, demonstrating the requirement for NOS2 activity in macrophages *in vivo* (Chan *et al.,* 1995; MacMicking *et al.,* 1997; Scanga *et al.,* 2001). Controversy notwithstanding, emerging evidence suggests that NOS2 may play a role in defense against *M. tuberculosis* in humans (Chan and Flynn, 1999; Nathan and Shiloh, 2000).

It stands to reason that some of the mechanisms important in control of initial *M. tuberculosis* infection may be required to maintain control of a persistent or latent infection. However, the role of and requirement for various cytokines and T cell subsets during persistent infection might be quite different than during acute infection. For example, while the relatively robust bacterial growth that occurs during initial acute infection may require an intense host inflammatory response for disease control, it might be more beneficial to the host to down-regulate that initial response during the subsequent controlled persistent phase of infection, as long as appropriate antimycobacterial mechanisms are in place to prevent reactivation. Murine models of latent infection, which involve establishment of a chronic, persistent *M. tuberculosis* infection, have been used to identify those immune components that are important in preventing reactivation once an immune response has been established. The genes for IFN-γ, TNF-α, and NOS2 continued to be expressed in the lungs of mice infected for at least a year, during the period of stable bacterial burdens, suggesting that these components might also be important in control of persistent infection (Flynn *et al.*, 1998).

A major antimycobacterial mechanism of macrophages is RNI production via NOS2, as mentioned above. The chronic infection and Cornell murine models of latent tuberculosis were used to assess the requirement for continued production of RNI during latent tuberculosis, using an inhibitor of NOS2 function, aminoguanidine (Flynn *et al.*, 1998). Bacterial numbers increased dramatically in the lungs of the aminoguanidine-treated mice, who succumbed to reactivation tuberculosis, while control mice maintained stable bacterial numbers and survived the length of the experiment. The effects of NOS2 on bacterial numbers were largely confined to the lungs until late in the treatment period; this was in contrast to the effects of NOS2 inhibition in acute infections, in which spleen and liver were as affected as lungs. This suggests that RNI production may be more important in control of pulmonary tuberculosis than for infection of other organs. Another study in which NIL [(N^6-(iminoethyl)lysine]was used to inhibit NOS2 activity also demonstrated the requirement for NOS2 in controlling chronic tuberculous infection (MacMicking *et al.*, 1997). In experiments using a modification of the Cornell model of latent tuberculosis, inhibition of NOS2 activity also led to reactivation, although unlike the original Cornell model, bacterial load was low but detectable in the mice prior to inhibition of NOS2 (Flynn *et al.*, 1998). These observations indicated that macrophage activation, and specifically the production of RNI, was necessary to maintain the infection in a quiescent state, and reactivation occurred when NOS2 was inhibited.

CD4 T cells are an important component of the tuberculous granuloma (Flynn *et al.*, 1992; Shen *et al.*, 1988; Law *et al.*, 1996) and produce IFN-γ at the site of infection for activation of macrophages (Fenhalls *et al.*, 2000). Depletion of CD4 T cells 6 months after infection with *M. tuberculosis* in the chronic persistent model resulted in a rapid and dramatic increase in bacterial numbers in all organs, indicating that control of mycobacterial growth is dependent on functioning CD4 T cells (Scanga *et al.*, 2000). Pathology consistent with high bacterial numbers was observed, with coalescing granulomata, necrosis, and cell infiltration. A surprising finding of this study was that the absence of CD4 T cells did not result in diminished overall expression of IFN-γ, and production of NOS2 in the granulomata was unimpaired. Apparently, other cellular sources of IFN-γ exist that are in proximity to the macrophage (and therefore induce NOS2). In fact, CD8 T cell numbers increased in CD4 T cell-depleted mice, and these cells appeared to be the source of IFN-γ. Thus, the increase in bacterial numbers occurred despite apparently wild type levels of IFN-γ and NOS2 in CD4 T cell-depleted mice, indicating that these functions are insufficient to control chronic tuberculosis, and that CD4 T cells play additional roles besides IFN-γ production and macrophage activation. Studies in viral systems have suggested that CD4 T cells are important for priming CD8 T cells and maintaining long term CD8 T cell (including CTL) function (Clarke, 2000; Kalams and Walker, 1998). Recent data indicates that CD8 CTL function in the lungs of acutely *M. tuberculosis*-infected mice is dependent on CD4 T cells (Serbina *et al.*, 2001). Clearly, the role of CD4 T cells in tuberculosis is complex and multifaceted, involving more than simply IFN-γ production for macrophage activation.

The inflammatory cytokine TNF-α is viewed as a classic double-edged sword in tuberculosis. On the one hand, TNF-α in acute infection limits *M. tuberculosis* growth, as well as participates in granuloma formation, and is essential in the protective response against this pathogen. However, many have speculated that TNF-α is detrimental to the host and responsible for much of the pathology and symptomatology in active tuberculosis (reviewed in Rook and Bloom, 1994; Tsenova *et al.*, 1999). Indeed, overproduction of TNF-α by recombinant BCG caused increased lung pathology in a murine model (Bekker *et al.*, 2000). It seems that TNF-α production must be carefully modulated to achieve control of *M. tuberculosis* infection, yet avoid causing excessive damage to the host tissues. In a chronic persistent tuberculosis model, expression of the TNF-α receptor by adenovirus in *M. tuberculosis*-infected mice resulted in reactivation of the infection (Adams *et al.*, 1995). Neutralization of TNF-α by antibody treatment also caused reactivation of the persistent infection in both a modified Cornell model (Scanga *et al.*, 1999) and the chronic persistent infection model (Mohan *et al.*, 2001). The clinical relevance of the TNF-α neutralization-based murine model of tuberculosis reactivation has been validated by a recent study reporting an increased rate of tuberculous infection in patients receiving anti-TNF therapy for the treatment of chronic inflammatory diseases such as rheumatoid arthritis (Keane *et al.*, 2001). Clinicoepidemiological evidence suggests that a significant proportion of these TNF-α neutralization-associated cases of tuberculosis is due to reactivation of a latent infection (Keane *et al.*, 2001). In murine studies, TNF-α neutralization during persistent infection resulted in an initial increase in bacterial numbers that stabilized at levels usually not associated with lethality in the mice. However, TNF-α neutralized mice succumbed to the infection and this appeared to be due to extensive, destructive immunopathology in the lungs. The granulomatous response became disorganized and there was an intense and unfocused infiltration of mononuclear cells into the lungs. In some cases, this intense inflammatory response progressed to squamous metaplasia, demonstrated by keratin deposition. Squamous metaplasia is usually observed in response to chronic inflammation of the lungs, suggesting an enhanced inflammatory state in the absence of TNF-α. These results suggested that TNF-α plays an important role in the regulation of the inflammatory response to *M. tuberculosis*, and may not only cause but also limit pathology during tuberculosis. Further exploration of the mechanisms by which TNF-α modulates pathology during tuberculosis is necessary. The ability of this cytokine to affect adhesion molecules, chemokines and chemokine receptor expression suggests that one important role for TNF-α during persistent infection is in directing cell migration to and within the lungs.

Given the key role of IFN-γ in controlling the initial acute infection with *M. tuberculosis*, it is generally believed that this cytokine is crucial to maintaining a latent *M. tuberculosis* infection. However, this has been difficult to conclusively demonstrate in murine latency models. Unlike TNF-α, it has been difficult to effectively neutralize this cytokine during chronic or acute *M. tuberculosis* infection using antibodies *in vivo* (Scanga and Flynn, unpublished). Close contact between T cells producing IFN-γ and the macrophages may limit the accessibility of this cytokine to a neutralizing antibody, particularly in the lungs of chronically infected mice, where pathology can be extensive. A thorough understanding of the role for IFN-γ in control of latent tuberculosis may await better antibodies or construction of mice with specifically regulated IFN-γ gene expression, such as conditional knockouts.

Other Animal Models

There are other experimental models of tuberculosis, including guinea pigs, rabbits and non-human primates. Guinea pigs are quite susceptible to *M. tuberculosis*, and infection in this species is progressively fatal, even with a very low dose infection (Smith *et al.*, 1966). This animal model was used by Robert Koch to establish Koch's postulates for identification of the causative agent of infections (Koch, 1882), and this model is still used to study *M. tuberculosis* pathogenesis. The lung pathology in infected guinea pigs shares some important characteristics with that of human tuberculosis, and these animals are frequently used in testing vaccines against tuberculosis (Wiegeshaus *et al.*, 1970; reviewed in McMurray, 1994). There is a wealth of historical and recent research on rabbits and tuberculosis, indicating that some strains of rabbits are relatively resistant to *M. tuberculosis* and may develop latent infection (Lurie, 1964; Dannenberg, 1994). Classic studies by Lurie demonstrated that infection with *M. tuberculosis* was contained within the lungs, while infection with virulent *M. bovis* was not controlled by the rabbits (Dannenberg, 1994). Pathology in the rabbits is similar to humans, with classic granuloma formation, including caseation and cavity formation (Lurie, 1964; Dannenberg, 1994). These factors make the rabbit an attractive model for studies of tuberculosis. However, the available molecular and immunologic tools to study this infection in rabbits are limited, and must be developed ad hoc. This, and the biosafety concerns of housing infectious rabbits, has somewhat limited the utility of this model, although recently there has been a renewed interest in use of this model to study latent tuberculosis.

Recently, an intriguing frog granuloma model has been developed that is potentially useful for the characterization of mechanisms by which tuberculous latency is established (Ramakrishnan *et al.*, 2000). This system is based on the ability of the mycobacterial pathogen *M. marinum* to establish a persistent infection in frogs. Rhamakrishna *et al.* (2000) have used this model and an *in vitro* culture system, in conjunction with differential fluorescence induction — a fluorescence—activated cell sorter-based method — to identify genes with enhanced expression *in vivo*. Detailed genetic analysis of *M.* marinum *mag 24-1*, a homolog of the *M. tuberculosis* PE-PGRS family, demonstrated that this gene is required for persistence in the frog granuloma model. The *M. tuberculosis* PE-PGRS genes are predicted to encode a family of glycine-rich proteins whose precise functions remain to be defined. While the utility of the *M. marinum*-frog model for the identification of *M. tuberculosis* genes that play a role in tuberculous latency has yet to be established, the chronic persistence and non-BSL3 nature of this system are desirable features for examining the pathogenic mechanisms of the tubercle bacillus.

Non-human primates have the potential to be a useful model for acute and latent tuberculosis studies. Non-human primates can contract tuberculosis upon exposure to humans with active disease, and this can have disastrous consequences for a primate colony. Early studies indicated that in rhesus macaques, the progression of *M. tuberculosis* infection was often rapid and fatal (Good, 1968). Primates with active disease are infectious to people and other animals. A recent study using cynomolgus macaques suggested that a very low dose of virulent *M. tuberculosis* delivered to the lungs could result in slowly progressive or subclinical disease in a percentage of animals tested (Walsh *et al.*, 1996). Use of non-human primates to study tuberculosis may provide a model closer to human tuberculosis than the other animal models currently in use. Historically, vaccines against tuberculosis were tested in macaque models of aerosol *M. tuberculosis* infection (Good, 1968; Ribi *et al.*, 1971). However, more recent development and use of a monkey model for studies of tuberculosis has been limited by the high cost of non-human primate housing and research and the necessity for BSL3 primate housing that would protect both humans and other primates in the colony. A small number of groups currently are developing this model in rhesus and cynomolgus macaques (Walsh *et al.,* 1996; Langermans *et al.,* 2001). An exciting development in the cynomolgus model is that ~40% of animals infected with a low dose of *M. tuberculosis* (<50 bacilli) in the lungs do not develop active disease, but appear to be latently infected (J. Flynn, unpublished). This would represent an animal model that may closely mimic the human latent infection, and provides a rich area for study of immunologic and microbiologic questions of latency. One distinct advantage to using macaques as a model system is the availability of immunologic, histological, cell biology, and molecular biology reagents, thanks in part to the use of macaques in Simian immunodeficieny virus research; in addition, antibodies against human cell surface molecules frequently crossreact with macaque proteins. This model could have a significant impact on our understanding of events in the lungs during acute and latent tuberculosis, as well as serve as a model for determining surrogate markers of protection and testing of vaccines, drugs, and immunotherapies prior to use in humans.

In vivo Identification of *M. tuberculosis* Genes Required for Tuberculous Persistence: Potential Use of the IVET and STM Technologies

The study of microbial pathogenesis and pathogen-host interactions have been facilitated by the development of technologies that allow direct identification of microbial genes preferentially expressed in the host *in vivo* (Mahan *et al.,* 1993; Hensel *et al.,* 1995). Two such systems, based on the *s*ignature-*t*agged *m*utagenesis (STM) and on the *in vivo e*xpression *t*echnology (IVET) technology, have been exploited to dissect the mechanisms by which *M. tuberculosis* bacilli survive within a host (Cox *et al.,* 1999; Camacho *et al.,* 1999; Dubnau *et al.,* 2002). The STM technology utilizes a set of transposons that carries a large number of degenerate "signature" tags. This set of transposons is introduced into recipient bacilli to generate transposon mutants such that each carries a unique signature tag. These unique tags allow mutants to be distinguished from one another. Tagged mutant strains are assembled in a multiwell microtiter plate and aliquots from each well are combined to form an "inoculum pool". This pool, comprised of a number of distinct mutants, is introduced into animals to screen for clones defective for growth *in vivo*. After an appropriate time, bacilli are harvested from infected animals and represent the "recovered pool". Tags from each pool are PCR-amplified and then used to probe filters arrayed with specific tags representing those harbored in each of the mutants in the inoculum pool. Mutants that are attenuated for persistence will be those bacteria containing tags present in the inoculum pool but under-represented or missing in the recovered pool. The STM technology has been adopted by at least two laboratories to define *M. tuberculosis* genes that are preferentially expressed and promote bacterial survival in the host (Cox *et al.,*

1999; Camacho *et al.,* 1999). Results of studies conducted independently by these two laboratories identified an *M. tuberculosis* region with genes predicted to function in phthiocerol dimycoserosate (PDIM; a mycobacterial cell wall—associated complex lipid) biosynthesis. Further biochemical and genetic analysis of this region revealed that the *pps* promoter and *fadD28* are essential for the production of PDIM. In addition, these studies have provided evidence that *mmpL7* encodes a product that participates in PDIM transport (Cox *et al.,* 1999). These data demonstrate the potential use of the STM technology for the identification of *M. tuberculosis* genes that are required for persistence of the tubercle bacillus in the host. Recently, Dubnau *et al.* have adopted the IVET system to identify *M. tuberculosis* genes that are preferentially expressed in macrophages (Dubnau *et al.,* 2002). This system exploits the recent characterization of the mechanisms that mediate resistance to the antituberculous agent isoniazid (Zhang and Telenti, 2000). In this system, a promoter library of H37Rv was generated by cloning random ~300-base-pair *M. tuberculosis* genomic fragments upstream of *inhA*, a mycobacterial gene that encodes INH resistance. Clones harboring promoters that are preferentially activated *in vivo* will lead to expression of InhA and can be selected for by the administration of INH. This system was tested in a murine acute tuberculosis model, and candidate genes potentially important for the survival in the host are being characterized (E. Dubnau, unpublished). These results suggest that the InhA-based IVET system can be potentially useful in trapping *M. tuberculosis* promoters whose activity *in vivo* is essential for the establishment of tuberculous latency.

Conclusion

Data generated by studies based on the described *in vitro* and *in vivo* models have shed considerable light on the physiology of *M. tuberculosis* and the host immune response during the persistent phase of tuberculous infection. The availability of the genome of the tubercle bacillus, together with the adaptation of microarray analysis, proteomics, STM, and IVET technologies for use in *M. tuberculosis* research will facilitate our understanding of tuberculous latency. Although the significance of data generated by the various models of tuberculous latency and reactivation remains to be determined, it appears that at least in certain systems, such as that based on CD4 depletion- and TNF-α neutralization-induced reactivation, the findings are clinically relevant. The recent resurgence in interest in non-human primates promises a model likely to resemble human latency more closely than the currently existing small animal systems. Finally, the use of human *M. tuberculosis*-infected tissues for the characterization of the host immune response to the tubercle bacillus, as well as the physiology of lesional bacilli, is likely to help advance our knowledge in tuberculous latency.

Acknowledgement

This work was supported in part by NIH grants HL71241, HL68526 and AI50732.

References

Adams, L.B., Mason, C.M., Kolls, J.K., Scollard, D., Krahenbuhl, J.L., and Nelson, S. 1995. Exacerbation of acute and chronic murine tuberculosis by adminstration of a tumor necrosis factor receptor-expressing adenovirus. J. Infect. Dis. 171: 400-405.

Altare, F., Durandy, A., Lammas, D., Emile, J.F., Lamhamedi, S., le Deist, F., Drysdale, P., Jouanguy, E., Doffinger, R., Bernaudin, F., Jeppsson, O., Gollob, J.A., Meinl, E., Segal, A.W., Fischer, A., Kumararatne, D., and Casanova, J.L. 1998. Impairment of mycobacterial immunity in human interleukin-12 receptor deficiency. Science 280: 1432-1435.

Bean, A.G.D., Roach, D.R., Briscoe, H., France, M.P., Korner, H., Sedgwick, J.D., and Britton, W.J. 1999. Structural deficiencies in granuloma formation in TNF gene-targeted mice underlie the heightened susceptibility to aerosol *Mycobacterium tuberculosis* infection, which is not compensated for by lymphotoxin. J. Immunol. 162: 3504-3511.

Behar, S.M., Dascher, C.C., Grusby, M.J., Wang, C.R., and Brenner, M.B. 1999. Susceptibility of mice deficient in CD1D or TAP1 to infection with *Mycobacterium tuberculosis*. J. Exp. Med. 189: 1973-1980.

Bekker, L.-G., Moreira, A.L., Bergtold, A., Freeman, S., Ryffel, B., and Kaplan, G. 2000. Immunopathologic effects of tumor necrosis factor alpha in murine mycobacterial infection are dose dependent. Infect. Immun. 68: 6954-6961.

Betts, J.C., Lukey, P.T., Robb, L.C., McAdam, R.A., and Duncan, K. 2002. Evaluation of a nutrient starvation model of *Mycobacterium tuberculosis* persistence by gene and protein expression profiling. Mol. Mirobiol. 43: 717-731.

Camacho, L.R., Ensergueix, D., Perez, E., Gicquel, B., and Guilhot, C. 1999. Identification of a virulence gene cluster of *Mycobacterium tuberculosis* by signature-tagged transposon mutagenesis. Mol. Microbiol. 34: 257-267.

Caruso, A.M., Serbina, N., Klein, E., Triebold, K., Bloom, B.R., and Flynn, J.L. 1999. Mice deficient in CD4 T cells have only transiently diminished levels of IFN-γ, yet succumb to tuberculosis. J. Immunol. 162: 5407-5416.

Chan, J., and Flynn, J. 1999. Nitric oxide in *Mycobacterium tuberculosis* infection. In: Nitric Oxide and Infection. F. Fang, ed. New York: Plenum Publishers. p. 281-310.

Chan, J., Tanaka, K., Carroll, D., Flynn, J.L., and Bloom, B.R. 1995. Effect of nitric oxide synthase inhibitors on murine infection with *Mycobacterium tuberculosis*. Infect. Immun. 63: 736-740.

Clarke, S.R.M. 2000. The critical role of CD40/CD40L in the CD4-dependent generation of CD8+ T cell immunity. J. Leuk. Biol. 67: 607-614.

Cole, S.T., Brosch, R., Parkhill, J., Garnier, T., Churcher, C., Harris, D., *et al.* 1998. Deciphering the biology of *Mycobacterium tuberculosis* from the complete genome sequence. Nature 393: 537–544.

Cooper, A.M., Dalton, D.K., Stewart, T.A., Griffen, J.P., Russell, D.G., and Orme, I.M. 1993. Disseminated tuberculosis in IFN-γ gene-disrupted mice. J. Exp. Med. 178: 2243-2248.

Cooper, A.M., Magram, J., Ferrante, J., and Orme, I.M. 1997. Interleukin 12 (IL-12) is crucial to the development of protective immunity in mice intravenously infected with *Mycobacterium tuberculosis.* J. Exp. Med. 186: 39-45.

Covert, B.A., Spencer, J.S., Orme, I.M., and Belisle, J.T. 2001. The application of proteomics in defining the T cell antigens of *Mycobacterium tuberculosis.* Proteomics 1: 574-586.

Cox, J.S., Chen, B., McNeil, M., and Jacobs, W.R. Jr. 1999. Complex lipid determines tissue-specific replication of *Mycobacterium tuberculosis* in mice. Nature 402: 79-83.

Cunningham, A.F., and Spreadbury, C.L. 1998. Mycobacterial stationary phase induced by low oxygen tension: cell wall thickening and localization of the 16-kilodalton α-crystallin homolog. J. Bacteriol. 180: 801-808.

Dannenberg, A.M. 1994. Rabbit model of tuberculosis. In: Tuberculosis: Pathogenesis, Protection, and Control. B. R. Bloom, ed. American Society for Microbiology, Washington, D.C. p. 149-56.

de Jong, R., Altare, F., Haagen, I.A., Elferink, D.G., Boer, T., van Breda Vriesman, P.J., Kabel, P.J., Draaisma, J.M., van Dissel, J.T., Kroon, F.P., Casanova, J.L., and Ottenhoff, T.H. 1998. Severe mycobacterial and Salmonella infections in interleukin-12 receptor-deficient patients. Science 280: 1435-1438.

Desjardin, L.E., Hayes, L.G., Sohaskey, C.D., Wayne, L.G., and Eisenach, K.D. 2001. Microaerophilic induction of the alpha-crystallin chaperone protein homologue (HspX) mRNA of *Mycobacterium tuberculosis.* J. Bacteriol. 183: 5311-5316.

Dubnau, E., Fontan, P., Manganelli, R., Soares-Appel, S., and Smith, I. 2002. *Mycobacterium tuberculosis* genes induced during infection of human macrophages. Infect. Immun. 70: 2787-2795.

Dye, C., Scheele, S., Dolin, R., Pathania, G., and Raviglione, M. 1999. Global burden of tuberculosis. Estimated incidence, prevalence, and mortality by country. J.A.M.A. 282: 677-686.

Feng, C.G., Bean, A.G.D., Hooi, H., Briscoe, H., and Britton, W.J. 1999. Increase in gamma interferon-secreting CD8+, as well as CD4+ T cells in lungs following aerosol infection with *Mycobacterium tuberculosis.* Infect. Immun. 67: 3242-3247.

Fenhalls, G., Wong, A., Bezuidenhout, J., van Helden, P., Bardin, P., and Lukey, P.T. 2000. *In situ* production of gamma interferon, interleukin-4, and tumor necrosis factor alpha mRNA in human lung tuberculous granuloma. Infect. Immun. 68: 2827-2836.

Fisher, M.A., Plikaytis, B.B., and Shinnick, T.M. 2002. Microarray analysis of the *Mycobacterium tuberculosis* transcriptional response to the acidic conditions found in phagosomes. J. Bacteriol. 184: 4025-4032.

Florczk, M.A., McCue, L.A., Stack, R.F., Hauer, C.R., and McDonough, K.A. 2001. Identification and characterization of mycobacterial proteins differentially expressed under standing and shaking culture conditions, including Rv2623c from a novel class of putative ATP-binding proteins. Infect. Immun. 69: 5777-5785.

Flynn, J., and Chan, J. 2001. Tuberculosis: Latency and Reactivation. Infect. Immun. 69: 4195-4201.

Flynn J.L., Scanga, C.A., Tanaka, K.E., and Chan, J. 1998. Effects of aminoguanidine on latent murine tuberculosis. J. Immunol. 160: 1796-1803.

Flynn, J., and Chan, J. 2001. Immunology of tuberculosis. Annu. Rev. Immunol. 19: 93-130.

Flynn, J.L., and Ernst J.D. 2000. Immune responses in tuberculosis. Curr. Opin. Immunol. 12: 432-436.

Flynn, J.L., Chan, J., Triebold, K.J., Dalton, D.K., Stewart, T.A., and Bloom, B.R. 1993. An essential role for Interferon-γ in resistance to *Mycobacterium tuberculosis* infection. J. Exp. Med. 178: 2249-2254.

Flynn, J.L., Goldstein, M.M., Triebold, K.J., Koller, B., and Bloom, B.R. 1992. Major histocompatibility complex class I-restricted T cells are required for resistance to *Mycobacterium tuberculosis* infection. Proc. Natl. Acad. Sci. U.S.A. 89: 12013-12017.

Flynn, J.L., Goldstein, M.M., Chan, J., Triebold, K.J., Pfeffer, K., Lowenstein, C.J., Schreiber, R., Mak, T.W., and Bloom, B.R. 1995. Tumor necrosis factor-α is required in the protective immune response against *M. tuberculosis* in mice. Immunity 2: 561-272.

Good, R.C. 1968. Biology of the mycobacterioses. Simian tuberculosis: immunologic aspects. Ann. N. Y. Acad. Sci. 154: 200-213.

Haas, D.W., and des Perez, R.M. 1995. *Mycobacterium tuberculosis*. In: Principles and Practice of Infectious Diseases. G.L. Mandell, J.E. Bennett, and R. Dolin, eds. Churchill Livingstone, New York. p. 389-415.

Hensel, M., Shea, J.E., Gleeson, C., Jones, M.D., Dalton, E., and Holden, D.W. 1995. Simultaneous identification of bacterial virulence genes by negative selection. Science 269: 400-403.

Ji, J., Chakraborty, A., Geng, M., Zhang, X., Amini, A., Bina, M., and Regnier, F.. 2000. Strategy for qualitative and quantitative analysis in proteomics based on signature peptides. J. Chromatogr. B744: 197-210.

Jouanguy, E., Altare, F., Lamhamedi, S., Revy, P., Emile, J.-F., Newport, M., Levin, M., Blanche, S., Seboun, E., Fischer, A., and Casanova, J.L. 1996. Interferon-γ receptor deficiency in an infant with fatal BacilleCalmette-Guerin infection. N. Engl. J. Med. 335: 1956-1961.

Jungblut, P.R., Muller, E.-C., Mattow, J., and Kaufmann, S.H.E. 2001. Proteomics reveals open reading frames in *Mycobacterium tuberculosis* H37Rv not predicted by genomics. Infect. Immun. 69: 5905-5907.

Kalams, S.A., and Walker, B.D. 1998. The critical need for CD4 help in maintaining effective cytotoxic T lymphocyte responses. J. Exp. Med. 188: 2199-2204.

Kaushal, D., Schroeder, B.G., Tyagi, S., Yoshimatsu, T., Scot, C., Ko, C., Carpenter, L., Mehrotra, J., Manabe, Y.C., Fleischmann, R.D., and Bishai, W.R. 2002. Reduced immunopathology and mortality despite tissue persistence in a *Mycobacterium tuberculosis* mutant lacking alternative σ factor, SigH. Proc. Natl. Acad. Sci. U.S.A. 99: 8330-8335.

Keane, J., Gershon, S., Wise, R.P., Mirabile-Levens, E., Kaszinica, J., Schwieterman, W.D., Siegel, J.N., and Braun, M.M. 2001. Tuberculosis associated with infliximab, a tumor necrosis factor a-neutralizing agent. N. Engl. J. Med. 345: 1098-1104.

Koch, R. 1882. Aeriologie der Tuberculose. Berlin Klin. Wochenschr. 19: 221-230.

Langermans, J.A.M., Anderson, P., van Soolingen, D., Vervenne R.A.W., Frost, P.A., van der Laan, T., van Pinsteren, L.A.H., van den Hombergh, J., Droom, S., Peekel, I., Florquin, S., and Thomas, A.W. 2001. Devergent effect of bacillus Calmette-Guerin (BCG) vaccination on *Mycobacterium tuberculosis* infection in highly related macaque species: Implications for primate models in tuberculosis vaccine research. Proc. Nat. Acad. Sci. U.S.A. 98: 11497-11502.

Law, K., Jagirdar, J., Weiden, M.D., Bodkin, M., and Rom, W.N. 1996. Tuberculosis in HIV-positive patients: cellular response and immune activation in the lung. Am. J. Respir. Crit. Care Med. 153: 1377-1384.

Lewinsohn. D., Alderson, M., Briden, A., Riddell, S., Reed, S., and Grabstein, K. 1998. Characterization of human CD8+ T cells reactive with Mycobacterium tuberculosis-infected antigen presenting cells. J. Exp. Med. 187: 1633-1640.

Loebel, R.O., Shorr, E., and Richardson, H.B. 1933. The influence of adverse conditions upon the respiratory metabolism and growth of human tubercle bacilli. J. Bacteriol. 26: 167-200.

Loebel, R.O., Shorr, E., and Richardson, H.B. 1933a. The influence of foodstuffs upon the respiratory metabolism and growth of human tubercle bacilli. J. Bacteriol. 26: 139-166.

Lurie, M.B. 1964. Resistance to tuberculosis: Experimental studies in native and acquired defense mechanisms. Harvard University Press, Cambridge, M. A.

MacMicking, J., North, R.J., LaCourse, R., Mudgett, J.S., Shah, S.K., and Nathan, C.F. 1997. Identification of nitric oxide synhthase as a protective locus against tuberculosis. Proc. Natl. Acad. Sci. U.S.A. 94: 5243-5248.

Mahan, M.J., Slauch, J.M., and Mekalanos, J.J. 1993. Selection of bacterial virulence genes that are specifically induced in host tissues. Science 259: 686-688.

Manganelli, R., Voskuil, M., Schoolnik, G.K., and Smith, I. 2001. The *Mycobacterium tuberculosis* ECF sigma factor σ^E: role in global gene expression and survival in macrophages. Mol. Microbiol. 41: 423-437.

Mattow, J., Jungblut, P.R., Muller, E.-C., and Kaufmann, S.H.E. 2001. Identification of acidic, low molecular mass proteins of *Mycobacterium tuberculosis* strain H37Rv by matrix-assisted laser desorption/ionization and electrospray ionization mass spectrometry. Proteomics 1: 494-507.

McCune, R.M., Feldmann, F.M., Lambert, H.P., and McDermott, W. 1966. Microbial persistence I. The capacity of tubercle bacilli to survive sterilization in mouse tissues. J. Exp. Med. 123: 445-468.

McCune, R.M., Feldmann, F.M., and McDermott, W. 1966a. Microbial persistence II. Characteristics of the sterile state of tubercle bacilli. J. Exp. Med. 123: 469-486.

McKinney, J.D., zu Bentrup, K.H., Munoz-Elias, E.J., Miczak, A., Chen, B., Chan, W.-T., Swenson, D., Sacchettini, J.C., Jacobs, W.R. Jr., and Russell, D.G. 2000. Persistence of *Mycobacterium tuberculosis* in macrophages and mice requires the glyoxylate shunt enzyme isocitrate lyase. Nature 406: 735-738.

McKinney, J.D., Jacobs, W.R. Jr, and Bloom, B.R. 1998. Persisting problems in tuberculosis. In: Emerging Infections. R.M. Krause RM, ed. Academic Press, San Diego, CA. p. 51-146.

McMurray, D.N. 1994. Guinea pig model of tuberculosis. In: Tuberculosis: Pathogenesis, Protection, and Control. B.R. Bloom, ed. American Society for Microbiology, Washington, D.C. 135-147.

Mohan, V.P., Scanga, C.A., Yu, K., Scott, H.M., Tanaka, K.E., Tsang, E., Flynn, J.L., and Chan, J. 2001. Tumor necrosis factor-α is required to prevent reactivation and limit pathology in latent murine tuberculosis. Infect. Immun. 69: 1847-1855.

Mollenkopf, H.-J., Jungblut, P.R., Raupach, B., Mattow, J., Lamer, S., Zimny-Arndt, U., Schaible, U.E., and Kaufmann, S.H.E. 1999. A dynamic two-dimensional polyacrylamide gel electrophoresis database: The mycobacterial proteome via Internet. Electrophoresis 20: 2172-2180.

Murthy, P.S., Sirsi, M., and Ramakrishnan, T. 1973. Effect of age on the enzymes of tricarboxylic acid and related cycles in *Mycobacterium tuberculosis* H37Rv. Am. Rev. Respir. Dis. 108: 689-690.

Nathan, C., and Shiloh, M.U. 2000. Reactive oxygen and nitrogen intermediates in the relationship between mammalian hosts and microbial pathogens. Proc. Natl. Acad. Sci. U.S.A. 97: 8841-8848.

Newport, M.J., Huxley, C., Huston, S., Hawrylowicz, C., Oostra, B., Williamson, R., and Levin, M. A. 1996. Mutation in the interferon-γ-receptor gene and susceptibility to mycobacterial infection. N. Engl. J. Med. 335: 1941-1949.

Orme, I.M. 1988. A mouse model of the recrudescence of latent tuberculosis in the elderly. Amer. Rev. Respir. Dis. 137: 716-718.

Orme, I., and Collins, F. 1984. Adoptive protection of the *Mycobacteria tuberculosis*-infected lung. Cell. Immun. 84: 113-120.

Parrish, N.M., Dick, J.D., and Bishai, W.R. 1998. Mechanisms of latency in *Mycobacterium tuberculosis*. Trends Microbiol. 6: 107-112.

Ramakrishnan, L., Federspiel, N.A., and Falkow, S. 2000. Granuloma-specific expression of mycobacterium virulence proteins from the glycine-rich PE-PGRS family. Science 288: 1436-1439.

Rees, R.J.W., and Hart, P.D. 1961. Analysis of the host-parasite equilibrium in chronic murine tuberculosis by total and viable bacillary counts. Brit. J. Exp. Pathol. 42: 83-88.

Ribi, E., Anacker, R.L., Barclay, W.R., Brehmer, W., Harris, S.C., Leif, W.R., and Simmons, J. 1971. Efficacy of mycobacterial cell wasll as a vaccine agaisnt airborne tuberculosis in the rhesus monkey. J. Infect. Dis. 123: 527-538.

Rook, G.A.W., and Bloom, B.R. 1994. Mechanisms of pathogenesis in tuberculosis. In: Tuberculosis: Pathogenesis, Protection and Control. B. R. Bloom, ed. American Society for Microbiology, Washington, D.C. 485-501.

Rosenkrands, I., Slayden, R.A., Crawford, J., Aagaard, C., Barry, C.E. III, and Anderson, P. 2002. Hypoxic response of *Mycobacterium tuberculosis* studied by metabolic labeling and proteome analysis of cellular and extracellular proteins. J. Bacteriol. 184: 3485-3491.

Rosenkrands, I., Weldingh, K., Hansen, C.V., Florio, W., Gianetri, I., and Andersen, P. 2000. Mapping and identification of *Mycobacterium tuberculosis* proteins by two-dimensional gel electrophoresis, microsequencing and immunodetection. Electrophoresis 21: 935-948.

Scanga, C.A., Mohan, V.P., Joseph, H., Yu, K., Chan, J., and Flynn J.L. 1999. Reactivation of latent tuberculosis: variations on the Cornell murine model. Infect. Immun. 67: 4531-4538.

Scanga, C.A., Mohan, V.P., Tanaka, K., Alland, D., Flynn, J.L., and Chan, J. 2001. The inducible nitric oxide synthase locus confers protection against aerogenic challenge of both clinical and laboratory strains of *Mycobacterium tuberculosis* in mice. Infect. Immun. 69: 7711-7717.

Scanga, C.A., Mohan, V.P., Yu, K., Joseph, H., Tanaka, K., Chan, J., and Flynn, J.L. 2000. Depletion of CD4+ T cells causes reactivation of murine persistent tuberculosis despite continued expression of IFN-γ and NOS2. J. Exp. Med. 192: 347-358.

Serbina, N.V., and Flynn, J.L. 1999. Early emergence of CD8+ T cells primed for production of Type 1 cytokines in the lungs of *Mycobacterium tuberculosis*-infected mice. Infect. Immun. 67: 3980-3988.

Serbina, N.V., Liu, C.-C., Scanga, C.A., and Flynn, J.L. 2000. CD8+ cytotoxic T lymphocytes from lungs of *M. tuberculosis* infected mice express perforin *in vivo* and lyse infected macrophages. J. Immunol. 165: 353-363.

Serbina, N.V., Lazarevic, V., and Flynn, J.L. 2001. CD4(+) T cells are required for the development of cytotoxic CD8(+) T cells during *Mycobacterium tuberculosis* infection. J. Immunol. 67: 6991-7000.

Shen, J., Barnes, P.F., Rea, T.H., and Meyer, P.R. 1988. Immunohistology of tuberculous adenitis in symptomatic HIV infection. Clin. Exp. Immunol. 72: 186-189.

Sherman, D.R., Voskuil, M., Schnappinger, D., Liao, R., Harrell, M.I., and Schoolnik, G.K. 2001. Regulation of the *Mycobacterium tuberculosis* hypoxic response gene encoding alpha-crystallin. Proc. Natl. Acad. Sci. U.S.A. 98: 7534-7539.

Smith, D.W., Wiegeshaus, E.H., Navalkar, R., and Grover, A.A. 1966. Host-parasite relationships in experimental airborne tuberculosis: I. Preliminary studies in BCG-vanncinated and nonvaccinated animals. J. Bacteriol. 91: 718-724.

Styblo, K. 1980. Recent advances in epidemiological research in tuberculosis. Adv. Tuberc. Res. 20: 1-63.

Tascon, R.E., Stavropoulos, E., Lukacs, K.V., and Colston, M.J. 1998. Protection against *Mycobacterium tuberculosis* infection by CD8 T cells requires production of gamma interferon. Infect. Immun. 66: 830-834.

Tsenova, L., Bergtold, A., Freedman, V.H., Young, R.A, and Kaplan, G. 1999. Tumor necrosis factor alpha is a determinant of pathogenesis and disease progressi.on in mycobacterial infection in the central nervous system. Proc. Natl. Acad. Sci. U.S.A. 96: 5657-5662.

Usha, V., Jayaraman, R., Toro, J.C., Hoffner, S.E., and Das, K.S. 2002. Glycine and alanine dehydrogenase activities are catalyzed by the same protein in *Mycobacterium smegmatis*: upregulation of both activities under microaerophilic adaptation. Can. J. Microbiol. 48: 7-13.

Walsh, G.P., Tan, E.V., de la Cruz, E.C., Abalos, R.M., Villhermonsa, L.G., Young, L.J., Cellona, R.V., Nazareno, J.B., and Horwitz, M.A. 1996. The Philappine cynomolgus monkey (*Macaca fascularis*) provides a new nonhuman primate model of tuberculosis that resembles human disease. Nature Med. 2: 430-436.

Wayne, L.G., and Salkin, D. 1956. The bacteriology of resected tuberculous pulmonary lesions. I. The effet of interval between reersal of infectiousness and subsequent surgery. Am. Rev. Tuberc. Pulm. Dis. 74: 376-387.

Wayne, L.G. 1976. Dynamics of submerged growth of *Mycobacterium tuberculosis* under aerobic and microaerophilic conditions. Am. Rev. Resp. Dis. 114: 807-811.

Wayne, L.G. 1977. Synchronized replication of *Mycobacterium tuberculosis*. Infect Immun 17: 528-530.

Wayne, L.G., and Sramek, H.A. 1979. Antigenic differences between extracts of actively replicating and synchronized resting cells of *Mycobacterium tuberculosis*. Infect. Immun. 24: 363-370.

Wayne, L.G., and Lin, K.-Y. 1982. Glyoxylate metabolism and adaptation of *Mycobacterium tuberculosis* to survival under anaerobic conditions. Infect Immun 37: 1042-1049.

Wayne, L.G., and Sramek, H.A. 1994. Metronidazole is bactericidal to dormant cells of *Mycobacterium tuberculosis*. Antiicrob. Agents Chemother. 38: 2054-2058.

Wayne, L.G., and Hayes, L.G. 1996. An *in vitro* model for sequential study of shiftdown of *Mycobacterium tuberculosis* through two stages of nonreplicating persistence. Infect. Immun. 64: 2062-2069.

Wiegeshaus, E., McMurray, D.N., Grover, A.A., Harding, G.E., and Simth, D.W. 1970. Host-parasite relationships in experimental airborne tuberculosis. III. Relevance to microbial enumeration to acquired resistance in guinea pigs. Am. Rev. Respir. Dis. 102: 422-429.

Wilson, M., DeRisi, J., Kristensen, H.-H., Imboden, P., Rane, S., Brown, P.O., and Schoolnik, G.K. 1999. Exploring drug-induced alterations in gene expression in *Mycobacterium tuberculosis* by microarray hybridization. Proc. Natl. Acad. Sci. U.S.A. 96: 12833-12838.

Yuan, Y., Crane, D.D., and Barry, C.E. 3rd. 1996. Stationary phase-associated protein expression in Mycobacterium tuberculosis: function of the mycobacterial alpha-crystallin homolog. J. Bacteriol. 178: 4484-4492.

Yuan, Y., Crane, D.D., Simpson, R.M., Zhu, Y.Q., Hickey, M.J., Sherman, D.R., and Barry, C.E. 3rd. 1998. The 16-kDa alpha-crystallin (Acr) protein of *Mycobacterium tuberculosis* is required for growth in macrophages. Proc. Natl. Acad. Sci. U.S.A. 95: 9578-9583.

Zhang, Y., and Telenti, A. 2000. Genetics of drug resistance in *Mycobacterium tuberculosis*. In: Molecular Genetics of Mycobacteria. G.F. Hatfull, W.R. Jacobs, Jr., eds. American Society of Microbiolgy Press, Washington D.C. p. 235-254.

Zu Bentrup, K.H., Miczak, A., Swenson, D.L., and Russell, D.G. 1999. Characterization of activity and expression of isocitrate lyase in *Mycobacterium avium* and *Mycobacterium tuberculosis*. J. Bacteriol. 181: 7161-7167.

From: Tuberculosis: The Microbe Host Interface
Edited by: Larry S. Schlesinger and Lucy E. DesJardin

Chapter 9

Molecular Epidemiology: Clinical Utility, Public Health Implications and Relevance To Pathogenesis

Peter F. Barnes and M. Donald Cave

Abstract

We review the advantages and disadvantages of the most widely used methods for genotyping *M. tuberculosis*. The current gold standard is based on the distribution of the insertion sequence IS*6110*, but genotyping based on polymorphism of mycobacterial interspersed repetitive units is a promising technology that provides automated, high-throughput results that can be expressed digitally and easily compared. Genotyping allows the clinician to identify cases of laboratory cross-contamination, and assists in treatment of patients with recurrent tuberculosis and those whose isolates show different drug susceptibility patterns. Population-based genotyping has major public health implications, demonstrating that recent transmission contributes significantly to tuberculosis morbidity and identifying subpopulations in which ongoing transmission is a problem. In locations where tuberculosis case rates are high, surveillance genotyping has the potential to identify and limit the spread of unsuspected tuberculosis outbreaks. Molecular epidemiologic studies have revealed several critical features about the pathogenesis of tuberculosis. First, some strains are much more widely distributed than others, suggesting that they have phenotypic characteristics that favor dissemination. Second, drug-resistant strains may be less transmissible than drug-susceptible ones, suggesting that some genes that encode for drug resistance also decreased fitness. Third, prior tuberculosis does not protect against new disease from exogenous reinfection, suggesting that effective vaccination must do more than mimic natural infection.

Introduction

Because of the inability to identify specific strains of *Mycobacterium tuberculosis* isolated from individual patients, knowledge of the transmission dynamics of tuberculosis during most of the twentieth century was based primarily on studies of tuberculosis outbreaks, and evaluation of changes in tuberculin skin test results and tuberculosis case rates in populations over time. In the early 1990s, the insertion element IS*6110* was first used to obtain a molecular "fingerprint" of *M. tuberculosis* isolates (Hermans *et al.*, 1990; Cave *et al.*, 1991). This breakthrough provided a quantum leap in our ability to trace the passage of individual *M. tuberculosis* strains through the population, and widespread use of this and related techniques has shattered previous myths and revolutionized our understanding of the transmission dynamics of tuberculosis. These studies have also begun to provide insight into the characteristics of the organism that favor transmission in humans. This chapter will review the genotyping methods used to identify specific *M. tuberculosis* strains, the clinical and public health applications of molecular epidemiology, and the relevance of molecular epidemiology to our understanding of the pathogenesis of tuberculosis.

Genotyping Methods

More than 20 methods of genotyping *M. tuberculosis* isolates have been described. Below we focus on the most widely used methods which yield reproducible results in different laboratories.

IS*6110*-Based Methods

IS*6110* is a 1365 bp insertion sequence with a 28 bp imperfect inverted repeat at each end (Thierry *et al.,* 1990). IS*6110* is present only in members of the *M. tuberculosis* complex and not in other mycobacteria (Cave *et al.,* 1991). More than 99.9% of *M. tuberculosis* strains have IS*6110*, inserted in 1 at 25 sites. A standard procedure for RFLP analysis is used worldwide, utilizing the restriction enzyme *PvuII*, which has only one cleavage site in IS*6110* (van Embden *et al.,* 1993). Southern blots of *M. tuberculosis* DNA are probed with a fragment of IS*6110* that lies to the right of the *PvuII* site, to generate RFLP patterns.

IS*6110*-based RFLP analysis is the most widely used method to separate *M. tuberculosis* strains. This method has two major advantages. First, large databases of IS*6110* RFLP patterns have been gathered over long periods in many parts of the world. Second, epidemiologic data has confirmed that, when isolates from different patients share IS*6110*-based RFLP patterns, the patients are frequently epidemiologically linked, and the RFLP results can therefore be used to evaluate the transmission dynamics of tuberculosis in a population (Table 1). This method is the gold standard for genotyping *M. tuberculosis* strains, and any new methods should be compared against this technique. Nevertheless, there are several significant disadvantages of IS*6110*-based RFLP analysis. First, many IS*6110* insertion sites are highly conserved in strains having less than 6 IS*6110* copies (Fomukong *et al.,* 1997), and IS*6110*-based polymorphism in low-copy strains is limited. To distinguish these low-copy strains, secondary typing procedures, based on other genetic elements, are often necessary, and are discussed below. Second, replicative transposition can lead to additional IS*6110* fragments, and approximately half of *M. tuberculosis* strains show a shift in one band of their IS*6110*-based RFLP pattern after 3-4 years (de Boer *et al.,* 1999). Multiple isolates from a patient may lose or gain a hybridizing fragment, and epidemiologic links have been frequently identified between patients whose isolates differ by 1 or 2 copies of IS*6110* (Mazurek *et al.,* 1991; Cave *et al.,* 1994). Third, IS*6110*-based RFLP analysis requires large amounts of DNA, requiring viable organisms and subculture of clinical isolates for several weeks. Finally, the technique is technically demanding and sophisticated software is required to compare large numbers of IS*6110*-based RFLP patterns.

A rapid technique, mixed-linker PCR, is based on PCR with one primer specific for IS*6110* and a second primer complementary to a linker ligated to the restricted genomic DNA (Haas *et al.,* 1993). The technique was automated by using a fluorescent-labeled IS*6110* primer and electrophoresis of the PCR products with internal lane size standards on a DNA sequencer (Butler *et al.,* 1996). RFLP patterns generated by this procedure parallel results obtained with standard IS*6110*-based RFLP analysis. The advantage of mixed linker

Table 1. Advantages and disadvantage of different genotyping methods for *M. tuberculosis*

Characteristic	IS6110	ML-PCR[1]	Spoligotype	VNTR[2]	MIRU[3]	PGRS[4]	SNP[5]	Microarray
Discriminatory power	+ + + +	+ + + +	+ +	+ +	+ + + +	+ + +	?	?
Test simplicity	+ +	+	+ + + +	+ + +	+ +	+ +	+ +	+
Automated	No	Yes	No	Yes	Yes	No	Yes	No
Data in digital format	No	No	Yes	Yes	Yes	No	Yes	No
Amount of DNA required	Large	Small	Small	Small	Small	Large	Small	Large
Simplicity of data analysis	+ +	+ +	+ + + +	+ + + +	+ + + +	+ +	+ + +	+
Large databases available	Yes	No	Yes	No	No	No	No	No

[1]Mixed-linker PCR
[2]Variable number of tandem repeats
[3]Spoligotyping based on polymorphism in the DR region
[4]Mycobacterial interspersed repetitive units

PCR-based genotyping is that it can be done rapidly on small amounts of DNA, including nonviable organisms. The disadvantage is that the procedure is technically demanding and costly (Table 1).

Direct Repeat Locus-Based Methods

The direct repeat (DR) locus is a single region of the *M. tuberculosis* chromosome that contains multiple copies of a 36 bp direct repeat that is reiterated between 10 and 50 times. Each repeat is separated from the next by non-repeated variable spacer DNA, each containing 37-41 bp (Hermans *et al.*, 1991). Although the overall arrangement of the spacers in the DR locus is conserved among strains, polymorphism results from deletion of segments of the DR locus. In its simplest format, the 36 bp repeat is used to probe blots of genomic DNA restricted with endonucleases for which there are no restriction sites in the 36 bp repeat (van Soolingen *et al.*, 1993). Polymorphism is generated by sequence differences in the spacer regions, and is independent of IS*6110*-based RFLP patterns.

Spacer oligonucleotide typing, or spoligotyping, involves PCR amplification of the DR locus using labeled primers which are directed outward from the 36 bp repeat, amplifying the variable spacers and repeats (Kamerbeek *et al.*, 1997). The labeled PCR products are then used to probe a membrane that contains covalently bound oligonucleotides corresponding to each of the 43 spacer sequences in *M. tuberculosis* strain H37Rv. Individual strains have different variable spacers and are distinguished by their positive and negative signals for each spacer. The addition of 51 additional spacer sequences to the 43 that are most commonly used did not improve discrimination of different *M. tuberculosis* strains (Sechi *et al.* 1998).

Spoligotyping has several advantages (Table 1). First, because small amounts of DNA are required, it can be performed on cultured cells or clinical samples (Kamerbeek *et al.*, 1997). Second, spoligotyping results, displayed as positive or negative for each spacer, can be readily expressed in digital format on a word processor (Dale *et al.*, 2001). The most significant disadvantage of spoligotyping is that it has lower discriminatory power to separate *M. tuberculosis* strains than most other genotyping methods (Kremer *et al.*, 1999; Yang *et al.*, 2000). This is in part because spoligotyping measures polymorphisms at a single genetic locus, whereas other methods evaluate multiple loci. Despite this shortcoming, spoligotyping may have a role in screening isolates, as strains with different spoligotypes generally have different IS*6110* patterns, obviating the need for more complex fingerprinting methods (Goguet de la Salmoniere *et al.*, 1997).

Tandem-Repeat Sequence-Based Methods

The genome of *M. tuberculosis* H37Rv contains 41 tandem-repeat loci, referred to as mycobacterial interspersed repeat units (MIRUs) (Supply *et al.*, 2000). Some tandem-repeat loci contain identical repeat units, and others contain repeats of variable sequence and length. Genotyping is based on the number and size of the repeats in each of these independent loci.

Identical repeat units are evaluated by variable number tandem repeat (VNTR) analysis, in which six different repeat loci are amplified in separate PCR reactions (Frothingham and Meeker-O'Connell, 1998). After gel electrophoresis, the size of the products and the number of repeat units at each locus can be calculated. Different strains have 1-7 repeat units at each locus. The advantage of this method is that number of repeat units at each of the six loci for a given strain can be expressed in digital format (Table 1). The major disadvantage of VNTR analysis is that it is less discriminatory than most other fingerprinting methods for separating different *M. tuberculosis* strains (Kremer *et al.*, 1999).

MIRU analysis employs the same principles as VNTR analysis to categorize 12 tandem repeat loci that are distributed around the chromosome. Different *M. tuberculosis* strains have variable numbers and sequences of tandem repeats (Supply *et al.*, 2000). Two to eight alleles were observed at each of the 12 loci, yielding approximately 20 million possible combinations of alleles (Mazars *et al.*, 2001; Supply *et al.*, 2001).

MIRU analysis has several advantages (Table 1). First, its ability to distinguish between strains is almost as great as that of IS*6110*. For 134 *M. tuberculosis* complex strains with no known epidemiologic links, MIRU analysis and IS*6110*-based RLFP analysis yielded 114 and 122 patterns, respectively (Mazars *et al.*, 2001; Supply *et al.*, 2001). Second, use of multiplex PCR, fluorescent-tagged PCR primers and an automated DNA sequencer allowed automated high throughput MIRU analysis and enabled establishment of a MIRU genotyping website that allows laboratories to compare their data and to generate a database for global epidemiologic surveillance of tuberculosis (Supply *et al.*, 2001). Third, the methods are relatively straightforward. Finally, the results are intrinsically digital and the analysis can be applied directly to cultures without DNA purification. There are two disadvantages of MIRU analysis. First, for *M. tuberculosis* strains with 5 or more copies of IS*6110*, MIRU analysis may be significantly less discriminatory than IS*6110*-based RFLP analysis. Second, a large database of MIRU results for different strains has not yet been accumulated. However, of all the currently available methods, MIRU is the most likely to replace IS*6110*-based RFLP analysis as an initial genotyping method in the future, particularly if evaluation of additional loci yields increased discriminatory power.

Polymorphic GC-Rich Sequence-Based Methods

The polymorphic GC-rich sequence (PGRS) is a repeated sequence that is present in 61 members of the PE family of genes that is distributed throughout the chromosome. PE proteins have a conserved 110 amino acid sequence that includes the unusual sequence Pro-Glu at the amino terminus. The PGRS subfamily consists of glycine-rich proteins, with multiple tandem repetitions of Gly-Gly-Ala and Gly-Gly-Asp. The 9 bp consensus sequence coding for these motifs is CGGCGGCAA (repeated 2-7 times in tandem) with most variation in bases 1 and 9 (Poulet and Cole, 1995).

PGRS-based RFLP analysis has been performed with several different restriction enzymes and 5 different probes have been used, including an RNA probe produced from transcription of an insert in the plasmid pTBN12 (Dwyer *et al.,* 1993), a DNA copy of this probe (Chaves *et al.* 1997), two other clones of the PGRS, pMBA1 and pMBA2, (Bigi *et al.,* 1995), and a 33 bp oligomer, based on the consensus sequence of the PGRS (Cousins *et al.,* 1998).

PGRS-based RFLP analysis has two major disadvantages, compared to IS*6110*-based RFLP analysis (Table 1). First, it is less discriminating. Second, the large number of hybridizing bands of variable intensity make it difficult to perform computerized comparison of RFLP patterns.

Genome-Based Analysis

Comparison of the genomic sequence of H37Rv and the clinical strain CDC1551 has revealed 1500 single nucleotide polymorphisms (SNPs), as well as deletions and duplications (Fraser *et al.,* 2000; Musser, 2001). Such differences can be exploited for genotyping by evaluation of SNPs (Sreevatsan *et al.,* 1997) and by microarray analysis (Kato-Maeda *et al.* 2001). Analysis of SNPs has to date been limited to those at codon 463 of the *katG* gene and codon 95 of the *gyrA* gene, which divides organisms into very large groups that are not sufficiently discriminating for molecular epidemiologic studies. Simultaneous evaluation of SNPs at multiple locations may allow this method to be used in the future.

Microarray analysis involves hybridization of genomic DNA to approximately 4,000 open reading frames in the *M. tuberculosis* genome, and preliminary studies have demonstrated that this method can distinguish different *M. tuberculosis* strains (Kato-Maeda *et al.* 2001). Microarray analysis is a theoretically attractive genotyping strategy, providing a comprehensive picture of all genes in an *M. tuberculosis* strain. However, this method has several disadvantages. First, the use of a microarray related to a specific strain such as H37Rv will only detect deletions relative to that strain and will not detect duplications or genes not present in H37Rv. Second, the methods used to obtain reproducible hybridization results and the data analysis are extremely complex and evolving, requiring expensive equipment and specialized software. Third, the capacity of this method to identify epidemiologically linked *M. tuberculosis* strains and to discriminate between unrelated strains remains uncertain. Therefore, much additional work is needed to determine if this method will be suitable for studies of molecular epidemiology.

Analysis of Population-based Genotyping Data

Early population-based genotyping studies defined clustered patients as those whose *M. tuberculosis* isolates shared an RFLP pattern, and most clustered patients were considered to have disease due to recent transmission of tuberculosis. Patients whose isolates had a distinctive RLFP pattern were considered to have disease due to reactivation of infection in the distant past (Alland *et al.*, 1994; Small *et al.*, 1994). To obtain an accurate estimate of the proportion of clustered cases in a population, it is important to evaluate as great a proportion of cases as possible. The percentage of clustered cases is underestimated if genotyping is not performed in a significant percentage of cases (Glynn *et al.*, 1999). In addition, if isolates are collected over a short period, some clustered cases will not be identified because infected persons will not yet have developed active tuberculosis.

Classification of patients as clustered or non-clustered was initially based on visual comparison of RFLP patterns of different isolates. With the establishment of very large IS*6110*-based RFLP databases and improved software technology, RFLP patterns are now evaluated by software programs, and sophisticated methods have been devised to adjust for the minor differences in apparent size of individual IS*6110* fragments, due to variability in electrophoresis on the gel (Salamon *et al.*, 1998).

To characterize the transmission dynamics of tuberculosis in a population, genotyping data must be interpreted in the context of detailed epidemiologic data on the patients. Many studies have shown that isolation of *M. tuberculosis* strains with identical RFLP patterns from different patients may represent recent transmission of tuberculosis. However, in some settings, identical RFLP patterns are due to transmission in the remote past (Braden *et al.*, 1997) or result from extensive circulation of a strain that was introduced into the community many years previously. Only epidemiologic data that permits identification of transmission links between patients can separate these possibilities.

Clinical Utility of Genotyping

Genotyping of isolates from individuals or from small groups of patients in the United States can be performed through regional laboratories that are supported by the Centers for Disease Control and Prevention. These studies are useful in the following situations:

1) Confirmation of laboratory cross-contamination. Of patients from whom *M. tuberculosis* is isolated in clinical laboratories, approximately 3% are due to cross-contamination and are therefore false-positive (Burman and Reeves, 2000). Positive cultures are most likely to be due to cross-contamination when all acid-fast smears are negative and only one specimen is culture-positive. When *M. tuberculosis* is isolated from samples obtained from a patient whose clinical findings are not suggestive of tuberculosis, genotyping of the patient isolates, isolates from other patients processed at the same time, and laboratory control strains can identify cases of cross-contamination and allow discontinuation of potentially toxic antituberculosis medications.

2) Evaluation of a patient's *M. tuberculosis* isolates with differing drug susceptibility patterns during or after antituberculosis therapy. This situation can arise because the original organism developed drug resistance during therapy, the patient was reinfected with a second *M. tuberculosis* strain, or is accounted for by laboratory cross-contamination. Genotyping of the patient's isolates and other isolates processed at the same time can distinguish these possibilities. If the original organism developed resistance during therapy, the possibility of non-adherence to therapy or reduction of antituberculosis drug concentrations due to malabsorption or drug interactions should be evaluated. On the other hand, if the patient was reinfected with a second *M. tuberculosis* strain, an epidemiologic investigation should be conducted to identify the source case.

3) Evaluation of second episodes of tuberculosis. In individual patients, it is important to know if a second episode is due to relapse of the original strain or to infection with another strain. This can be determined by genotyping of the strains from both episodes. If the second episode is due to a relapse, the drug susceptibility of the original isolate can be used to guide therapy, and reasons for treatment failure such as malabsorption of antituberculosis medications and nonadherence should be considered. If the second episode is due to exogenous infection, a search should be made for the new source case. In the setting of antituberculosis treatment trials, it is important to determine if second episodes of tuberculosis are due to relapse or exogenous infection, as only the former result from failure of therapy.

Public Health Implications of RFLP Analysis

Investigation of Suspected Outbreaks.

If epidemiologic data suggest a tuberculosis outbreak in a community, genotyping of the *M. tuberculosis* isolates, in combination with epidemiologic investigation of the cases and contacts, can separate coincidental occurrence of an unexpectedly high number of cases from an outbreak involving multiple patients. This strategy can identify the source case and delineate the extent of the outbreak, and has been used to characterize outbreaks in various settings, including bars, jails and complex social networks (Kline *et al.*, 1995; Jones *et al.*, 1999; Fitzpatrick *et al.*, 2001). Although genotyping has been used most frequently for outbreak investigation, population-based molecular epidemiology has yielded the most insight into critical aspects of the transmission dynamics of tuberculosis, providing guidance for the development of effective public health policies.

Population-based Molecular Epidemiology

The validity of population-based molecular epidemiology depends on genotyping the vast majority of *M. tuberculosis* isolates in the target population, combined with detailed epidemiologic information, usually gained by prolonged interviews, preferably at the time of diagnosis of tuberculosis. In the absence of epidemiologic data, genotyping results are difficult to interpret. For example, in urban areas, persons whose *M. tuberculosis* isolates share RFLP patterns are generally linked by recent transmission (Small *et al.*, 1994; Barnes *et al.*, 1997) whereas shared RFLP patterns in rural locations are often not due to recent transmission (Braden *et al.*, 1997).

Recent Transmission Contributes Significantly to Tuberculosis Morbidity

Prior to the advent of genotyping, it was widely believed that the vast majority of cases of tuberculosis in the United States and other industrialized nations were due to reactivation of infection acquired in the distant past. However, population-based molecular epidemiologic studies in multiple locations indicate that 20-50% of tuberculosis cases in urban areas are due to recent transmission of disease (Small *et al.*, 1994; Barnes *et al.*, 1997; Bishai *et al.* 1998; Weis *et al.*, 2002). This high frequency of ongoing transmission is probably due to two factors whose importance was underappreciated prior to the availability of genotyping:

1) The vast majority of transmission of tuberculosis probably precedes the initiation of therapy. Therefore, even in locations where universal directly observed therapy and high rates of treatment completion have been achieved for more than a decade, the percentage of tuberculosis cases due to recent transmission remains 32-36% (Bishai *et al.*, 1998; Weis *et al.*, 2002). Furthermore, failure of tuberculosis patients to seek medical care for more than 60 days after the development of symptoms contributed significantly to spread of tuberculosis in a community that followed the recommended elements of tuberculosis control (Chin *et al.*, 2000).

2) There is substantial transmission of tuberculosis to casual contacts in urban areas where tuberculosis is prevalent. Extensive unsuspected transmission of tuberculosis to casual contacts has been documented in workplaces, homeless shelters, bars and other social settings (Kline *et al.*, 1995; Barnes *et al.*, 1997; Valway *et al.*, 1998; Golub *et al.*, 2001), and tuberculosis patients often have limited contact with other patients whose isolates share the same genotype (Barnes *et al.*, 1997; Weis *et al.*, 2002). Much of this transmission is unsuspected by public health authorities because these casual contacts are not identified during routine public health investigations that focus primarily on household and workplace contacts, although these contacts can be elicited in retrospect (Small *et al.*, 1994; Barnes *et al.*, 1997; Chin *et al.*, 2000; Weis *et al.*, 2002).

Variability in Local Transmission Dynamics

Population-based molecular epidemiology has revealed that the transmission dynamics of tuberculosis vary greatly in different locations. In cities where homelessness is common, homeless shelters are often foci of tuberculosis transmission (Barnes *et al.*, 1997; Weis *et al.*, 2002). In other locations, health care facilities and bars have been some of the dominant sites of tuberculosis transmission (Alland *et al.*, 1994; Small *et al.*, 1994; Yagenehdoost *et al.*, 1999). Most population-based studies have been performed in large cities, and it will be important to characterize the transmission dynamics of tuberculosis in small cities and rural areas. Because the patterns of tuberculosis transmission vary by location, local efforts to identify high-risk populations and tuberculosis transmission sites are paramount for development of effective tuberculosis control strategies.

Evaluation of Tuberculosis Control Programs and Identification of High-Risk Populations

One potential use of population-based molecular epidemiology is to determine the efficacy of ongoing tuberculosis control efforts by measuring changes in the annual rate of clustered tuberculosis cases over a prolonged period. This strategy can also identify subpopulations of patients for whom tuberculosis control efforts should be intensified. In San Francisco, public health interventions aimed at reducing tuberculosis transmission were instituted in the early 1990s, and over a 6-year period, these interventions were associated with a significant reduction in the rate of clustered tuberculosis cases diagnosed, both in the entire population and in high-risk populations such as HIV-infected persons (Jasmer *et al.,* 1999).

The identification of subpopulations where transmission rates are highest has been facilitated by the development of analytical methods to calculate the transmission index, defined as the number of tuberculosis cases generated by a single source case (Borgdorff *et al.,* 1998: Borgdorff *et al.,* 2000). These analyses have demonstrated that transmission indices vary significantly by nationality and ethnicity. In Holland, the transmission index was 3-5-fold higher among Moroccan, Somalian and Surinamese than among Dutch nationals (Borgdorff *et al.,* 1998). In San Francisco, the transmission index was 7-fold higher among U.S.-born African Americans than among U.S.-born whites or Hispanics (Borgdorff *et al.,* 2000). This increase was not due to an increased prevalence of HIV infection or of sputum acid-fast smear-positive disease among African Americans. The reasons for an increased transmission index in certain ethnic groups remain speculative but may be related to delays in seeking medical attention, increased susceptibility to infection or progression of infection to disease as a result of genetic factors, or social factors such as overcrowding that facilitate disease transmission. Regardless of the reasons, these findings suggest that if public health resources are limited, focusing these resources on specific subpopulations in which transmission indices are highest can most efficiently reduce tuberculosis transmission.

Interventions to Reduce Transmission of Tuberculosis

Based on the traditional view that most tuberculosis cases in industrialized nations were due to reactivation of infection acquired in the distant past and that most transmission occurred to close contacts of tuberculosis patients, public health resources were directed primarily toward effective therapy of tuberculosis cases and evaluation of their close contacts for tuberculosis infection and disease. Molecular epidemiologic studies have demonstrated that an unexpectedly high percentage of tuberculosis cases are due to recent transmission from casual contact with patients with undiagnosed tuberculosis.

Therefore, a significant percentage of tuberculosis cases are potentially preventable through focused application of public health resources to break the cycle of transmission.

The importance of casual contact makes it imperative for public health contact investigators to question tuberculosis patients closely about their contacts outside the traditional home and workplace, and to identify locations where patients spend significant periods of time, as they may infect persons at these locations despite minimal personal contact. A complementary strategy is to screen high-risk populations for tuberculosis to detect cases before they routinely seek medical attention. Molecular epidemiologic studies have demonstrated that a limited number of tuberculosis transmission sites can contribute a significant proportion of the tuberculosis morbidity in a population (Jones *et al.*, 1999; Barnes *et al.*, 1999), and screening at these sites is likely to be particularly rewarding. Chest radiography was found to be an excellent screening tool to detect tuberculosis in incarcerated persons, those who frequent shelters for the homeless, and indigent persons seeking social services in New York City (Barnes, 1998; Jones *et al.*, 1999; Schluger *et al.*, 1999). Obtaining sputum smears for acid-fast bacilli from homeless shelter occupants has also been highly effective (Kimmerling *et al.*, 1999). Screening high-risk populations for latent tuberculosis infection, followed by treatment to prevent development of disease is theoretically attractive, but low rates of prescription of isoniazid and completion of prescribed therapy reduce the practical value of this approach (Schluger *et al.*, 1999).

In locations where tuberculosis case rates are high and ongoing tuberculosis transmission has been documented, prospective population-based surveillance genotyping has the potential to identify and limit the spread of unsuspected tuberculosis outbreaks. IS*6110*-based genotyping is not suitable for this purpose because results are not rapidly available and because comparison of RFLP patterns of different isolates is relatively complex. Surveillance genotyping hinges on the availability of reproducible, relatively simple methods to analyze small amounts of DNA in clinical samples and *M. tuberculosis* cultures, and to compare isolate genotypes rapidly, using a digital code or other methods. MIRU analysis is the most promising candidate for population-based genotyping, and could identify clustered cases several days after obtaining an acid-fast-positive clinical specimen or a positive culture for *M. tuberculosis*. Patients infected with the same *M. tuberculosis* strain could then be interviewed soon after diagnosis, and tuberculosis transmission sites could be rapidly identified for intensive targeted interventions. The success of this strategy will depend on commitment of resources for surveillance genotyping and close communication between laboratory facilities and public health care systems. If this approach significantly reduces tuberculosis transmission, expansion of these efforts to other locations can then be considered.

Relevance of Molecular Epidemiology to the Pathogenesis of Tuberculosis

Microbial Factors That Favor Transmission of Tuberculosis

The traditional dogma is that all *M. tuberculosis* strains are equally virulent and that a single tubercle bacillus can cause disease in humans. However, most population-based molecular epidemiologic studies in different countries have demonstrated that a small percentage of *M. tuberculosis* strains cause the majority of cases (Small *et al.*, 1994; Barnes *et al.*, 1997; van Soolingen *et al.*, 1999), implying that some strains are better equipped to spread in the population. The widespread C strain in New York City was more resistant to reactive nitrogen intermediates than other *M. tuberculosis* strains, perhaps favoring its survival in human macrophages (Friedman *et al.*, 1997). The CDC1551 strain caused a large outbreak in Kentucky and Tennessee (Valway *et al.*, 1998), and induced greater production of proinflammatory monokines such as TNF-α, as well as the type 1 cytokines IL-12 and IFN-γ, in infected mice, compared to mice infected with other *M. tuberculosis* strains. Lipids extracted from CDC1551 also induced production of higher concentrations of TNF-α and IL-12 by human monocytes (Manca *et al.*, 1999). It has been hypothesized that the marked immune response elicited by CDC1551 may account for the unusually large tuberculin skin test responses observed in persons infected with this strain (Valway *et al.*, 1998). Another study evaluated two *M. tuberculosis* strains that caused large clusters in Texas, and found that one strain killed mice much more rapidly than another strain, rapid death being associated with failure to mount a type 1 response (Manca *et al.*, 2001). The sum of these studies suggests that the phenotypes of *M. tuberculosis* strains can markedly affect their interactions with the host, and that different strains may use distinct mechanisms that favor their capacity to spread in the population.

The best example of a successful group of *M. tuberculosis* strains is the Beijing family, which has caused outbreaks in many parts of the world (Yang *et al.*, 1998; Bifani *et al.*, 1999; Caminero *et al.*, 2001a) and constitutes the dominant strain in multiple locations throughout Asia and the United States (van Soolingen *et al.*, 1995; Soini *et al.*, 2000; Bifani *et al.*, 1999; Ahn *et al.*, 2001; Bifani *et al.*, 2002). All members of the Beijing genotype have the following characteristics: (1) high number of copies of IS*6110*; (2) an IS*6110* insertion in the origin or replication; (3) they belong to genetic group 1 (Sreevatsan *et al.*, 1997), characterized by a CTG at codon 463 of the *katG* gene and an ACC at codon 95 of the *gyrA* gene; (4) an identical spoligotype pattern that shows only spacers 35-43 (Bifani *et al.*, 2002). Two prominent members of the Beijing family are the W strain that caused large outbreaks of multidrug-resistant and drug-susceptible tuberculosis in the north-east United States (Bifani *et al.*, 2002), and the 210 strain, which is widely distributed in

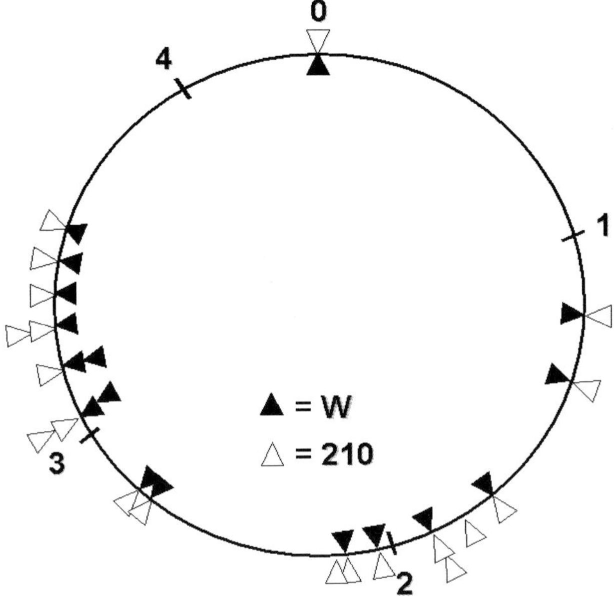

Figure 1. Locations of IS*6110* in the W and 210 strains, indicated on the genomic map of *M. tuberculosis* strain H37Rv. The numbers indicate the map positions in megabases on the *M. tuberculosis* chromosome. Sixteen IS*6110* insertion sites are shared by the two strains.

the south-west and south-central United States (Barnes *et al.*, 1997; Yang *et al.*, 1998; Weis *et al.*, 2002). These strains share 16 insertion sites of IS*6110* (Figure 1), suggesting that the position of these elements may play a role in their capacity to spread in the population.

It is possible that the Beijing family strains are widely distributed because they were introduced into multiple geographic locations long before other strains, and have had more time to spread in the community. However, we believe that it is more likely that these strains have an enhanced capacity to spread in human populations, either because they are more readily aerosolized, are more effective at establishing infection, or are more likely to progress from infection to disease. This hypothesis is supported by three findings. First, isolates of the Beijing genotype were significantly more common in young patients than in older patients in Vietnam, suggesting that spread of these strains was due to transmission recently, rather than in the remote past (Ahn *et al.*, 2000). Second, after introduction of a single Beijing family isolate to a European island, it rapidly spread and accounted for 27% of the tuberculosis cases reported on the island after three years (Caminero *et al.*, 2001a). Third, *M. tuberculosis* isolates of the 210 strain grow more rapidly in human macrophages than other strains, suggesting that it is more effective at avoiding host defenses (Zhang *et al.*, 1999).

It has been postulated that the Beijing genotype is widespread because people vaccinated with BCG remain more susceptible to Beijing family strains than to other *M. tuberculosis* strains (van Soolingen *et al.*, 1995). However, this is unlikely because the frequency of Beijing family isolates are similar in BCG-vaccinated and non-vaccinated persons (Ahn *et al.*, 2000) and the Beijing genotype is dominant in countries such as the United States, where widespread BCG vaccination has not been given for decades. It is intriguing to speculate that one or more IS*6110* elements in the Beijing genotype may have a promoter element that upregulates expression of downstream virulence genes (Beggs *et al.*, 2000). An alternative possibility is that the DR region in *M. tuberculosis* is involved in regulating replication (van Embden *et al.*, 2000) and that the pattern of repeats in this region that characterizes the Beijing family may enhance bacterial division (Beggs *et al.*, 2000). Identification of the genetic and phenotypic characteristics that favor the spread of the Beijing genotype will provide important insight into our understanding of the population biology of *M. tuberculosis* and its relationship with the human host.

Relationship of Drug Resistance to Disease Transmission

Studies in the 1950s showed that isoniazid-resistant *M. tuberculosis* strains were less virulent for guinea pigs than drug-susceptible isolates (Middlebrook and Cohn, 1953), but studies in humans have yielded conflicting results (Siminel *et al.*, 1979; Snider *et al.*, 1985), in part because transmission was assessed predominantly by results of tuberculin skin testing, which do not provide definitive information on the source of infection. Molecular epidemiologic studies have provided additional information on the transmission potential of drug-susceptible and drug-resistant strains.

A population-based study in Holland showed that isoniazid-resistant isolates were 0.4-0.7 times as likely to be clustered as drug-susceptible isolates, whereas clustering was similar in streptomycin-resistant and in drug-susceptible isolates (van Soolingen *et al.*, 1999). In Mexico and in South Africa, multidrug-resistant isolates were 0.2-0.3 times as likely to be clustered as drug-susceptible organisms (de Lourdes Garcia-Garcia *et al.*, 2000; Godfrey-Fausett *et al.*, 2000). The most comprehensive evaluation of this question to date was a population-based molecular epidemiologic study in San Francisco that included 1520 cases of drug-susceptible and 280 cases of drug-resistant tuberculosis. Preliminary results of this study show that the transmission indices of isoniazid-, streptomycin-, and rifampin-resistant organisms were 0.2, 0.5 and 0.7 times that of drug-susceptible strains, respectively (Burgos *et al.*, 2001). In Los Angeles, RFLP analysis of isolates from 102 patients with multidrug-resistant tuberculosis, most with acid-fast smear-positive cavitary disease, revealed only 5 clusters of two patients each, suggesting limited transmission of these strains (Nitta *et al.*, 2001). The sum of these studies suggests that drug-resistant isolates, particularly isoniazid-resistant ones, are less likely to cause secondary cases than drug-susceptible isolates. If additional studies confirm that these findings are not due to confounding variables, this would indicate that some genes that confer drug resistance may also reduce mycobacterial virulence. For example, isoniazid-resistant organisms have an increased frequency of *katG* mutations, and some *katG* mutations attenuate virulence of *M. bovis* and H37Rv in animals (Wilson *et al.*, 1995; Li *et al.*, 1998). Altered *katG* function might reduce catalase-peroxidase activity and inhibit the ability of *M. tuberculosis* to resist the antimycobacterial effects of macrophages.

Although the reduced transmission potential of drug-resistant strains may have significant implications for our understanding of virulence factors of *M. tuberculosis*, it is unlikely to affect the clinical management of patients with drug-resistant tuberculosis. Such patients are likely to be infectious for more prolonged periods than those with drug-susceptible disease. Although drug-resistant strains may be less transmissible on average than drug-susceptible ones, large outbreaks of multidrug-resistant tuberculosis have been documented (Friedman *et al.*, 1995; Kruuner *et al.*, 2001) and it is not possible to identify more transmissible strains. Furthermore, the public health consequences of transmission of drug-resistant tuberculosis are so much more serious than those of drug-susceptible disease that precautions to minimize transmission should not be relaxed for patients with drug-resistant tuberculosis, even if their transmission potential is relatively low.

The Role of Reinfection

It has traditionally been believed that when immunocompetent persons become infected with or develop disease from *M. tuberculosis*, they develop a substantial degree of protection against infection with a second *M. tuberculosis* strain. Therefore, it is recommended that an immunocompetent person with a history of a positive tuberculin skin test should not receive isoniazid after exposure to a new case of tuberculosis. Similarly, second episodes of tuberculosis are treated as reactivation of disease due to the organism causing the first episode. These assumptions have been challenged by three molecular epidemiologic studies in South Africa and Europe, which found second episodes of tuberculosis to be due to exogenous reinfection in 16-75% of cases (van Rie *et al.*, 1999; Bandera *et al.*, 2001; Caminero *et al.*, 2001b). The percentage of cases due to reinfection increased as the incidence of tuberculosis rose in the population. In populations where the prevalence of tuberculosis is low, patients are unlikely to encounter a tuberculosis patient and be reinfected, whereas reinfection is more common in high-prevalence populations.

These findings have important implications for our understanding of the nature of immunologic protection against tuberculosis. If natural infection does not protect against reinfection, it would explain in part the relative ineffectiveness of vaccination with *M. bovis* BCG, and imply that an effective vaccine must do more than mimic natural infection with *M. tuberculosis*. From a public health perspective, frequent reinfection would mandate treatment of persons recently exposed to tuberculosis, regardless of prior tuberculin skin test status. Similarly, patients with second episodes of tuberculosis should be treated with a regimen targeted toward the drug resistance patterns of current tuberculosis patients in the community, rather than that of the patient's original strain, particularly in locations where the prevalence of tuberculosis is high.

References

Ahn, D.D., Borgdorff, M.W., Van, L.N., Lan, N.T.N., van Gorkom, T., Kremer, K., and van Soolingen, D. 2000. *Mycobacterium tuberculosis* Beijing genotype emerging in Vietnam. Emerg. Infect. Dis. 6: 302-305.

Alland, D., Kalkut, G.E., Moss, A.R., McAdam, R.A., Hahn J.A., Bosworth, W., Drucker, E., and Bloom, B.R. 1994. Transmission of tuberculosis in New York City. An analysis by DNA fingerprinting and conventional epidemiologic methods. N. Engl. J. Med. 330: 1710-1716.

Bandera, A., Gori, A., Catozzi, L., Espost, A.D., Marchetti, G., Molteni, C., Ferrario, G., Codecasa, L., Penati, V., Matteelli, A., and Franzetti, F. 2001. Molecular epidemiology study of exogenous reinfection in an area with a low incidence of tuberculosis. J. Clin. Microbiol. 39: 2213-2218.

Barnes, P.F., Yang, Z., Preston-Martin, S., Pogoda, J.M., Jones, B.E., Otaya, M., Eisenach, K.D., Knowles, L., Harvey, S., and Cave, M.D. 1997. Patterns of tuberculosis transmission in central Los Angeles. JAMA. 278: 1159-1163.

Barnes, P. 1998. Reducing ongoing transmission of tuberculosis. JAMA 280: 1702-1703.

Barnes, P.F., Yang, Z., Pogoda, J.M., Preston-Martin, S., Jones, B.E., Otaya, M., Knowles, L., Harvey, S., Eisenach, K.D., and Cave, M.D. 1999. Foci of tuberculosis transmission in central Los Angeles. Am. J. Respir. Crit. Care Med. 159: 1081-1086.

Beggs, M. L., Eisenach, K. D. and Cave, M. D. 2000. Mapping of IS*6110* insertion sites in two epidemic strains of *Mycobacterium tuberculosis*. J. Clin Microbiol. 38: 2923-2928.

Bifani, P.J., Mathema, B., Liu, Z., Moghazeh, S.L., Shopsin, B., Tempalski, B., Driscoll, J., Frothingham, R., Musser, J.M., Alcabes, P., and Kreiswirth, B.N. 1999. Identification of a W variant outbreak of *Mycobacterium tuberculosis* via population-based molecular epidemiology. JAMA 282: 2321-2327.

Bifani, P.J., Mathema, B., Kurepina, N.E., and Kreiswirth, B.N. 2002. Global dissemination of the *Mycobacterium tuberculosis* W-Beijing family strains. Trends. Microbiol. 10: 45-52.

Bigi, F., Romano, M.I., Alito, A., and Cataldi, A. 1995. Cloning of a novel polymorphic GC-rich repetitive DNA from *Mycobacterium bovis*. Res. Microbiol. 146: 341-348.

Bishai, W.R., Graham, N.M.H., Harrington, S., Pope D., Hooper, N., Astemborski, J., Sheely L., Vlahov, D., Glass G.E., and Chaisson R.E. 1998. Molecular and geographic patterns of tuberculosis transmission after 15 years of directly observed therapy. JAMA. 280: 1679-1684.

Borgdorff, M.W., Nagelkerke, N., van Soolingen, D., de Haas, P.E.W., Veen, J., and van Embden, J.D.A. 1998. Analysis of tuberculosis transmission between nationalities in the Netherlands in the period 1993-1995 using DNA fingerprinting. Am. J. Epidemiol. 147: 187-195.

Borgdorff, M.W., Behr, M.A., Nagelkerke, N.J.D., Hopewell, P.C., and Small, P.M. 2000. Transmission of tuberculosis in San Francisco and its association with immigration and ethnicity. Int. J. Tuberc. Lung Dis. 4: 287-294.

Braden, C.R., Templeton, G.L., Cave, M.D., Valway, S., Onorato, I.M., Castro, K.G., Moers, D., Yang, Z., Stead, W.W., and Bates, J.H. 1997. Interpretation of restriction fragment length polymorphism analysis of *Mycobacterium tuberculosis* isolates from a state with a large rural population. J. Infect. Dis. 175: 1446-1452.

Burgos, M., DeReimer, K., Small, P., Hopewell, P., and Daley, C. 2001. Differential transmission of drug-resistant and drug-susceptible *Mycobacterium tuberculosis*. Program and Abstracts of the U.S.-Japan Thirty-Sixth Tuberculosis and Leprosy Research Conference: 206-211.

Burman, W.J., and Reeves, R.R. 2000. Review of false-positive cultures for *Mycobacterium tuberculosi*s and recommendations for avoiding unnecessary treatment. Clin. Infect. Dis. 31: 1390-1395.

Butler W.R., Hass, W.H., and Crawford J.T. 1996. Automated DNA fingerprinting analysis of *Mycobacterium tuberculosis* using fluorescent detection of PCR products. J Clin Microbiol. 34: 1801-1803.

Caminero, J.A., Pena, M.J., Campos-Herrero, M.I., Rodriguez, J.C., Garcia, I., Cabrera, P., Lafoz, C., Samper, S., Takiff, H., Afonso, O., Pavon, J.M., Torres, M.J., Van Soolingen, D., Enarson, D.A., and Martin, C. 2001. Epidemiologic evidence of the spread of a *Mycobacterium tuberculosis* strain of the Beijing genotype on Gran Canaria Island. Am. J. Respir. Crit. Care Med. 164: 1165-1170.

Caminero, J.A., Pena, M.J., Campos-Herrero, M.I., Rodriguez, J.C., Afonso, O., Martin, C., Pavon, J.M., Torres, M.J., Burgos, M., Cabrera, P., Small, P.M., and Enarson, D.A. 2001. Exogenous reinfection with tuberculosis on a European island with a moderate incidence of disease. Am. J. Respir. Crit. Care Med. 163: 717-720.

Cave, M.D., Eisenach, K.D., McDermott, P.F., Bates, J.J., and Crawford, J.T. 1991. IS*6110*: conservation of sequence in the *Mycobacterium tuberculosis* complex and its utilization in DNA fingerprinting. Mol. Cell. Probes 5: 73-80.

Cave, M.D., Eisenach, K.D., Templeton, G., Salfinger, M., Mazurek, G., Bates, J. and Crawford, J.T. 1994. Stability of DNA fingerprint pattern produced with IS*6110* strains of *Mycobacterium tuberculosis*. J. Clin. Microbiol. 32: 262-266.

Chaves, F., Dronda, F., Cave, M.D., Alonso, M., Amparo, G., Eisenach, K.D., Oretega, M.D., Lopez Cubero, L., Fernandez, I., Catalan, S., and Bates, J.H. 1997. A longitudinal study of transmission of tuberculosis in a large prison population. Am. J. Respir. Crit. Care Med. 155: 719 -725.

Chin, D.P., Crane, C.M., Diul, M.Y., Sun, S.J., Agraz, R., Taylor, S., Desmond, E., and Wise, F. 2000. Spread of *Mycobacterium tuberculosis* in a community implementing recommended elements of tuberculosis control. JAMA 283: 2968-2974.

Cousins, D., Williams, S., Liebana, E., Aranaz, A., Bunschoten, A., Van Embden, J., and Ellis, T. 1998. Evaluation of four DNA typing techniques in epidemiological investigations of bovine tuberculosis. J. Clin. Microbiol. 36: 168-178.

Dale, J.W., Brittain, D., Cataldi, A.A., Cousins, D., Crawford, J.T., Driscoll, J., Heersma, H., Lilleback, T., Quitugua, T., Rastogi, N., Skuce, R.A., Sola, C., Van Soolingen, D., and Vincent, V. 2001. Spacer oligonucleotide typing of bacteria of the *Mycobacterium tuberculosis* complex: recommendations for standardized nomenclature. Int. J. Tuberc. Lung Dis. 5: 216-219.

De Boer, A.S., Borgdorff, M.W., de Haas, P.E.W., Nagelkerke, N.J.D., van Embden, J.D.A., and van Soolingen, D. 1999. Analysis of rate of change of IS*6110* RFLP patterns of *Mycobacterium tuberculosis* based on serial patient isolates. J. Infect. Dis. 180: 1238-1244.

de Lourdes Garcia-Garcia, M., Ponce-de-Leon, A., Jimenez-Corona, M.E., Jimenez-Corona, A., Palacios,-Martinez, M., Balandrano-Campos, S., Ferreyra-Reyes, L., Juarez-Sandino, L., Sifuentes-Osornio, J., Olivera-Diaz, H., Valdespino-Gomez, J.L., and Small, P.M. 2000. Clinical consequences and transmissibility of drug-resistant tuberculosis in southern Mexico. Arch. Intern. Med. 160: 630-636.

Dwyer, B., Jackson, K., Raios, K., Sievers, A., Wilshire, E., and Ross, B. 1993. DNA restriction fragment analysis to define an extended cluster of tuberculosis in homeless men and their associates. J. Infect. Dis. 167: 490-494.

Fitzpatrick, L.K., Hardacker, J.A., Heirendt, W., Agerton, T., Streicher, A., Melnyk, H., Ridzon, R., Valway, S., and Onorato, I. 2001. A preventable outbreak of tuberculosis investigated through an intricate social network. Clin. Infect. Dis. 33 : 1801-1806.

Fomukong, N., Beggs, M., el Hajj, H., Templeton, G., Eisenach, K., and Cave, M.D. 1998. Differences in the prevalence of IS*6110* insertion sites in *Mycobacterium tuberculosis* strains: low and high copy number of IS*6110*. Tuberc. Lung Dis. 78: 109-116.

Fraser, C.M., Eisen, J., Fleischmann, R.D., Ketchum, K.A., and Peterson, S. 2000. Comparative genomics and understanding of microbial biology. Emerg. Infect. Dis. 6: 505-512.

Friedman, C.R., Stoeckle, M.Y., Kreiswirth, B.N., Johnson Jr., W.D., Manoach, K.M., Berger, J., Sathianathan, K., Hafner, A., and Riley, L.W. 1995.

Transmission of multidrug-resistant tuberculosis in a large urban setting. Am. J. Respir. Care Med. 152: 355-359.

Friedman, C.R., Quinn, G.C., Kreiswirth, B.N., Perlman, D.C., Salomon, N., Schluger, N.W., Lutfey, M., Berger, J., Poltoratskaia, N., and Riley, L.W. 1997. Widespread dissemination of a drug-susceptible strain of *Mycobacterium tuberculosis*. J. Infect. Dis. 176: 478-484.

Frothingham, R., and Meeker-O'Connell, W.A. 1998. Genetic diversity in the *Mycobacterium tuberculosis* complex based on variable numbers of tandem DNA repeats. Microbiology. 144: 1189-96.

Glynn, J.R., Vynnycky, E., and Fine, P.E.M. 1999. Influence of sampling on estimates of clustering and recent transmission of *Mycobacterium tuberculosis* derived from DNA fingerprinting techniques. Am. J. Epidemiol. 149: 366-371.

Godfrey-Faussett, P., Sonnenberg, P., Shearer, S., Bruce, M.C., Mee, C., Morris, L., and Murray, J. 2000. Tuberculosis control and molecular epidemiology in a South African gold-mining community. Lancet 356: 1066-1071.

Goguet de la Salmoniere, Y-O., Li H.M., Torrea, G., Bunschoten, A., van Embden, J., and Gicquel, B. 1997. Evaluation of spoligotyping in a study of the transmission of *Mycobacterium tuberculosis*. J. Clin. Microbiol. 35: 2210-2214.

Golub, J.E., Cronin, W.A., Obasanjo, O.O., Coggin, W., Moore, K., Pope, D.S., Thompson, D., Sterling, T.R., Harrington, S., Bishai, W.R., and Chaisson, R.E. 2001. Transmission of *Mycobacterium tuberculosis* through casual contact with an infectious case. Arch. Intern. Med. 161: 2254-2258.

Haas, W.H., Butler, W.R., Woodley, C.L., and Crawford, J.T. 1993. Mixed-linker polymerase chain reaction: a new method for rapid fingerprinting of isolates of the *Mycobacterium tuberculosis* comples. J. Clin. Microbiol. 31: 1293-1298

Hermans, P.W., van Soolingen, D., Dale, J.W., Schuitema, A.R., McAdam, R.A., Catty, D., and van Embden, J.D. 1990. Insertion element IS986 from *Mycobacterium tuberculosis*: a useful tool for diagnosis and epidemiology of tuberculosis. J. Clin. Microbiol. 28: 2051-2058.

Jasmer, R.M., Hahn, J.A., Small, P.M., Daley, C.L., Behr, M.A., Moss, A.R., Creasman, J.M., Schecter, G.F., Paz, E.A., and Hopewell, P.C. 1999. A molecular epidemiologic analysis of tuberculosis trends in San Francisco, 1991-1997. Ann. Intern. Med. 130: 971-978.

Jones, T.F., Craig, A.S., Valway, S., Woodley, C.L., and Schaffner, W. 1999. Transmission of tuberculosis in a jail. Ann. Intern. Med. 131: 557-563.

Kamerbeek, J., Schouls, L., Kolk, A., van Agtervveld, M. van Soolingen, D., Kuijper, S., Bunschoten, A., Molhuizen, H., Shaw, R., Goyal, G. and van Embden, J.D.A. 1997. Simultaneous detection and strain differentiation of *Mycobacterium tuberculosis* for diagnosis and epidemiology. J. Clin. Microbiol. 35: 907-914.

Kato-Maeda, M., Rhee, J.T., Gingeras, T.R., Salamon, H., Drenkow, J., Smittipat, N., and Small, P.M. 2001. Comparing genomes within the species *Mycobacterium tuberculosis*. Genome Res. 11: 547-554.

Kimmerling, M.E., Shakes, C.F., Carlisle, R., Lok, K.H., Benjamin, W.H., and Dunlap, N.E. 999. Spot sputum screening: evaluation of an intervention in two homeless shelters. Int. J. Tuberc. Lung Dis. 3: 613-619.

Kline, S.E., Hedemark, L.L., and Davies, S.F. 1995. Outbreak of tuberculosis among regular patrons of a neighborhood bar. N. Engl. J. Med. 333: 222-227.

Kremer, K., Van Soolingen, D., Frothingham, R., Haas, W.H., Hermans, P.W.M., Martin, C., Palittapongarnpim, P., Plikaytis, B.B., Riley, L.W., Yakrus, M.A., Musser, J.M., and van Embden, J.D.A. 1999. Comparison of methods based on different molecular epidemiological markers for typing of *Mycobacterium tuberculosis* complex strains: interlaboratory study of discriminatory power and reproducibility. J. Clin. Microbiol. 37: 2607-2618.

Kruuner, A., Hoffner, S.E., Sillastu, H., Danilovits, M., Levina, K., Svenson, S.B., Ghebremichael, S., Koivula, T., and Kallenius, G. 2001. Spread of drug-resistant pulmonary tuberculosis in Estonia. J. Clin. Microbiol. 39: 3339-3345.

Li, Z., Kelley, C., Collins, F., Rouse, D., and Morris, S. 1998. Expression of katG in *Mycobacterium tuberculosis* is associated with its growth and persistence in mice and guinea pigs. J. Infect. Dis. 177: 1030-1035.

Manca, C., Tsenova, L., Barry III, C.E., Bergtold, A., Freeman, S., Haslett, P.A.J., Musser, J.M., Freedman, V.H., and Kaplan, G. 1999. *Mycobacterium tuberculosis* CDC1551 induces a more vigorous host response *in vivo* and *in vitro*, but is not more virulent than other clinical isolates. J. Immunol. 162: 6740-6746.

Manca, C., Tsenova, L., Bergtold, A., Freeman, S., Tovey, M., Musser, J.M., Barry III, C.E., Freedman, V.H., and Kaplan, G. 2001. Virulence of a *Mycobacterium tuberculosis* clinical isolate is determined by failure to induce Th1 type immunity and is associated with induction of IFN-α/β. Proc. Natl. Acad. Sci. USA. 98: 5752-5757.

Mazurek, G.H., Cave, M.D., Eisenach, K.D., Wallace, R.J. Jr., Bates, J.H., and Crawford, J.T. 1991. Chromosomal DNA fingerprint patterns produced with IS*6110* as strain-specific markers for epidemiologic study of tuberculosis. J. Clin. Microbiol. 29: 2030-2033.

Mazars, E., Lesjean, S., Banuls, A-L., Gilbert, M., Vincent, V., Gicquel, B., Tibayrenc, M., Locht, C., and Supply, P. 2001. High-resolution minisatellite-based typing as a portable approach to global analysis of *Mycobacteruim tuberculosis* molecular epidemiology. Proc. Nat. Acad. Sci. U.S.A. 98: 1901-1906.

Middlebrook, G., and Cohn, M.L. 1953. Some observations on the pathogenicity of isoniazid-resistant variants of tubercle bacilli. Science 118: 297-299.

Musser, J.M. 2001. Single nucleotide polymorphisms in *Mycobacterium tuberculosis* structural genes. Emerg. Infect. Dis. 7: 486-487.

Nitta, A.T., Knowles, L.S., Kim, J., Lehnkering, E.L., Borenstein, L.A., Davidson, P.T., Harvey, S.M., and De Koning, M.L. 2002. Limited transmission of multidrug-resistant tuberculosis despite a high proportion of infectious cases in Los Angeles County, California. Am. J. Respir. Dis. Crit. Care Med. 165: 812-817.

Poulet, S., and Cole, S.T. 1995. Characterization of the highly abundant polymorphic GC-rich-repetitive sequence (PGRS) present in *Mycobacterium tuberculosis*. Arch. Microbiol. 163: 87-95.

Salamon, H., Segal, M.R., Ponce de Leon, A., and Small, P.M. 1998. Accommodating error analysis in comparison and clustering of molecular fingerprints. Emerg. Infect. Dis. 4 : 159-168.

Schluger, N.W., Huberman, R., Holzman, R., Rom, W.N., and Cohen, D.I. 1999. Screening for infection and disease as a tuberculosis control measure among indigents in New York City, 1994-1997. Int. J. Tuberc. Lung Dis. 3: 281-286.

Sechi, L.A., Zanetti, S., Dupre, I., Delogu, G., and Fadda, G. 1998. Enterobacterial repetitive intergenic consensus sequences as molecular targets for typing of *Mycobacterium tuberculosis* strains. J. Clin. Microbiol. 36: 128-132.

Siminel, M., Bungezianu, G., and Anastasatu, C. 1979. The risk of infection and disease incontacts with patients excreting *Mycobacterium tuberculosis* sensitive and resistant to isoniazid. Bull. Int .Union Tuberc. Lung Dis. 54: 263 (abstract).

Small, P.M., Hopewell, P.C., Singh, S.P., Paz, A., Parsonnet, J., Ruston, D.C., Schecter, G.F., Daley, C.L., and Schoolnik, G.K. 1994. The epidemiology of tuberculosis in San Francisco. A population-based study using conventional and molecular methods. N. Engl. J. Med. 330: 1703-1709.

Snider, D.E., Kelly, G.D., Cauthen, G.M., Thompson, N.J., and Kilburn, J.O. 1985. Infection and disease among contacts of tuberculosis cases with drug-resistant and drug-susceptible bacilli. *Am. Rev. Respir .Dis.* 132: 125-132.

Soini, H., Pan, X., Amin, A., Graviss, E.A., Siddiqui, A., and Musser, J.M. 2000. Characterization of *Mycobacterium tuberculosis* isolates from patients in Houston, Texas, by spoligotyping. Infect. Immun. 38: 669-676.

Sreevatsan, S., Pan, X., Stockbauer, K.E., Connell, N.D., Kreiswirth, B.N., Whittam, T.S., and Musser, J.M. 1997. Restricted structural gene polymorphism in the *Mycobacterium tuberculosis* complex indicates evolutionary recent global dissemination. Proc. Natl. Acad. Sci. USA. 94: 9869-9874.

Supply, P., Mazars, E., Lesjean, S., Vincent, V., Gicquel, B., and Locht, C. 2000. Variable human minisatellite-like regions in the *Mycobacterium tuberculosis* genome. Mol. Microbiol. 36: 762-771.

Supply, P., Lesjean, S., Savine, E., Kremer, K., Van Soolingen, D., and Locht, C. 2001. Automated high-throughput genotyping for study of global epidemiology of *Mycobacterium tuberculosis* based on mycobacterial interspersed repetitive units. J. Clin. Microbiol. 39: 3563-3571.

Thierry, D., Cave, M.D., Eisenach, K.D., Crawford, J.C., Bates, J.H., Gicquel, B., and Guesdon, J.L. 1990. IS*6110*, an IS-like element of the *Mycobacterium tuberculosis* complex. Nuc. Acid Res. 18: 188.

Valway, S.E., Sanchez, M.P.C., Shinnick, T.M., Orme, I., Agerton, T., Hoy, D., Jones, J.S., Westmoreland, H., and Onorato, I.M. 1998. An outbreak involving extensive transmission of a virulent strain of *Mycobacterium tuberculosis*. N. Engl. J. Med. 338: 633-639.

van Embden, J.D.A, Cave, M.D., Crawford, J.T. Dale, J.W., Eisenach, K.D., Gicquel, B., Hermans, P., Martin, C., McAdam, R., Shinnick, T.M., and Small, P.M.. 1993. Strain identification of *Mycobacterium tuberculosis* by DNA fingerprinting: recommendation for a standardized methodology. J. Clin. Microbiol. 31: 406-409.

van Embden, J.D., van Gorkom, T., Kremer, K., Jansen, R., van Der Zeijst, B.A., and Schouls, L.M. 2000. Genetic variation and evolutionary origin of the direct repeat locus of *Mycobacterium tuberculosis* complex bacteria. J. Bacteriol. 182: 2393-2401.

van Rie, A., Warren, R., Richardson, M., Victor, T., Gie, R.P., Enarson, D.A., Beyers, N., and van Helden, P. 1999. Exogenous reinfection as a cause of recurrent tuberculosis after curative treatment. N. Engl. J. Med. 341: 1174-1179.

van Soolingen, D., Petra, E.W., de Haas, P., Hermans, P., Groenen, P., and van Embden, J.D. 1993. Comparison of various repetitive DNA elements as genetic markers for strain differentiation and epidemiology of *Mycobacterium tuberculosis*. J. Clin. Microbiol. 31: 1987-1995.

van Soolingen, D., Qian, L., de Haas, P.E., Douglas, J.T., Traore, H., Portaels, F., Qing, H.Z., Enkhsaikan, D., Nymadawa, P., and van Embden, J.D. 1995. Predominance of a single genotype of *Mycobacterium tuberculosis* in countries of east Asia. J. Clin. Microbiol. 33: 3234-3238.

van Soolingen, D., Borgdorff, M.W., de Haas, P.E.W., Sebek, M.M.G.G., Veen, J., Dessens, M., Kremer, K., and van Embden, J.D.A. 1999. Molecular epidemiology of tuberculosis in the Netherlands: a nationwide study from 1993 through 1997. J. Infect. Dis. 180: 726-736.

Weis, S.E., Pogoda, J.M., Yang, Z., Cave, M.D., Wallace, C., Kelley, M., and Barnes, P.F. 2002. Transmission dynamics of tuberculosis in Tarrant County, Texas. Am. J. Respir. Crit. Care Med. 166: 36-42.

Wilson, T.M., de Lisle, G.W., and Collins, D.M. 1995. Effect of inhA and katG on isoniazid resistance and virulence of *Mycobacterium bovis*. Mol. Microbiol. 15: 1009-1015.

Yaganehdoost, A., Graviss, E.A., Ross, M.W., Adams, G.J., Ramaswamy, S., Wanger, A., Frothingham, R., Soini, H., and Musser, J.M. 1998. Complex transmission dynamics of clonally related virulent *Mycobacterium tuberculosis* associated with barhopping by predominantly human immunodeficiency virus—positive gay men. J. Infect. Dis. 180: 1245-1251.

Yang, Z., Barnes, P.F., Chaves, F., Eisenach, K.D., Weis, S., Bates, J.H., and Cave, M.D. 1998. Diversity of DNA fingerprints of *Mycobacterium tuberculosis* isolates in the United States. J. Clin. Microbiol. 36: 1003-1007.

Yang, Z. H., Ijaz, K., Bates, J.H., Eisenach, K.D, and Cave, M. D. 2000. Spoligotyping and PGRS fingerprinting of *Mycobacterium tuberculosis* strains having few copies of IS*6110*. J. Clin. Microbiol. 38: 3572-3576.

Zhang, M., Gong, J., Yang, Z., Samten, B., Cave, M.D., and Barnes, P.F. 1999. Enhanced capacity of a widespread strain of *Mycobacterium tuberculosis* to grow in macrophages. J. Infect. Dis. 179: 1213-1217.

Index

S

T

V